THERMODYNAMIC AND TRANSPORT
PROPERTIES OF ORGANIC SALTS

IUPAC CHEMICAL DATA SERIES

Number 1	Y. MARCUS	Critical Evaluation of Some Equilibrium Constants involving Organophosphorus Extractants
Number 2	A. S. KERTES	Critical Evaluation of Some Equilibrium Constants involving Alkylammonium Extractants
Number 3	Y. MARCUS	Equilibrium Constants of Liquid-Liquid Distribution Reactions— Organophosphorus Extractants
Number 4	Y. MARCUS	Equilibrium Constants of Liquid-Liquid Distribution Reactions—Alkylammonium Salt Extractants
Number 5	S. ANGUS et al.	International Thermodynamic Tables of the Fluid State—Argon
Number 6	S. ANGUS et al.	International Thermodynamic Tables of the Fluid State—Ethylene
Number 7	S. ANGUS et al.	International Thermodynamic Tables of the Fluid State—Carbon Dioxide
Number 8	S. ANGUS et al.	International Thermodynamic Tables of the Fluid State—Helium
Number 9	A. R. H. COLE	Tables of Wavenumbers for the Calibration of Infrared Spectrometers, 2nd edition
Number 10	Y. MARCUS & D. G. HOWERY	Ion Exchange Equilibrium Constants
Number 11	R. TAMAMUSHI	Kinetic Parameters of Electrode Reactions of Metallic Compounds
Number 12	D. D. PERRIN	Dissociation Constants of Organic Bases in Aqueous Solution
Number 13	G. CHARLOT, A. COLUMEAU & M. J. C. MARCHON	Selected Constants: Oxidation-Reduction Potentials of Inorganic Substances in Aqueous Solutions
Number 14	C. ANDEREGG	Critical Survey of Stability Constants of EDTA Complexes
Number 15	Y. MARCUS et al.	Equilibrium Constants of Liquid-Liquid Distribution Reactions—Compound Forming Extractants, Solvating Solvents and Inert Solvents
Number 16	S. ANGUS et al.	International Thermodynamic Tables of the Fluid State—Methane
Number 17	W. A. E. McBRYDE	A Critical Review of Equilibrium Data for Proton- and Metal Complexes of 1,10-Phenanthroline, 2,2,'-Bipyridyl and Related Compounds
Number 18	J. STARY & H. FREISER	Equilibrium Constants of Liquid-Liquid Distribution Reactions—Chelating Extractants
Number 19	D. D. PERRIN	Dissociation Constants of Inorganic Acids and Bases in Aqueous Solution
Number 20	S. ANGUS et al.	International Thermodynamic Tables of the Fluid State—Nitrogen
Number 21	E. HÖGFELDT	Stability Constants of Metal-Ion Complexes: Part A—Inorganic Ligands
Number 22	D. D. PERRIN	Stability Constants of Metal-Ion Complexes: Part B—Organic Ligands
Number 23	E. P. SERJEANT & B. DEMPSEY	Ionization Constants of Organic Acids in Aqueous Solution
Number 24	J. STARY et al.	Critical Evaluation of Equilibrium Constants involving 8-Hydroxyquinoline and its Metal Chelates
Number 25	S. ANGUS et al.	International Thermodynamic Tables of the Fluid State—Propylene (Propene)
Number 26	Z. KOLARIK	Critical Evaluation of Equilibrium Constants involving Acidic Organophosphorus Extractants
Number 27	A. M. BOND & G. T. HEFTER	Critical Survey of Stability Constants and Related Thermodynamic Data of Fluoride Complexes in Aqueous Solution
Number 28	P. FRANZOSINI & M. SANESI	Thermodynamic and Transport Properties of Organic Salts

NOTICE TO READERS

Dear Reader

If your library is not already a standing/continuation order customer to this series, may we recommend that you place a standing/continuation order to receive immediately upon publication all new volumes. Should you find that these volumes no longer serve your needs, your order can be cancelled at any time without notice.

ROBERT MAXWELL
Publisher at Pergamon Press

INTERNATIONAL UNION OF PURE AND APPLIED CHEMISTRY
(PHYSICAL CHEMISTRY DIVISION, COMMISSION ON THERMODYNAMICS)

THERMODYNAMIC AND TRANSPORT PROPERTIES OF ORGANIC SALTS

Prepared for publication by
PAOLO FRANZOSINI and MANLIO SANESI
Institute of Physical Chemistry and Electrochemistry
University of Pavia, Italy

IUPAC Chemical Data Series, No. 28

PERGAMON PRESS

OXFORD · NEW YORK · TORONTO · SYDNEY · PARIS · FRANKFURT

U.K.	Pergamon Press Ltd., Headington Hill Hall, Oxford OX3 0BW, England
U.S.A.	Pergamon Press Inc., Maxwell House, Fairview Park, Elmsford, New York 10523, U.S.A.
CANADA	Pergamon of Canada, Suite 104, 150 Consumers Road, Willowdale, Ontario M2J 1P9, Canada
AUSTRALIA	Pergamon Press (Aust.) Pty. Ltd., P.O. Box 544, Potts Point, N.S.W. 2011, Australia
FRANCE	Pergamon Press SARL, 24 rue des Ecoles, 75240 Paris, Cedex 05, France
FEDERAL REPUBLIC OF GERMANY	Pergamon Press GmbH, 6242 Kronberg-Taunus, Hammerweg 6, Federal Republic of Germany

First edition 1980

British Library Cataloguing in Publication Data
International Union of Pure and Applied Chemistry.
Commission on Thermodynamics
Thermodynamic and transport properties of organic salts.
-(International Union of Pure and Applied Chemistry. Chemical data series; no. 28).
1. Chemistry, Physical organic
2. Salts - congresses   3. Thermochemistry
4. Transport theory
I. Title   II. Franzosini, P   III. Sanesi, M
IV. Series
547   QD476   80-40689

ISBN 0-08-022378-8

Printed in Great Britain by A. Wheaton & Co. Ltd., Exeter

CONTENTS

Commission on Thermodynamics   vi
Foreword   vii
Acknowledgements   viii
Contributors   ix
Introduction   x

Part 1—Pure Salts   1
Chapter 1.1—PVT Relationships
K. TÖDHEIDE   3
Chapter 1.2—Thermal Properties
M. SANESI, A. CINGOLANI, P. L. TONELLI and P. FRANZOSINI   29
Chapter 1.3—Melting Mechanism of Pure Salts
A. R. UBBELOHDE   119
Chapter 1.4—Transport Properties of Pure Molten Salts with Organic Ions
A. KISZA   125

Part 2—Salt Mixtures   179
Chapter 2.1—Phase Diagrams
P. FRANZOSINI, P. FERLONI and M. SANESI   181
Chapter 2.2—Transport Properties of Binary Mixtures of Molten Salts with Organic Ions
A. KISZA   243

Part 3—Solutions   273
Chapter 3.1—Transport Properties of Organic Salts in Aqueous Solutions
P. STENIUS, S. BACKLUND and P. EKWALL   275
Chapter 3.2—Thermodynamic Quantities of Micelle Formation
P. STENIUS, S. BACKLUND and P. EKWALL   295
Chapter 3.3—Formation of Lyotropic Liquid Crystals by Organic Salts
P. STENIUS and P. EKWALL   321

Appendixes   341
Appendix 1—Aspects of the Structure of Pure Solid Organic Salts
K. FONTELL   343
Appendix 2—Electrochemical Studies In Molten Organic Salts
M. FIORANI   359

COMMISSION ON THERMODYNAMICS
1977-1979

Titular Members

M. Laffitte (Chairman)
G. M. Schneider (Secretary)
S. Angus, G. T. Armstrong, V. A. Medvedev
Y. Takahashi, I. Wadsö, W. Zielenkiewicz

Associate Members

H. Chihara, J. F. Counsell, M. Diaz-Peña
P. Franzosini, R. D. Freeman, V. A. Levitskii
G. Somsen, C. E. Vanderzee

National Representatives

J. Pick (Czechoslovakia)
M. Rätzsch (German Democratic Republic)
J. C. G. Calado (Portugal)

INTERNATIONAL UNION OF PURE AND APPLIED CHEMISTRY

IUPAC Secretariat: Bank Court Chambers, 2-3 Pound Way
Cowley Centre, Oxford OX4 3YF, UK

FOREWORD

One of the aims of the Commission on Thermodynamics of the International Union of Pure and Applied Chemistry is to enhance the quality of publications in the field of chemical thermodynamics.

The work of the Commission is concerned by accurate definitions of words used in the field (what does mean "standard"? for example), by nomenclature and units (shall we say Gibbs energy or Gibbs function, in accordance with ISO or IUPAP recommandations?), by the edition of books whose aim is to supply experimentalists with authoritative guidance (the last one, published in 1979 is entitled "Combustion Calorimetry").

These books may give advices on the "State of the Art" or may give the best values of thermodynamic quantities, selected by international expertise, which means the best experts.

The present volume belongs to this type. The principle of such a book has been proposed to our Commission by Dr Paolo Franzosini in 1977 in Warsaw. We agreed upon and two years later, at the following meeting at Davos, the work was nearly completed, under the guidance of Dr Franzosini and Dr Manlio Sanesi, by a group of well known scientists, who have made the work at a speed which may be considered as a record. If we had Olympic Games in this type of sport, they would surely win a Gold Medal!

All editors and authors must be thanked for their valuable work.

Professor Marc Laffitte
Chairman
IUPAC Commission on Thermodynamics

ACKNOWLEDGEMENTS

First of all, the most hearty thanks are due to all contributors, whose co-operative attitude and punctuality in following the planned timetable greatly facilitated the editorial job.

The Editors also feel deeply indebted on one hand to the several IUPAC Officers who took interest in the preparation of the book, in particular to Professor S. Sunner (Past-President of the Physical Chemistry Division), Professor E.F. Westrum, Jr., and Professor M. Laffitte (Past-Chairman and Chairman, respectively, of the Commission on Thermodynamics), Dr. M. Williams (Executive Secretary of the Union) and Mr. P.D. Gujral (Assistant Secretary for Publications); and on the other hand both to the Publishers, Pergamon Press, for their assistance, and to a number of copyright holders for permission to reproduce figures.

Finally, as the co-authors of Chapters 1.2 and 2.1, we wish to express sincere thanks to Professor C.A. Angell (Purdue University, West Lafayette, Indiana), to Dr. M. Blander (Argonne National Laboratory, Argonne, Illinois) and again to Professor Westrum, who were so kind as to read carefully through the text and to give very useful suggestions.

The Editors

CONTRIBUTORS

S. Backlund
> Department of Physical Chemistry, Åbo Akademi, Turku (Finland)

A. Cingolani
> *Centro di Studio per la Termodinamica ed Elettrochimica dei Sistemi Salini Fusi e Solidi del CNR* c/o Institute of Physical Chemistry and Electrochemistry of the University, Pavia (Italy)

P. Ekwall
> Gråhundsvägen 134, Farsta (Sweden)

P. Ferloni
> *Centro di Studio per la Termodinamica ed Elettrochimica dei Sistemi Salini Fusi e Solidi del CNR* c/o Institute of Physical Chemistry and Electrochemistry of the University, Pavia (Italy)

M. Fiorani
> Institute of Analytical Chemistry of the University, Padova (Italy)

K. Fontell
> Chemical Center, University of Lund, Lund (Sweden)

P. Franzosini
> *Centro di Studio per la Termodinamica ed Elettrochimica dei Sistemi Salini Fusi e Solidi del CNR* c/o Institute of Physical Chemistry and Electrochemistry of the University, Pavia (Italy)

A. Kisza
> Institute of Chemistry, University of Wroclaw, Wroclaw (Poland)

M. Sanesi
> *Centro di Studio per la Termodinamica ed Elettrochimica dei Sistemi Salini Fusi e Solidi del CNR* c/o Institute of Physical Chemistry and Electrochemistry of the University, Pavia (Italy)

P. Stenius
> The Swedish Institute for Surface Chemistry, Stockholm (Sweden)

K. Tödheide
> Institute of Physical Chemistry and Electrochemistry of the University, Karlsruhe (FRG)

P. L. Tonelli
> *Centro di Studio per la Termodinamica ed Elettrochimica dei Sistemi Salini Fusi e Solidi del CNR* c/o Institute of Physical Chemistry and Electrochemistry of the University, Pavia (Italy)

A. R. Ubbelohde
> Dept. of Chemical Engineering and Chemical Technology, Imperial College, London (UK)

INTRODUCTION

A remarkable amount of physico-chemical work was carried out during the last few decades on salts with organic anion and/or cation: valuable reviews have been already published, e.g., by J.E. Gordon (*"Applications of Fused Salts in Organic Chemistry"*, in *"Techniques and Methods of Organic and Organometallic Chemistry"*, Vol. 1, ed. by D.B. Denner, M. Dekker, New York, 1969, pp. 51 - 188) and by J.E. Lind, Jr. (*"Molten Organic Salts - Physical Properties"*, in *"Advances in Molten Salt Chemistry"*, Vol. 2, ed. by J. Braunstein, G. Mamantov and G.P. Smith, Plenum Press, New York, 1973, pp. 1 - 26).

Still, since the interest in this research field proved recently to be in progressive expansion - due *inter alia* to the fact that, as stressed by Duruz, Michels and Ubbelohde {*Proc. Roy. Soc.*, <u>A322</u>, 282 (1971)}, "the ease with which the cation radii and the detailed configurations of the anions can be varied at will makes organic ionic melts into a unique class of model ionic liquids, potentially suitable for very diverse scientific and technological applications" - it seemed that drawing together in a new updated report the available material on thermodynamic and transport properties of organic salts would be desirable.

The report, which was intended to be neither a true handbook (i.e., a simple collection of tabulated data) nor a suite of review articles but something intermediate, was realized through the co-operative work of an international staff under the auspices of the IUPAC Commission on Thermodynamics.

The material was arranged into three parts, dealing with pure salts, mixtures and solutions, respectively. Additional information, thought to be of interest for a better general knowledge of these salts, has been given in two appendixes concerning respectively the structure of the pure solids and the use of the melts in electrochemical studies.

Although the initial aim at a picture as comprehensive as possible had to be somewhat reduced through several cuts in order to maintain the size of the volume within reasonable limits, it is hoped that the result may still be of some usefulness for a number of colleagues active in this field.

Part 1
PURE SALTS

Editors' note.

Throughout Part 1 (except for Chapter 1.1), Part 2 and Part 3 complete literature references are given at the end of each Chapter in alphabetical order: for papers by the same author(s) a chronological sequence is followed. In the text and/or at the end of the Tables abridged references (i.e., limited to the name(s) of the author(s), and to the year of publication when necessary) are also given: papers by the same author(s) published in the same year are distinguished with the letters a, b, ..., _according to the sequence mentioned above._

PVT RELATIONSHIPS

K. Tödheide

INTRODUCTION

Interest in accurate PVT data mainly originates in two facts:

1. PVT data enable the separation of the density and temperature dependences of macroscopic properties measured as functions of pressure and temperature. Consequently, their knowledge greatly facilitates the interpretation of macroscopic properties of physical systems in terms of the microscopic properties of their constituents.

2. PVT data enable the construction of an equation of state from which the thermodynamic properties of the system may be calculated as functions of temperature and pressure or density.

In principle, PVT data may be obtained in quite different ways:

a) from statistical mechanics by purely analytical methods;
b) from computer simulation "experiments";
c) from experimental determinations;
d) from estimates using empirical equations of state.

One aim of this chapter is to discuss briefly the present state of the statistical treatment and of computer simulation methods for molten salts as well as their results for PVT data. Another goal is to present and discuss the PVT data for organic molten salts determined experimentally and to describe a method by which PVT data at high pressure can be estimated from those at low pressure with sufficiently high accuracy.

PVT DATA FOR MOLTEN ORGANIC SALTS

CALCULATION FROM STATISTICAL MECHANICS

The equation of state of a system of N_α particles of species α ($\alpha = 1, 2, ..., \nu$) in a volume V at temperature T may be written in the form

$$P = P(V,T,N_\alpha) \tag{1}$$

and can be calculated from the relation

$$P(V,T,N_\alpha) = -\{\partial A(V,T,N_\alpha)/\partial V\}_{T,N_\alpha} \tag{2}$$

if the Helmholtz free energy, $A(V,T,N_\alpha)$, which is a Gibbs function of the system, is known. $A(V,T,N_\alpha)$ is connected with the properties of the constituents of the system by the following equations derived from statistical mechanics:

$$A(V,T,N_\alpha) = -kT\ln Q \tag{3}$$

$$Q = (\prod_{\alpha=1}^{\nu} N_\alpha! \ \Lambda_\alpha^{3N_\alpha})^{-1} Q_{int} \ Z \tag{4}$$

$$Z = \int\int \prod_{i\alpha} \Omega_\alpha^{-1} \exp(-U/kT) dr_{i\alpha} d\omega_{i\alpha} \tag{5}$$

with $\Lambda_\alpha = (2\pi m_\alpha kT/h^2)^{-\frac{1}{2}}$ and $\Omega_\alpha = \int d\omega_{i\alpha}$.

$\vec{r}_{i\alpha}$ is the position of the i-th particle of species α and $\omega_{i\alpha}$ its orientation. U is the potential energy of the system and depends on the positions and orientations of all particles. In

Eq. (4) it has been assumed that the internal partition function Q_{int} is only dependent on the intramolecular or intraionic structure and is thus independent of volume. It is obvious that this assumption is questionable for the more complex organic salts.

Two difficulties are encountered in calculating the configuration integral Z from Eq. (5): a suitable expression is needed for the potential energy U and the integral must be evaluated using this expression for U. The progress so far achieved in this direction was reviewed for liquids in general by Barker & Henderson [1] and for molten salts in particular by Gillan [2]. Usually the total potential energy is written as a sum of effective pair potentials

$$U = \sum_{\alpha,\beta} \sum'_{i,j} \phi_{\alpha\beta} \{|\vec{r}_{i\alpha} - \vec{r}_{j\beta}|, (\omega_{i\alpha} - \omega_{j\beta})\} \tag{6}$$

where $i \neq j$ if $\alpha = \beta$. At present the statistical treatment of molten salts is still restricted to ions with spherical symmetry. Thus Eq. (6) reduces to

$$U = \sum_{\alpha,\beta} \sum'_{i,j} \phi_{\alpha\beta} (|\vec{r}_{i\alpha} - \vec{r}_{j\beta}|) \tag{7}$$

Even for the description of the simplest pure molten salts three different pair potentials of this kind are required which usually are represented as sums of coulombic, exponential repulsion and dispersion terms:

$$\phi_{\alpha\beta} = z_\alpha z_\beta r^{-1} + B_{\alpha\beta} \exp(-A_{\alpha\beta} r) - C_{\alpha\beta} r^{-6} - D_{\alpha\beta} r^{-8} \tag{8}$$

Sangster & Dixon [3] reviewed the pair potentials used for alkali halides and the methods for the determination of the coefficients in the potentials.

For the evaluation of the multi-dimensional integral in Eq. (5) perturbation methods appear to be most promising. In order to apply those methods, the properties of the actual system are related to those of a reference system. In particular, the pair potential is written as a sum of the pair potential of the reference system ϕ_0 and a perturbative term ϕ_1:

$$\phi(r) = \phi_0(r) + \phi_1(r) \tag{9}$$

It is supposed that the integral in Eq. (5) can be solved for $\phi_0(r)$ and the influence of $\phi_1(r)$ on the free energy can be determined by a perturbation calculation. For simple liquids the hard spheres system usually is chosen as reference system. For molten salts the so-called Restricted Primitive Model (RPM) has been suggested as reference system. In the RPM the ions are charged spheres all having the same diameter. For the RPM Eq. (5) has so far only been solved approximately in the so-called Mean Spherical Approximation (MSA) [4,5]. The thermodynamic properties and the equation of state of the RPM were calculated by Larsen [6,7] by computer simulation using the Monte Carlo method.

At present, the construction of a suitable reference system for molten salts has not yet been completed. No successful attempts have been made to establish the relations between this reference system and real molten salts. It is, therefore, obvious that the calculation of PVT data by purely analytical methods will not be feasible in the near future, especially not for more complex molten salts, among them the molten organic salts.

It should be mentioned that in order to circumvent the evaluation of the integral in Eq. (5) a number of models have been developed for molten salts {e.g., Quasi-lattice [8,9], Hole [10,11] Free Volume [12,13], Significant Structures [14,15]} which either give an expression for the partition function of the system or enable the direct calculation of some macroscopic properties. None of these models yielded results from which an equation of state could be constructed with sufficient accuracy. Future progress in this direction is not expected. Consequently, models for molten salts will not be discussed here.

COMPUTER SIMULATION STUDIES

The application of computer simulation studies to molten salts was extensively reviewed by Woodcock [16] and by Sangster & Dixon [3]. Most of the studies were performed on molten alkali halides. Among other properties PVT data were obtained from the calculations. A comparison with experimental results over a broad pressure range became possible when Goldmann & Tödheide [17] published PVT data for molten KCl to 1320 K and 6 kbar with an accuracy of 0.4 % in density. Woodcock & Singer [18] employed the Monte Carlo (MC) method to calculate PVT data for potassium chloride using the Huggins–Mayer potential {Eq. (8)} with coefficients determined by Fumi & Tosi [19] from the properties of solid potassium chloride at 298 K. Five of their calculated PVT points fall into the density range of the experiments. At a constant temperature of 1033 K the deviation of the calculated densities from the experimental values range from +1.3 to −2.0 percent in the pressure range from 1 kbar to 4 kbar. This corresponds to pressure deviations of −225 to +854 bar, if temperature and density are chosen as independent variables. MC computations by Larsen et al. [20] using the same type of pair potential yielded PVT data with density values which are low by about 4 %. Densities obtained by Lewis [21] from Molecular Dynamics (MD) calculations also using the same pair potential fall about 7 % below the experimental results. Other MC calculations on molten potassium chloride performed at constant temperature and constant pressure rather than constant volume using the Pauling potential [22,23] or including polarization effects [24] yielded even less satisfactory results. The calculated densities at the melting temperature at 1 bar are low by 8 to 15 %. MC and MD calculations [3,25] on other molten alkali halides using the same types of pair potential generally led to larger discrepancies with experimental data than for potassium chloride.

PVT data for a molten salt with non-spherical ions were calculated by Van Wechem [26]. The pressures calculated for potassium nitrate at various temperatures for experimental densities at 1 bar range from 1 kbar to 1.8 kbar.

It may be concluded that computer simulation studies are a powerful tool to investigate the relative contributions of different parts of the pair potential to thermodynamic, structural (radial distribution functions) and dynamic properties of molten salts rather than to evaluate accurate numerical results for PVT data and related properties. Even in favourable cases, i.e., for salts with two spherical ions not too different in size in the range of high temperatures and high pressures, PVT calculations by computer simulation methods do not attain accuracies comparable to those which can be reached in careful experimental investigations. In many cases the errors are beyond the limits tolerable for practical uses of the data. For molten organic salts the situation is even less favourable, because non-spherical ions with a complex internal structure as well as covalency effects may be involved. For these cases adequate pair potentials are not yet available which are a pre-requisite for computer simulation calculations.

EXPERIMENTAL RESULTS

As pointed out in the preceding sections, PVT data for molten organic salts can only be obtained from experimental determinations at the present time. In this section experimental results on PVT data and related properties are presented. Table 1 shows a list of organic salts for which information on PVT data in the liquid phase is available. The numbers in the last column of Table 1 refer to the numbers of those tables in which information on the respective salt is presented.

TABLE 1. List of salts

No.	Substance	Formula	Table No.
	I. Salts with organic cation		
	A. Primary ammonium salts		
	1. Chlorides		
1	Methylammonium chloride	$CH_3NH_3^+Cl^-$	2
2	Ethylammonium chloride	$C_2H_5NH_3^+Cl^-$	2
3	*iso*Propylammonium chloride	$i.C_3H_7NH_3^+Cl^-$	2
4	Butylammonium chloride	$C_4H_9NH_3^+Cl^-$	2
	2. Bromides		
5	Ethylammonium bromide	$C_2H_5NH_3^+Br^-$	2

TABLE 1 (Continued)

3. Fluoborates

6	Methylammonium tetrafluoborate	$CH_3NH_3^+BF_4^-$	2
7	Ethylammonium tetrafluoborate	$C_2H_5NH_3^+BF_4^-$	2
8	Propylammonium tetrafluoborate	$C_3H_7NH_3^+BF_4^-$	2
9	*iso*Propylammonium tetrafluoborate	$i.C_3H_7NH_3^+BF_4^-$	2
10	Butylammonium tetrafluoborate	$C_4H_9NH_3^+BF_4^-$	2
11	*iso*Butylammonium tetrafluoborate	$i.C_4H_9NH_3^+BF_4^-$	2
12	Cyclohexylammonium tetrafluoborate	$C_6H_{11}NH_3^+BF_4^-$	2

4. Others

13	Methylammonium nitrate	$CH_3NH_3^+NO_3^-$	2
14	Ethylammonium nitrate	$C_2H_5NH_3^+NO_3^-$	2

B. Secondary ammonium salts

15	Dimethylammonium chloride	$(CH_3)_2NH_2^+Cl^-$	2
16	Diethylammonium chloride	$(C_2H_5)_2NH_2^+Cl^-$	2
17	Methylphenylammonium chloride	$(CH_3)(C_6H_5)NH_2^+Cl^-$	2
18	Ethylphenylammonium chloride	$(C_2H_5)(C_6H_5)NH_2^+Cl^-$	2
19	Methylphenylammonium bromide	$(CH_3)(C_6H_5)NH_2^+Br^-$	2
20	Dimethylammonium nitrate	$(CH_3)_2NH_2^+NO_3^-$	2
21	Diethylammonium nitrate	$(C_2H_5)_2NH_2^+NO_3^-$	2
22	Dimethylammonium tetrafluoborate	$(CH_3)_2NH_2^+BF_4^-$	2
23	Diethylammonium tetrafluoborate	$(C_2H_5)_2NH_2^+BF_4^-$	2

C. Tertiary ammonium salts

24	Dimethylphenylammonium bromide	$(CH_3)_2(C_6H_5)NH^+Br^-$	2
25	Tri*iso*pentylammonium iodide	$(i.C_5H_{11})_3NH^+I^-$	2
26	Tri*iso*pentylammonium thiocyanate	$(i.C_5H_{11})_3NH^+SCN^-$	2
27	Dimethylphenylammonium hydrogensulfate	$(CH_3)_2(C_6H_5)NH^+HSO_4^-$	2
28	Trimethylammonium tetrafluoborate	$(CH_3)_3NH^+BF_4^-$	2
29	Triethylammonium tetrafluoborate	$(C_2H_5)_3NH^+BF_4^-$	2

D. Quaternary ammonium salts
1. Halides

30	Tetrabutylammonium bromide	$(C_4H_9)_4N^+Br^-$	2
31	Tetrabutylammonium iodide	$(C_4H_9)_4N^+I^-$	2
32	Tetra*iso*pentylammonium iodide	$(i.C_5H_{11})_4N^+I^-$	2
33	Tetrahexylammonium iodide	$(C_6H_{13})_4N^+I^-$	2

2. Perchlorates

34	Tetrabutylammonium perchlorate	$(C_4H_9)_4N^+ClO_4^-$	2
35	Tetrapentylammonium perchlorate	$(C_5H_{11})_4N^+ClO_4^-$	2
36	Tetra*iso*pentylammonium perchlorate	$(i.C_5H_{11})_4N^+ClO_4^-$	2
37	Hexyldipentylbutylammonium perchlorate	$(C_6H_{13})(C_5H_{11})_2(C_4H_9)N^+ClO_4^-$	2

TABLE 1 (Continued)

38	Dihexyldibutylammonium perchlorate	$(C_6H_{13})_2(C_4H_9)_2N^+ClO_4^-$	2
39	Octyltributylammonium perchlorate	$(C_8H_{17})(C_4H_9)_3N^+ClO_4^-$	2
40	Trihexylethylammonium perchlorate	$(C_6H_{13})_3(C_2H_5)N^+ClO_4^-$	2
41	Undecyltripropylammonium perchlorate	$(C_{11}H_{23})(C_3H_7)_3N^+ClO_4^-$	2
42	Dioctyldiethylammonium perchlorate	$(C_8H_{17})_2(C_2H_5)_2N^+ClO_4^-$	2
43	Tetradecyltriethylammonium perchlorate	$(C_{14}H_{29})(C_2H_5)_3N^+ClO_4^-$	2
44	Hexadecylethyldimethylammonium perchlorate	$(C_{16}H_{33})(C_2H_5)(CH_3)_2N^+ClO_4^-$	2

3. Fluoborates

45	Tetrapropylammonium tetrafluoborate	$(C_3H_7)_4N^+BF_4^-$	2
46	Tetrabutylammonium tetrafluoborate	$(C_4H_9)_4N^+BF_4^-$	2,3
47	Tetrapentylammonium tetrafluoborate	$(C_5H_{11})_4N^+BF_4^-$	4
48	Tetrahexylammonium tetrafluoborate	$(C_6H_{13})_4N^+BF_4^-$	2,5,12
49	Tetraheptylammonium tetrafluoborate	$(C_7H_{15})_4N^+BF_4^-$	6

4. Others

50	Tetrapropylammonium hexafluorophosphate	$(C_3H_7)_4N^+PF_6^-$	2
51	Tetrabutylammonium hexafluorophosphate	$(C_4H_9)_4N^+PF_6^-$	2
52	Tetrapentylammonium thiocyanate	$(C_5H_{11})_4N^+SCN^-$	2
53	Tetrabutylammonium tetrachloromanganate	$\{(C_4H_9)_4N^+\}_2MnCl_4^{2-}$	2
54	Tetrabutylammonium tetrabromomanganate	$\{(C_4H_9)_4N^+\}_2MnBr_4^{2-}$	2

E. Phosphonium and arsonium salts

55	Tributylbenzylphosphonium chloride	$(C_4H_9)_3(CH_2C_6H_5)P^+Cl^-$	2
56	Tributyl-2,4-dichlorobenzyl-phosphonium chloride	$(C_4H_9)_3(CH_2C_6H_3Cl_2)P^+Cl^-$	2
57	Tetrapropylphosphonium tetrafluoborate	$(C_3H_7)_4P^+BF_4^-$	2,7
58	Tetrapropylarsonium tetrafluoborate	$(C_3H_7)_4As^+BF_4^-$	2,8

F. Pyridinium salts

59	Pyridinium chloride	$(C_5H_5NH)^+Cl^-$	2,12
60	Pyridinium bromide	$(C_5H_5NH)^+Br^-$	2
61	N-Methylpyridinium chloride	$(C_5H_5NCH_3)^+Cl^-$	2
62	N-Methylpyridinium bromide	$(C_5H_5NCH_3)^+Br^-$	2
63	N-Methylpyridinium iodide	$(C_5H_5NCH_3)^+I^-$	2
64	4-Methylpyridinium chloride	$(CH_3C_5H_4NH)^+Cl^-$	2
65	4-Methylpyridinium bromide	$(CH_3C_5H_4NH)^+Br^-$	2
66	4-Methyl-N-methyl-pyridinium bromide	$(CH_3C_5H_4NCH_3)^+Br^-$	2

II. Salts with organic anion
A. Alkanoates
1. Alkali alkanoates

67	Lithium ethanoate	$CH_3COO^-Li^+$	2
68	Sodium ethanoate	$CH_3COO^-Na^+$	2
69	Potassium ethanoate	$CH_3COO^-K^+$	2
70	Rubidium ethanoate	$CH_3COO^-Rb^+$	2

TABLE 1 (Continued)

71	Cesium ethanoate	$CH_3COO^-Cs^+$	2
72	Sodium propanoate	$C_2H_5COO^-Na^+$	2
73	Sodium butanoate	$C_3H_7COO^-Na^+$	2
74	Sodium *iso*butanoate	$i.C_3H_7COO^-Na^+$	2
75	Sodium *iso*pentanoate	$i.C_4H_9COO^-Na^+$	2
	2. Other alkanoates		
76	Zinc hexanoate	$(C_5H_{11}COO^-)_2Zn^{2+}$	2
77	Lead hexanoate	$(C_5H_{11}COO^-)_2Pb^{2+}$	2
78	Zinc octanoate	$(C_7H_{15}COO^-)_2Zn^{2+}$	2
79	Lead octanoate	$(C_7H_{15}COO^-)_2Pb^{2+}$	2
80	Zinc decanoate	$(C_9H_{19}COO^-)_2Zn^{2+}$	2
81	Cadmium decanoate	$(C_9H_{19}COO^-)_2Cd^{2+}$	2
82	Lead decanoate	$(C_9H_{19}COO^-)_2Pb^{2+}$	2
83	Zinc dodecanoate	$(C_{11}H_{23}COO^-)_2Zn^{2+}$	2
84	Cadmium dodecanoate	$(C_{11}H_{23}COO^-)_2Cd^{2+}$	2
85	Lead dodecanoate	$(C_{11}H_{23}COO^-)_2Pb^{2+}$	2
86	Zinc tetradecanoate	$(C_{13}H_{27}COO^-)_2Zn^{2+}$	2
87	Cadmium tetradecanoate	$(C_{13}H_{27}COO^-)_2Cd^{2+}$	2
88	Lead tetradecanoate	$(C_{13}H_{27}COO^-)_2Pb^{2+}$	2
89	Zinc hexadecanoate	$(C_{15}H_{31}COO^-)_2Zn^{2+}$	2
90	Cadmium hexadecanoate	$(C_{15}H_{31}COO^-)_2Cd^{2+}$	2
91	Lead hexadecanoate	$(C_{15}H_{31}COO^-)_2Pb^{2+}$	2
92	Zinc octadecanoate	$(C_{17}H_{35}COO^-)_2Zn^{2+}$	2
93	Cadmium octadecanoate	$(C_{17}H_{35}COO^-)_2Cd^{2+}$	2
94	Lead octadecanoate	$(C_{17}H_{35}COO^-)_2Pb^{2+}$	2
	B. Benzenesulfonates		
95	Potassium benzenesulfonate	$C_6H_5SO_3^-K^+$	2
96	Rubidium benzenesulfonate	$C_6H_5SO_3^-Rb^+$	2
97	Cesium benzenesulfonate	$C_6H_5SO_3^-Cs^+$	2

III. Salts with organic cation and organic anion

A. Tetraalkyl- and tetraphenyl-borates

98	Tetrapropylammonium tetraethylborate	$(C_3H_7)_4N^+(C_2H_5)_4B^-$	2,9
99	Tetraethylammonium tetrapropylborate	$(C_2H_5)_4N^+(C_3H_7)_4B^-$	2,10
100	Tetrabutylammonium tetrabutylborate	$(C_4H_9)_4N^+(C_4H_9)_4B^-$	2,11
101	Tetrapropylammonium tetraphenylborate	$(C_3H_7)_4N^+(C_6H_5)_4B^-$	2
102	Tetrabutylammonium tetraphenylborate	$(C_4H_9)_4N^+(C_6H_5)_4B^-$	2
103	Propylbutylpentyloctylammonium tetraphenyl=borate	$(C_3H_7)(C_4H_9)(C_5H_{11})(C_8H_{17})N^+$ $(C_6H_5)_4B^-$	2
104	Dimethyldinonylammonium tetraphenylborate	$(CH_3)_2(C_9H_{19})_2N^+(C_6H_5)_4B^-$	2

B. Ammonium picrates

1. Primary ammonium picrates

105	Ethylammonium picrate	$C_2H_5NH_3^+C_6H_2(NO_2)_3O^-$	2

TABLE 1 (Continued)

106	Propylammonium picrate	$C_3H_7NH_3^+C_6H_2(NO_2)_3O^-$	2
107	Butylammonium picrate	$C_4H_9NH_3^+C_6H_2(NO_2)_3O^-$	2
108	*iso*Butylammonium picrate	$i.C_4H_9NH_3^+C_6H_2(NO_2)_3O^-$	2
109	Pentylammonium picrate	$C_5H_{11}NH_3^+C_6H_2(NO_2)_3O^-$	2
110	*iso*Pentylammonium picrate	$i.C_5H_{11}NH_3^+C_6H_2(NO_2)_3O^-$	2
111	Heptylammonium picrate	$C_7H_{15}NH_3^+C_6H_2(NO_2)_3O^-$	2
112	Hexadecylammonium picrate	$C_{16}H_{33}NH_3^+C_6H_2(NO_2)_3O^-$	2

2. Secondary ammonium picrates

113	Dimethylammonium picrate	$(CH_3)_2NH_2^+C_6H_2(NO_2)_3O^-$	2
114	Methylethylammonium picrate	$(CH_3)(C_2H_5)NH_2^+C_6H_2(NO_2)_3O^-$	2
115	Diethylammonium picrate	$(C_2H_5)_2NH_2^+C_6H_2(NO_2)_3O^-$	2
116	Dipropylammonium picrate	$(C_3H_7)_2NH_2^+C_6H_2(NO_2)_3O^-$	2
117	Dibutylammonium picrate	$(C_4H_9)_2NH_2^+C_6H_2(NO_2)_3O^-$	2
118	Di*iso*pentylammonium picrate	$(i.C_5H_{11})_2NH_2^+C_6H_2(NO_2)_3O^-$	2
119	Dihexadecylammonium picrate	$(C_{16}H_{33})_2NH_2^+C_6H_2(NO_2)_3O^-$	2

3. Tertiary ammonium picrates

120	Triethylammonium picrate	$(C_2H_5)_3NH^+C_6H_2(NO_2)_3O^-$	2
121	Tripropylammonium picrate	$(C_3H_7)_3NH^+C_6H_2(NO_2)_3O^-$	2
122	Tributylammonium picrate	$(C_4H_9)_3NH^+C_6H_2(NO_2)_3O^-$	2
123	Tri*iso*pentylammonium picrate	$(i.C_5H_{11})_3NH^+C_6H_2(NO_2)_3O^-$	2

4. Quaternary ammonium picrates

124	Dimethyldipropylammonium picrate	$(CH_3)_2(C_3H_7)_2N^+C_6H_2(NO_2)_3O^-$	2
125	Triethylpropylammonium picrate	$(C_2H_5)_3(C_3H_7)N^+C_6H_2(NO_2)_3O^-$	2
126	Methyltripropylammonium picrate	$(CH_3)(C_3H_7)_3N^+C_6H_2(NO_2)_3O^-$	2
127	Diethyldipropylammonium picrate	$(C_2H_5)_2(C_3H_7)_2N^+C_6H_2(NO_2)_3O^-$	2
128	Ethyltripropylammonium picrate	$(C_2H_5)(C_3H_7)_3N^+C_6H_2(NO_2)_3O^-$	2
129	Tetrapropylammonium picrate	$(C_3H_7)_4N^+C_6H_2(NO_2)_3O^-$	2
130	Tetrabutylammonium picrate	$(C_4H_9)_4N^+C_6H_2(NO_2)_3O^-$	2
131	Tetrapentylammonium picrate	$(C_5H_{11})_4N^+C_6H_2(NO_2)_3O^-$	2
132	Tetra*iso*pentylammonium picrate	$(i.C_5H_{11})_4N^+C_6H_2(NO_2)_3O^-$	2

Molar volume and density at normal pressure

Most of the PVT measurements on molten organic salts have been performed at normal (standard) pressure of 1 atm (1 atm = 1.01325 bar = 101325 Pa) in order to determine the temperature dependences of molar volume, V_m, or density, ρ. Standard experimental methods for density measurements, e.g., buoyancy, pycnometric and dilatometric techniques, can be used for molten organic salts, since the upper temperature limit of the experiments exceeds 700 K only in exceptional cases. This is due to the relatively low melting temperature of organic salts and to their tendency to undergo thermal decomposition at moderate temperatures.

The accuracy claimed in the original papers for the published density values varies from about 0.1 to 1.0 percent. Where measurements for the same salt have been carried out in different laboratories using different techniques, the disagreement can be as high as 2 to 3 percent, but is in most cases of the order of about 1 percent. Part of the differences may be due to chemical impurities in the salts, since the melting temperatures also differ in these cases. It is therefore assumed that the values calculated with the data given in Table 2 are accurate to about 1 percent. If a higher accuracy is needed, it is recommended to go back to the original literature for a careful examination of the error limits in a particular study.

It was found that the molar volumes as well as the densities at normal pressure can be represented as linear functions of temperature. This, of course, is due to the small relative variation of both quantities (about $5 \cdot 10^{-4}$/K) with temperature in combination with the narrow temperature intervals over which measurements could be taken. Consequently the data found in the literature were fitted to linear functions of temperature

$$V_m = a + b\,T \tag{10}$$

and

$$\rho = c - d\,T \tag{11}$$

The coefficients a, b, c, d were determined by linear regression and are presented in Table 2. The deviation of the experimental points from the straight lines is generally much less than 1 percent, the standard deviation being of the order of 0.1 percent. In the second column of Table 2 the temperature range is given for which the representation of the data by Eq.s (10) and (11) is valid. However, a slight extrapolation down to the melting temperature should not result in substantially larger errors than those inside the interval.

In column 7 of Table 2 the thermal expansion coefficient defined as

$$\alpha = (1/V)(\partial V/\partial T)_P \tag{12}$$

is given. The listed values for α were calculated from the relation

$$\alpha = -(1/\rho)(\partial\rho/\partial T)_P = d/(c - d\,T) \tag{13}$$

at a temperature given in column 8 which is $T_0 = 1.05\,T_F$ (T_F: melting temperature) unless the temperature value is marked by a superscript (+). The accuracy of the expansion coefficients quoted is estimated to be about ±5 percent. Column 9 gives the references from which the data were taken.

TABLE 2. Coefficients for the calculation of molar volumes, V_m, and densities, ρ, according to Eq.s (10) and (11), respectively. Thermal expansion coefficient, α. (The units for the coefficients are: $a/10^{-6}m^3mol^{-1}$; $b/10^{-9}m^3mol^{-1}K^{-1}$; $c/10^3kg\ m^{-3}$; $d/10^{-1}kg\ m^{-3}K^{-1}$; $\alpha/10^{-4}K^{-1}$).

No.	$\dfrac{T\ \text{range}}{K}$	a	b	c	d	α	$\dfrac{T_0}{K}$	References
1	498–533	47.17	37.00	1.313	5.700	5.56	523	27
2	383–493	66.69	36.69	1.176	4.343	4.51	401	27
3	427–493	84.07	40.00	1.105	3.742	3.99	447	27
4	490–508	103.1	34.00	1.052	2.800	2.75	,500[+]	27
5	445–490	41.66	100.5	2.176	16.12	11.2	455	28
6	468–498	65.02	47.71	1.703	7.307	5.44	491	29
7	423–503	77.19	66.15	1.584	7.575	6.07	444	29
8	443–493	89.06	82.83	1.498	7.454	6.45	459	29
9	393–493	95.91	70.10	1.441	6.394	5.43	441	29
10	473–513	102.9	95.70	1.410	6.840	6.37	491	29
11	433–493	104.0	92.10	1.417	6.900	6.20	442	29

TABLE 2 (Continued)

12	473–503	120.0	95.80	1.436	6.440	5.70	475	29
13	373.8	74.60 (V_m)		1.261 (ρ)				30
14	290–343	76.05	44.46	1.385	5.853	4.97	300	30, 31
15	440–503	65.96	41.50	1.171	4.618	4.82	462	27
16	497–523	91.01	60.00	1.120	4.280	4.69	522	27
17	400–440	107.1	68.68	1.276	5.229	4.94	416	31
18	455–476	120.7	78.05	1.233	4.927	4.93	475	31
19	377.9	137.0 (V_m)		1.358 (ρ)				30
20	348–391	76.64	45.84	1.364	5.662	4.92	365	30
21	372–390	86.95	111.6	1.403	9.210	8.55	390	30
22	373–493	79.75	63.70	1.557	7.346	5.74	378	29
23	453–493	108.0	87.50	1.372	6.230	5.76	467	29
24	355–415	103.5	133.1	1.742	11.32	8.59	375	30, 31
25	373–423	236.6	206.3	1.399	7.143	6.50	392	32
26	353–403	230.0	261.1	1.132	6.891	8.10	353	30
27	384–410	137.6	90.20	1.526	6.578	5.16	380	31
28	493–523	93.03	76.70	1.441	6.460	5.82	513	29
29	393–493	133.7	103.6	1.322	6.070	5.65	408	29
30	393–419	231.9	224.7	1.280	6.950	6.98	412	33, 34
31	420–442	228.8	259.5	1.446	8.388	7.58	438	33, 34
32	423–443	267.1	335.0	1.367	7.750	8.08	441	30, 32
33	395–423	338.7	356.7	1.294	7.350	7.43	396	34
34	483–493	236.5	260.0	1.228	5.900	7.14	490[+]	32
35	393–433	311.4	261.6	1.192	5.887	6.12	411	35
36	393–433	310.3	237.0	1.216	5.870	5.81	411	32
37	358–393	337.3	201.2	1.140	4.692	4.87	375	35
38	358–393	325.8	236.5	1.166	5.481	5.70	374	35
39	343–383	327.7	235.9	1.164	5.540	5.73	357	35
40	333–363	324.3	246.2	1.174	5.830	5.96	336	35
41	343–373	321.2	256.8	1.183	6.150	6.38	356	35
42	343–373	324.2	248.1	1.170	5.776	5.98	352	35
43	428–443	291.3	320.9	1.221	6.840	7.47	446	35
44	428–443	306.2	296.3	1.185	6.226	6.88	450	35
45	525–547	183.8	221.6	1.257	6.611	7.02	547	33
46	436–539	249.5	233.1	1.191	5.812	6.55	457	33
48	375–551	353.2	344.6	1.130	5.772	6.35	382	33

TABLE 2 (Continued)

50	513–545	259.3	93.57	1.243	3.224	3.01	536	33
51	529–548	252.2	271.8	1.325	6.557	6.78	546	33
52	324–383	316.1	242.7	1.078	5.385	6.09	339	36,37,38
53	402–463	536.8	379.2	1.201	5.268	5.44	421	39
54	395–430	548.7	391.6	1.508	7.243	5.57	393	40
55	451–513	255.9	221.7	1.171	5.480	6.20	458	34
56	428–494	266.1	256.8	1.353	6.900	6.91	410	34
57	517–547	209.9	189.3	1.238	5.716	6.07	539	41
58	488–530	224.1	171.8	1.394	6.349	5.52	508	41
59	417–471	78.66	51.72	1.395	5.798	5.10	438	42,43,44
60	497–537	80.16	59.00	1.810	7.055	5.33	518	42
61	423–433	93.87	47.00	1.349	4.964	4.12	430	42
62	425–468	97.04	54.90	1.723	6.528	4.52	446	42
63	390–460	110.3	55.60	1.943	6.880	4.14	410	45
64	440–473	91.51	63.32	1.339	5.755	5.24	462	42
65	435–518	96.50	64.24	1.703	7.004	5.10	457	42
66	429–498	112.4	59.10	1.610	5.701	4.25	450	42
67	553–603	36.16	38.49	1.558	7.414	6.58	581	46,47
68	603–673	45.56	31.50	1.643	6.170	4.93	633	46,47,48,49
69	575–663	52.67	31.40	1.733	6.000	4.38	604	46,47,49
70	514–653	54.34	34.92	2.460	8.983	4.55	540	46,47,48
71	470–633	64.51	31.31	2.843	8.975	3.74	494	46,47,48
72	561–623	59.45	44.71	1.463	5.823	5.20	589	50
73	597–623	66.37	65.15	1.427	6.388	6.16	610	50
74	527–623	67.78	69.21	1.403	6.589	6.34	553	50
75	553–623	85.50	70.76	1.296	5.432	5.54	581	50
76	438–472	176.6	217.7	1.459	8.460	7.77	437	51
77	375–460	183.8	197.0	2.152	12.17	7.12	364	51
78	419–467	225.5	267.6	1.370	7.80	7.54	431	51
79	400–450	222.0	271.0	1.975	12.00	7.88	376	51
80	427–467	275.0	329.5	1.291	7.27	7.39	423	51
81	423–466	247.9	343.0	1.571	9.78	8.44	421	52
82	400–450	272.2	331.0	1.788	10.73	7.87	396	51
83	414–467	327.8	377.6	1.250	7.06	7.42	423	51
84	414–484	314.2	366.0	1.435	8.17	7.44	413	52
85	390–455	283.3	456.0	1.787	12.16	9.34	397	51
86	410–434	386.7	407.4	1.211	6.64	7.13	421	51
87	408–472	365.3	424.0	1.376	7.91	7.49	404	52
88	390–450	359.0	457.0	1.619	9.93	8.12	399	51
89	408–476	405.3	518.8	1.233	7.35	7.96	421	51

TABLE 2 (Continued)

90	401–476	414.3	430.0	1.338	7.60	7.34	398	52
91	390–450	407.1	517.0	1.550	9.51	8.16	405	51
92	432–462	456.2	564.8	1.206	7.02	7.70	419	51
93	401–478	427.0	620.0	1.361	8.87	8.94	416	52
94	400–450	444.3	589.0	1.515	9.42	8.33	408	51
95	648–703	67.39	104.6	2.146	10.71	7.55	680	46
96	598–753	112.0	56.40	2.048	6.393	3.88	628	46
97	505–773	114.0	64.13	2.371	7.780	3.97	530	46
98	393–423	301.5	238.0	0.9767	4.666	5.91	401	53
99	363–403	306.9	230.1	0.9704	4.620	4.76	371	53
100	393–433	455.2	398.9	0.9835	4.999	6.39	402	53,54
101	482–512	380.5	332.1	1.207	5.638	6.11	503	33
102	513–543	448.1	356.1	1.144	4.945	5.63	536	33
103	388–418	531.1	392.1	1.101	5.096	5.68	401	35
104	383–413	548.7	362.4	1.077	4.660	5.16	374	35
105	443–453	159.1	100.0	1.645	6.700	4.99	453[+]	55
106	418–433	169.9	105.8	1.621	6.574	4.92	433[+]	55
107	413–423	181.3	120.0	1.585	6.675	5.12	423[+]	55
108	413–423[++]	179.8	120.0	1.610	7.080	5.41	424[+]	55
109	398–423[++]	197.1	129.1	1.535	6.594	5.25	423[+]	55
110	403–423[++]	179.2	167.5	1.627	8.560	6.77	423[+]	55
111	388–423[++]	221.5	154.8	1.477	6.555	5.45	417	55
112	388–433	329.5	280.7	1.331	6.648	6.27	407	56
113	434–443	156.8	100.0	1.667	6.860	5.03	443[+]	55
114	370–443	170.1	106.0	1.627	6.773	4.97	389	55
115	348–463	182.1	123.1	1.584	6.913	5.19	365	56
116	373–463	208.2	151.8	1.500	6.798	5.51	391	55
117	373–463	217.4	242.6	1.480	8.505	7.40	391	56
118	368–463	251.0	228.4	1.424	7.372	6.47	386	55
119	333–443	566.0	478.1	1.152	5.867	6.17	344	56
120	448–468	207.0	147.4	1.500	6.486	5.43	470	55
121	388–438	240.6	204.2	1.444	7.220	6.28	407	55
122	383–443	281.2	231.1	1.381	6.796	6.12	398	56
123	398–458	312.7	300.4	1.337	7.063	6.78	418	55
124	373–473	231.6	171.0	1.460	6.629	5.50	384	55
125	418–483	238.0	186.4	1.460	6.721	5.77	438	55
126	363–473	258.1	187.7	1.416	6.408	5.44	373	55
127	353–473	253.7	180.8	1.445	6.492	5.39	371	55

TABLE 2 (Continued)

128	383–473	266.3	204.9	1.412	6.559	5.70	399	55
129	393–493	281.1	213.1	1.382	6.270	5.58	413	31,55,57,59
130	366–493	334.7	254.8	1.322	6.098	5.61	384	56,57,58,59
131[+++]	353–450	406.6	297.1	1.230	5.673	5.54	364	59
132	361–493	391.4	304.5	1.264	5.931	5.71	379	55,59

[+]: temperature lower than 1.05 T_F (T_F: melting temperature)

[++]: supercooled

[+++]: for No.s 129, 130 and 132 the values of Ref. [59] differ from other measurements; therefore, most probably, the molar volume is about 2 % high and the density about 2 % low

PVT data at high pressures

Information on high-pressure PVT data for molten organic salts is scarce. Measurements have been performed on nine salts in two laboratories up to a maximum pressure of 5200 bar. In both cases a piezometer technique was applied using mercury as separating fluid. The position of the mercury level inside the piezometer, which determines the volume of the salt at a given temperature and pressure was monitored either by a conductance method (Barton & Speedy [60]) or by a stainless steel float the position of which was measured by means of a linear variable transformer (Grindley & Lind [53]). The accuracy achieved in these careful studies is of the order of 0.1 percent in density.

Also the high-pressure isobars of molar volume and density are straight lines within the limits of error. The coefficients of Eq.s (10) and (11) were calculated from the experimental data points at constant pressures by linear regression and are presented in Tables 3–11 together with thermal expansion coefficients calculated according to Eq. (13).

An interpretation of the high-pressure PVT data of molten tetrabutylammonium tetrabutylborate, tetraethylammonium tetrapropylborate and tetrapropylammonium tetraethylborate in relation to the PVT data of non-ionic fluids containing molecules isoelectronic to the ions in the salts was given by Lind ([63]). Pressures of the order of 1 to 2 kbar are required to bring the volume of the non-electrolyte to that of the corresponding salt at the same temperature. The compression factor $Z = PV_m/RT$ is about eight units less in the salt system than in the non-electrolyte at all temperatures and volumes. This difference is interpreted in terms of the thermodynamic equation of state

$$P = -(\partial A/\partial V)_T = T(\partial S/\partial V)_T - (\partial U/\partial V)_T \tag{14}$$

The examination of this equation showed that the major contribution to the lower pressure of the salts arises from the second term in Eq. (14) which increases considerably because of the coulombic interactions. $(\partial S/\partial V)_T$ is only slightly lower for the salts than for non-electrolytes.

The results obtained were compared with those derived from the Restricted Primitive Model in the Mean Spherical Approximation ([64]; see also the section on Calculation from Statistical Mechanics). It turned out that the theory predicts the correct sign of all the changes in the thermodynamic properties upon charging the system from a non-electrolyte to a molten salt, but it underestimates the effects. The theory predicts a decrease of one or two in the compression factor Z upon charging, compared to eight in the experiments, and the calculated change in $(\partial U/\partial V)_T$ amounts to only 20 to 35 percent of that observed experimentally. These underestimates arise from effects peculiar to the real system, e.g., deformability of the ions and ion-induced dipole effects. The comparison shows that even at high densities, where the individual differences in the attractive parts of the pair potentials become less important, the RPM does not yield quantitatively the thermodynamic properties of these molten organic salts.

In Table 12 a few experimental values are given for the isothermal compressibility, $\kappa_T = -(1/V)(\partial V/\partial P)_T$ and for the thermal pressure coefficient $\beta_V = (\partial P/\partial T)_V$. They show considerable scatter. The error limits of these values are estimated to be about ±10 percent.

TABLE 3. No. 46: Tetrabutylammonium tetrafluoborate, $(C_4H_9)_4N^+BF_4^-$ (Ref.[60])
Values for α given at 457 K; units for a, b, c, d given in Table 2

P bar	T range K	a	b	c	d	$\dfrac{\alpha}{10^{-4}K^{-1}}$
10	416–480	254.7	221.4	1.189	5.772	6.24
250	416–480	263.3	187.5	1.176	5.080	5.38
500	420–480	266.4	169.3	1.174	4.727	4.93
750	440–480	269.8	152.5	1.170	4.366	4.50
1000	440–473	270.5	142.7	1.172	4.178	4.26

TABLE 4. No. 47: Tetrapentylammonium tetrafluoborate, $(C_5H_{11})_4N^+BF_4^-$ (Ref.[60])
Values for α given at 413 K; units for a, b, c, d given in Table 2

P bar	T range K	a	b	c	d	$\dfrac{\alpha}{10^{-4}K^{-1}}$
10	393–486	309.8	268.1	1.145	5.543	6.05
250	397–486	321.5	222.2	1.130	4.794	5.14
500	402–494	325.4	198.3	1.129	4.425	4.68
1000	413–494	328.4	168.3	1.132	3.955	4.08
1500	433–494	329.7	147.5	1.136	3.611	3.66
2000	443–494	330.3	131.6	1.139	3.332	3.33

TABLE 5. No. 48: Tetrahexylammonium tetrafluoborate, $(C_6H_{13})_4N^+BF_4^-$ (Ref.[60])
Values for α given at 382 K; units for a, b, c, d given in Table 2

P bar	T range K	a	b	c	d	$\dfrac{\alpha}{10^{-4}K^{-1}}$
10	368–473	355.9	336.2	1.129	5.733	6.30
500	394–473	371.9	257.0	1.119	4.749	5.06
1000	394–473	375.8	219.2	1.123	4.280	4.46
1500	403–473	377.1	194.3	1.129	3.959	4.05
2000	413–473	378.5	173.4	1.132	3.658	3.69

TABLE 6. No. 49: Tetraheptylammonium tetrafluoborate, $(C_7H_{15})_4N^+BF_4^-$ (Ref.[60])
Values for α given at 420 K; units for a, b, c, d given in Table 2

P bar	T range K	a	b	c	d	$\dfrac{\alpha}{10^{-4}K^{-1}}$
10	400–503	409.8	377.6	1.109	5.555	6.35
500	407–503	432.2	278.0	1.093	4.432	4.89
1000	417–503	437.6	232.9	1.094	3.923	4.22

TABLE 6 (Continued)

| 1500 | 426–503 | 439.3 | 204.7 | 1.098 | 3.593 | 3.79 |
| 2000 | 434–503 | 440.3 | 182.8 | 1.102 | 3.319 | 3.45 |

TABLE 7. No. 57: Tetrapropylphosphonium tetrafluoborate, $(C_3H_7)_4P^+BF_4^-$ (Ref.[41])
Units for a, b, c, d given in Table 2

P bar	T range K	a	b	c	d
10		229.0	152.1	1.184	4.665
250		190.8	213.0	1.310	6.694
500		201.6	182.6	1.290	5.984
1000	521–531	225.4	121.7	1.226	4.260
1500		219.0	121.7	1.260	4.462
2000		213.9	121.7	1.284	4.564

TABLE 8. No. 58: Tetrapropylarsonium tetrafluoborate, $(C_3H_7)_4As^+BF_4^-$ (Ref.[41])
Units for a, b, c, d given in Table 2

P bar	T range K	a	b	c	d
10		217.8	183.9	1.401	6.437
250		217.1	172.4	1.412	6.207
500		211.6	172.4	1.444	6.434
1000	503–513	197.9	183.9	1.515	7.241
1500		185.9	195.4	1.580	8.046
2000		192.4	172.4	1.567	7.356
2500		205.8	137.9	1.520	6.092

TABLE 9. No. 98: Tetrapropylammonium tetraethylborate, $(C_3H_7)_4N^+(C_2H_5)_4B^-$ (Ref.[53])
Values for α given at 401 K; units for a, b, c, d given in Table 2

P bar	T range K	a	b	c	d	$\dfrac{\alpha}{10^{-4}K^{-1}}$
200		305.8	215	0.9718	4.30	5.38
400		308.4	198	0.9705	4.05	5.01
600		309.8	185	0.9717	3.88	4.75
800	393–433	310.8	174	0.9734	3.74	4.54
1000		311.8	164	0.9745	3.60	4.34
1200		310.8	160	0.9748	3.45	4.12
1400		309.7	155	0.9767	3.35	3.98

TABLE 10. No. 99 : Tetraethylammonium tetrapropylborate, $(C_2H_5)_4N^+(C_3H_7)_4B^-$ (Ref.[53])
Values for α given at 371 K; units for a, b, c, d given in Table 2

P bar	T range K	a	b	c	d	$\dfrac{\alpha}{10^{-4}K^{-1}}$
200		308.7	213	0.9696	4.35	5.38
400		310.8	197	0.9697	4.13	5.06
600		311.7	185	0.9706	3.95	4.79
800		312.2	175	0.9716	3.79	4.56
1000		312.3	167	0.9728	3.65	4.36
1200		313.3	157	0.9752	3.55	4.21
1400		313.4	150	0.9768	3.44	4.05
1600		313.3	144	0.9793	3.36	3.93
1800		312.2	141	0.9810	3.27	3.80
2000		312.4	135	0.9841	3.22	3.72
2200	363–403	311.5	132	0.9858	3.14	3.61
2400		311.6	127	0.9885	3.09	3.54
2600		310.8	124	0.9898	3.01	3.43
2800		310.7	120	0.9932	2.99	3.39
3000		309.2	120	0.9958	2.95	3.33
3200		309.8	114	0.9967	2.87	3.22
3400		308.6	113	0.9990	2.83	3.17
3600		308.3	110	1.0016	2.80	3.12
3800		306.9	110	1.0033	2.75	3.05
4000		307.5	105	1.0068	2.75	3.04
4200		308.1	100	1.0083	2.70	2.97

TABLE 11. No. 100: Tetrabutylammonium tetrabutylborate, $(C_4H_9)_4N^+(C_4H_9)_4B^-$ (Ref.[53])
Values for α given at 402 K; units for a, b, c, d given in Table 2

P bar	T range K	a	b	c	d	$\dfrac{\alpha}{10^{-4}K^{-1}}$
200		463.2	355	0.9799	4.60	5.79
400		468.3	322	0.9787	4.30	5.34
600		471.0	298	0.9790	4.07	4.99
800		473.3	277	0.9791	3.86	4.68
1000		473.9	262	0.9812	3.72	4.47
1200		473.7	250	0.9832	3.59	4.28
1400	393–433	475.3	235	0.9850	3.47	4.10
1600		475.1	225	0.9868	3.36	3.94
1800		475.3	215	0.9895	3.28	3.82
2000		474.0	209	0.9918	3.20	3.71
2200		474.3	200	0.9939	3.12	3.59
2400		473.4	194	0.9961	3.05	3.49

TABLE 11 (Continued)

2600		472.0	190	0.9989	3.00	3.42
2800		471.1	185	1.0016	2.95	3.34
3000		471.7	177	1.0036	2.89	3.26
3200		471.1	172	1.0055	2.83	3.17
3400		468.9	171	1.0076	2.78	3.10
3600		469.4	164	1.0104	2.75	3.06
3800		468.7	160	1.0123	2.70	2.99
4000	393–433	467.2	158	1.0153	2.68	2.95
4200		465.9	156	1.0182	2.66	2.92
4400		465.0	153	1.0201	2.62	2.86
4600		464.2	150	1.0224	2.59	2.82
4800		462.7	149	1.0258	2.59	2.81
5000		462.5	145	1.0275	2.55	2.76
5200		462.8	140	1.0286	2.50	2.69

TABLE 12. Compressibility, κ_T, and thermal pressure coefficient, β_V

No.	Formula	$\dfrac{T}{K}$	$\dfrac{\kappa_T}{10^{-6} \text{bar}^{-1}}$	$\dfrac{\beta_V}{\text{bar K}^{-1}}$	References
48	$(C_6H_{13})N^+BF_4^-$	375	69.9	9.05	61
48		399	85.0	7.55	61
48		403	78.0	8.2	61
48		435	89.8	7.3	61
59	$(C_5H_5NH)^+Cl^-$	428	34.7		62

Estimates of high-pressure PVT data

Since experimental information on PVT data of molten organic salts at high pressures is very limited, it might be desirable to have an empirical equation of state from which estimated high-pressure data can be obtained from those at normal pressure. The empirical equation derived by Tait [85] has proven to describe excellently the variation of state along isotherms using only two adjustable parameters for a large number of liquids, although it yields an unphysical result at the limit of infinite pressure. The equation in its integrated form reads

$$\{\rho(P) - \rho(P_0)\}/\rho(P) = \{V(P_0) - V(P)\}/V(P_0) = A \ln\{(B + P)/(B + P_0)\} \qquad (15)$$

where A and B are constants and the subscript "₀" denotes a reference point, usually chosen at $P_0 = 1$ bar. A varies only slightly from one substance to another, e.g., from 0.09 for liquid hydrocarbons to 0.15 for water. The B parameter is usually of the order 10^2 to 10^3 bar. Thus P_0 in Eq. (15) may be neglected in most cases. It has been shown [65] that Eq. (15) is also applicable to molten salts and that it can be made an equation of state by permitting the parameters A and B to become simple temperature functions. Kuss & Taslimi [66] found that the PVT data of a large number of organic liquids could be well described by the Tait equation, when A was held constant and the temperature dependence of B was calculated from the compressibility at normal pressure. Derivation of Eq. (15) with respect to pressure yields

$$-\{\partial V(P)/\partial P\}_T = A\, V(P_o)/(B + P) \tag{16}$$

For the compressibility at $P_o = 1$ bar one obtains

$$\kappa_T(P_o,T) = A/(B + P_o) \tag{17}$$

where P_o may be neglected with respect to B. For molten salts the value A = 0.1 has been introduced which seems to be the best choice [65]. This results in the simple relation

$$B/\text{bar} = 0.1\, \kappa_T^{-1}(P_o,T)/\text{bar} \tag{18}$$

A combination of Eq. (18) with Eq. (15) leads to

$$\{\rho(P,T) - \rho(P_o,T)\}/\rho(P,T) = \{V(P_o,T) - V(P,T)\}/V(P_o,T) =$$

$$= 0.1\, \ln\left[\, \{0.1\, \kappa_T^{-1}(P_o,T) + P\}/0.1\, \kappa_T^{-1}(P_o,T)\,\right] \tag{19}$$

Densities calculated from Eq. (19) using densities and compressibilities at 1 bar agree to better than one percent with experimental values for potassium chloride [17] to 6 kbar, for potassium and sodium nitrates [67,68] to 5 kbar, and also for tetrahexylammonium tetrafluoborate [60] up to 2 kbar. It thus appears that Eq. (19) enables an estimate of high-pressure PVT data for molten organic salts with tolerable error limits.

If no experimental values for the compressibility are available at normal pressure, they may be calculated from other properties according to one of the following relations [61,65,69]

$$\kappa_T = \alpha/\beta_V \tag{20}$$

$$\kappa_T = V_m/(u^2\, M) + \alpha(\partial T/\partial P)_S \tag{21}$$

$$\kappa_T = \alpha^2 T V_m/C_P + V/(u^2\, M) \tag{22}$$

$$\kappa_T = \alpha^2 T V_m/(C_P - C_V) \tag{23}$$

where u is the velocity of sound and all other symbols have their usual meaning.

For some simple inorganic molten salts it was possible [70] to calculate the compressibility from the thermal expansion coefficient or from the surface tension of the salt using the relations derived from Scaled Particle Theory [71]. For molten organic salts the results were unsatisfactory (see also Ref. [63]). This supports the observation that the hard spheres system does not lend itself as a model system for organic molten salts.

PVT DATA FOR SOLID ORGANIC SALTS

No effort has been made to collect molar volume or density data for crystalline organic salts at 298 K and normal pressure which may be found in standard data compilations [72,73] or may be estimated from a summation of empirical volume increments of the constituent atomic and ionic species (e.g., [74]). Most of the organic salts undergo several phase transitions at normal pressure between room temperature and their melting temperature (see Chapter 1.2). A high percentage of the phase transitions are rather sluggish and show hysteresis effects. Consequently, the properties of the salts are dependent on their thermal history. This was shown for quaternary ammonium salts [38,75] as well as for alkali alkanoates [50]. In addition, most of the experimental results on the temperature dependence of densities or molar volumes for solid organic salts are published in graphical rather than numerical or analytical form and in some publications relative rather than absolute values of density or molar volume are given. A complete graphical reproduction of the published material would certainly go beyond the scope of this chapter. Instead a reference list is given in Table 13.

TABLE 13. References for the temperature dependence of density and molar volume of solid or=
          ganic salts

Formula	References	Formula	References
Primary ammonium salts		$C_{11}H_{23}COONa$	78,80
		$C_{12}H_{25}COONa$	78
$(C_{10}H_{21}NH_3)_2MnCl_4$	76	$C_{13}H_{27}COONa$	78,80
$(C_{12}H_{25}NH_3)_2MnCl_4$	76	$C_{14}H_{29}COONa$	78
$(C_{14}H_{29}NH_3)_2MnCl_4$	76	$C_{15}H_{31}COONa$	78,80
$(C_{16}H_{33}NH_3)_2MnCl_4$	76	$C_{16}H_{33}COONa$	78
		$C_{17}H_{35}COONa$	78,79,80
Quaternary ammonium salts		$C_{18}H_{37}COONa$	78
		$C_{19}H_{39}COONa$	80
$(C_4H_9)_4NBr$	75	$C_{21}H_{43}COONa$	80
$(C_5H_{11})_4NBr$	75	$C_{25}H_{51}COONa$	80
$(C_6H_{13})_4NBr$	75		
$(C_4H_9)_4NI$	75	Potassium alkanoates	
$(C_5H_{11})_4NI$	75		
$(i.C_5H_{11})_4NI$	75	CH_3COOK	49,81,83
$(C_6H_{13})_4NI$	75	$C_7H_{15}COOK$	82
$(C_7H_{15})_4NI$	75	$C_9H_{19}COOK$	82
$(C_5H_{11})_4NSCN$	38,75	$C_{11}H_{23}COOK$	82
$(C_5H_{11})_4NNO_3$	75	$C_{13}H_{27}COOK$	82
$(C_6H_{13})_4NClO_4$	75	$C_{15}H_{31}COOK$	82,83
$(C_6H_{13})_4NBF_4$	75	$C_{17}H_{35}COOK$	79,80,82,83
Lithium alkanoates		Rubidium alkanoates	
$C_9H_{19}COOLi$	77	$C_{17}H_{35}COORb$	79
$C_{11}H_{23}COOLi$	77		
$C_{13}H_{27}COOLi$	77,78	Cesium alkanoates	
$C_{14}H_{29}COOLi$	78		
$C_{15}H_{31}COOLi$	77,78	$C_{11}H_{23}COOCs$	84
$C_{16}H_{33}COOLi$	78	$C_{13}H_{27}COOCs$	84
$C_{17}H_{35}COOLi$	77,78,79	$C_{15}H_{31}COOCs$	84
		$C_{17}H_{35}COOCs$	79,84
Sodium alkanoates			
		Silver alkanoates	
CH_3COONa	49		
C_2H_5COONa	50	CH_3COOAg	83
C_3H_7COONa	50	C_2H_5COOAg	83
$C_5H_{11}COONa$	80	C_3H_7COOAg	83
$C_7H_{15}COONa$	80	C_4H_9COOAg	83
$C_9H_{19}COONa$	80	$C_8H_{17}COOAg$	83
$C_{10}H_{21}COONa$	78	$C_9H_{19}COOAg$	83

TABLE 13 (Continued)

$C_{11}H_{23}COOAg$	83	Sodium 2-methyltetradecanoate	78
$C_{15}H_{31}COOAg$	83	Sodium 2-methylhexadecanoate	78
$C_{17}H_{35}COOAg$	83	Sodium 10-methyloctadecanoate	79
		Sodium 12-hydroxyoctadecanoate	79
Branched and substituted sodium alkanoates		Sodium 9-keto-10-methyloctadecanoate	79
		Sodium 9-phenyloctadecanoate	79
Sodium isoButanoate	50		
Sodium isoPentanoate	50	Sodium oleate	80

P.W. Bridgman measured linear compressions along the different crystal axes as well as volume compressions for a number of crystalline organic salts at or near room temperature up to 12 kbar [86], 30 kbar [88], and 40 kbar [87] (volume effect only) using different techniques. His results are presented in Tables 14—24 for those substances which, according to Bridgman, do not undergo phase transitions in the reported pressure range. The values of the most precise determination [88] were incorporated in the tables in those cases where the compression of a substance was studied by more than one technique. Since it was found earlier that Bridgman's pressure scales differ slightly from the now accepted pressure scale, the original pressures were corrected and converted into bar as pressure unit. The corrected pressures are given in addition to Bridgman's original values in the tables. The correction in the 12 kbar range was made by multiplying Bridgman's pressure values by f = 7569/(7640 0.98067) = 1.0102 which is the ratio of what is now regarded to be the best value for the melting pressure of mercury at 273.15 K [89] and the value Bridgman used for his pressure calibration. In the 30 kbar range corrections to Bridgman's manganin pressure scale were applied which were evaluated and published by Babb [90].

TABLE 14. Ethylammonium hexachlorostannate, $(C_2H_5NH_3)_2SnCl_6$ (Ref. [86])
Pressure range: 12 kbar
T = 303.15 K
Relative compression: $-100 \, \Delta x/x_\circ = a \, P$

	original values $a/10^{-4}kg^{-1}cm^2$	corrected values (see text) $a/10^{-4}bar^{-1}$
‖ hexagonal axis	5.75	5.70
⊥ hexagonal axis	3.95	3.91
volume	12.9	12.8

TABLE 15. Dextrose sodium chloride, $C_6H_{12}O_6NaCl$ (Ref. [88]; see also Ref. [87])
T = 298.15 K

Pressure		Relative compression		
original	corrected	x direction	z direction	volume
$kg \, cm^{-2}$	bar	$-100 \, \Delta l/l_\circ$	$-100 \, \Delta l/l_\circ$	$-100 \, \Delta V/V_\circ$
2500	2478	0.506	0.286	1.290
5000	4958	1.028	0.555	2.590
10000	9910	1.972	1.089	4.932
15000	14857	2.661	1.573	6.740
20000	19801	3.352	2.000	8.460

TABLE 15 (Continued)

25000	24731	3.963	2.443	10.026
30000	29626	4.557	2.828	11.484

TABLE 16. Dextrose sodium iodide, $C_6H_{12}O_6NaI$ (Ref. [88]; see also Ref. [87])
T = 298.15 K

Pressure		Relative compression		
original	corrected	x direction	z direction	volume
kg cm^{-2}	bar	$-100\ \Delta l/l_0$	$-100\ \Delta l/l_0$	$-100\ \Delta V/V_0$
2500	2478	0.522	0.388	1.424
5000	4958	0.992	0.760	2.718
10000	9910	1.828	1.468	5.036
15000	14857	2.566	2.135	7.093
20000	19801	3.209	2.707	8.854
25000	24731	3.806	3.233	10.456
30000	29626	4.308	3.714	11.834

TABLE 17. Sodium xylenesulfonate, ortho-, meta-, and para.1H$_2$O-, $C_8H_{11}O_4SNa$ (Ref. [87])
T = 298.15 K
Molar volumes at 1 bar, $V_m/10^{-6}m^3mol^{-1}$: ortho: 153.3; meta: 151.6; para.1H$_2$O: 163.0

Pressure		Relative compression		
original	corrected	ortho-	meta-	para.1H$_2$O-
kg cm^{-2}	bar	$-100\ \Delta V/V_0$	$-100\ \Delta V/V_0$	$-100\ \Delta V/V_0$
2500	2478	2.06	2.36	1.95
5000	4958	3.91	4.40	3.54
10000	9910	7.06	7.75	6.09
15000	14857	9.59	10.29	8.14
20000	19801	11.65	12.29	9.87
25000	24731	13.42	13.90	11.37
30000	29626	14.92	15.24	12.61

TABLE 18. Strontium methanoate, $(HCOO)_2Sr$ (Ref. [88]; see also Ref. [87])
T = 298.15 K

Pressure		Relative compression			
original	corrected	a direction	b direction	c direction	volume
kg cm^{-2}	bar	$-100\ \Delta l/l_0$	$-100\ \Delta l/l_0$	$-100\ \Delta l/l_0$	$-100\ \Delta V/V_0$
2500	2478	0.330	0.198	0.383	0.908
5000	4958	0.610	0.390	0.729	1.722

TABLE 18 (Continued)

10000	9910	1.130	0.764	1.371	3.230
15000	14857	1.613	1.114	1.976	4.632
20000	19801	2.068	1.434	2.548	5.932
25000	24731	2.484	1.740	3.077	7.130
30000	29626	2.857	2.035	3.574	8.236

TABLE 19. Sodium potassium 2,3-dihydroxybutanedioate (tartrate) tetrahydrate, $C_4H_4O_6NaK \cdot 4H_2O$
(Ref. [88]; see also Ref.s [86],[87])
T = 298.15 K

Pressure		Relative compression			
original	corrected	a direction	b direction	c direction	volume
kg cm^{-2}	bar	$-100\ \Delta l/l_0$	$-100\ \Delta l/l_0$	$-100\ \Delta l/l_0$	$-100\ \Delta V/V_0$
2500	2478	0.618	0.257	0.387	1.258
5000	4958	1.211	0.495	0.740	2.430
10000	9910	2.227	0.957	1.374	4.496
15000	14857	3.132	1.372	1.921	6.296
20000	19801	3.968	1.730	2.412	7.904
25000	24731	4.730	2.025	2.849	9.316
30000	29626	5.403	2.278	3.272	10.580

TABLE 20. Sodium ammonium 2,3-dihydroxybutanedioate (tartrate), $C_4H_4O_6NaNH_4$ (Ref. [88]; see also Ref. [87])
T = 298.15 K

Pressure		Relative compression			
original	corrected	x direction	y direction	z direction	volume
kg cm^{-2}	bar	$-100\ \Delta l/l_0$	$-100\ \Delta l/l_0$	$-100\ \Delta l/l_0$	$-100\ \Delta V/V_0$
2500	2478	0.668	0.248	0.378	1.288
5000	4958	1.307	0.477	0.737	2.502
10000	9910	2.444	0.892	1.388	4.658
15000	14857	3.425	1.261	2.006	6.556
20000	19801	4.299	1.597	2.532	8.212
25000	24731	5.093	1.918	3.007	9.712
30000	29626	5.716	2.216	3.436	10.972

TABLE 21. Rubidium 2,3-dihydroxybutanedioate (tartrate), $C_4H_4O_6Rb_2$ (Ref. [88]; see also Ref. [87])
$T = 298.15$ K

Pressure		Relative compression		
original	corrected	x direction	z direction	volume
kg cm^{-2}	bar	$-100\ \Delta l/l_0$	$-100\ \Delta l/l_0$	$-100\ \Delta V/V_0$
2500	2478	0.431	0.179	1.035
5000	4958	0.849	0.329	2.015
10000	9910	1.638	0.611	3.838
15000	14857	2.344	0.856	5.448
20000	19801	3.008	1.051	6.910
25000	24731	3.610	1.224	8.228
30000	29626	4.142	1.375	9.376

TABLE 22. Ammonium 2,3-dihydroxybutanedioate (tartrate), $C_4H_4O_6(NH_4)_2$ (Ref. [86])
Pressure range: 12 kbar
Relative compression: $-100\ \Delta x/x_0 = a\ P - b\ P^2$

$T = 303.15$ K	original values		corrected values	
	a/10^{-5}kg^{-1}cm^2	b/10^{-10}kg^{-2}cm^4	a/10^{-5}bar^{-1}	b/10^{-10}bar^{-2}
a direction	8.47	16.8	8.39	16.5
b direction	13.71	9.3	13.58	9.1
c direction	26.85	46.1	26.60	45.2
volume	49.03	79.3	48.57	77.8
$T = 348.15$ K				
a direction	8.75	19.3	8.67	18.9
b direction	14.51	11.8	14.37	11.6
c direction	27.48	48.2	27.22	47.3
volume	50.84	87.0	50.37	85.4

TABLE 23. Ethylenediamine 2,3-dihydroxybutanedioate (tartrate), $C_2H_8N_2 \cdot C_4H_6O_6$ (Ref. [87])
$T = 298.15$ K

Pressure		Relative compression
original	corrected	volume
kg cm^{-2}	bar	$-100\ \Delta V/V_0$
5000	4958	2.90
10000	9910	5.23
15000	14857	7.18
20000	19801	8.86
25000	24731	10.33
30000	29626	11.61

TABLE 24. Morpholine hydrogentartrate, $C_4H_9ON \cdot C_4H_6O_6$ (Ref. [88]; see also Ref. [87])
$T = 298.15$ K

Pressure		Relative compression			
original	corrected	x direction	y direction	z direction	volume
kg cm^{-2}	bar	$-100\ \Delta l/l_0$	$-100\ \Delta l/l_0$	$-100\ \Delta l/l_0$	$-100\ \Delta V/V_0$
2500	2478	0.809	0.214	0.654	1.676
5000	4958	1.479	0.415	1.622	3.484
10000	9910	2.655	0.797	2.798	6.134
15000	14857	3.598	1.144	3.711	8.236
20000	19801	4.427	1.468	4.445	10.018
25000	24731	5.150	1.771	5.112	11.592
30000	29626	5.809	2.046	5.616	12.920

REFERENCES

1. BARKER, J.A., and HENDERSON, D., *Rev. Mod. Phys.*, **48**, 587 (1976)

2. GILLAN, M.J., *Phys. Chem. Liq.*, **8**, 121 (1978)

3. SANGSTER, M.J.L., and DIXON, M., *Adv. Phys.*, **25**, 247 (1976)

4. WAISMAN, E., and LEBOWITZ, J.L., *J. Chem. Phys.*, **56**, 3086 (1972)

5. WAISMAN, E., and LEBOWITZ, J.L., *J. Chem. Phys.*, **56**, 3093 (1972)

6. LARSEN, B., *Chem. Phys. Lett.*, **27**, 47 (1974)

7. LARSEN, B., *J. Chem. Phys.*, **65**, 3431 (1976)

8. FRENKEL, Ya.I., *Acta Physicochim. URSS*, **3**, 633, 913 (1935)

9. BRESLER, S.E., *Acta Physicochim. URSS*, **10**, 491 (1939)

10. ALTAR, W., *J. Chem. Phys.*, **5**, 577 (1937)

11. FÜRTH, R., *Proc. Cambridge Phil. Soc.*, **37**, 252 (1941)

12. TURNBULL, D., and COHEN, M.H., *J. Chem. Phys.*, **29**, 1049 (1958)

13. COHEN, M.H., and TURNBULL, D., *J. Chem. Phys.*, **31**, 1164 (1959)

14. EYRING, H., REE, T., and HIRAI, N., *Proc. Nat. Acad. Sci. U.S.*, **44**, 683 (1958)

15. EYRING, H.., and JHON, M.S., *"Significant Liquid Structures"*, Wiley, New York, 1969

16. WOODCOCK, L.V., in *"Advances in Molten Salt Chemistry"*, ed. by Braunstein, J., Mamantov,
     G., and Smith, G.P., Vol. 3, 1-74, Plenum Press, New York, 1975

17. GOLDMANN, G., and TÖDHEIDE, K., *Z. Naturforsch.*, **31a**, 656 (1976)

18. WOODCOCK, L.V., and SINGER, K., *Trans. Faraday Soc.*, **67**, 12 (1971)

19. FUMI, F.G., and TOSI, M.P., *J. Phys. Chem. Solids*, **25**, 31 (1964)

20. LARSEN, B., FØRLAND, T., and SINGER, K., *Mol. Phys.*, **26**, 1521 (1973)

21. LEWIS, J.W.E., "Atlas Computer Laboratory", Chilton, Didcot, Berkshire, private communica-
     tion, 1974

22. ADAMS, D.J., and McDONALD, I.R., *J. Phys.:C*, **7**, 2761 (1974)

23. ROMANO, S., and McDONALD, I.R., *Physica*, **76**, 625 (1973)

24. ROMANO, S., and MARGHERITIS, C., *Physica*, **77**, 557 (1974)

25. LEWIS, J.W.E., and SINGER, K., *JCS Faraday Trans. II*, 71, 41 (1975)

26. VAN WECHEM, H., *Thesis*, Amsterdam, 1976

27. KISZA, A., and HAWRANEK, J., *Z. physik. Chem. (Leipzig)*, 237, 210 (1968)

28. ZABINSKA, G., and KISZA, A., *Proc. III Int. Conf. Molten Salt Chem.*, Wroclaw-Karpacz
        (Poland), 347 (1979)

29. GATNER, K., *Pol. J. Chem.*, in press (1979)

30. WALDEN, P., *Bull. Acad. Imp. Sci. Petersbourg*, 6 (8), 405 (1914)

31. SUGDEN, S., and WILKINS, A., *J. Chem. Soc.*, 1929, 1291

32. WALDEN, P., and BIRR, E.J., *Z. physik. Chem.*, A160, 57 (1932)

33. LIND, J.E., Jr., ABDEL-REHIM, H.A.A., and RUDICH, S.W., *J. Phys. Chem.*, 70, 3610 (1966)

34. GRIFFITH, T.R., *J. Chem. Eng. Data*, 8, 568 (1963)

35. GORDON, J.E., and SUBBA RAO, G.N., *J. Amer. Chem. Soc.*, 100, 7445 (1978)

36. KENAUSIS, L.C., EVERS, E.C., and KRAUS, C.A., *Proc. Nat. Acad. Sci. U.S.*, 48, 121 (1962)

37. KENAUSIS, L.C., EVERS, E.C., and KRAUS, C.A., *Proc. Nat. Acad. Sci. U.S.*, 49, 141 (1963)

38. COKER, T.G., WUNDERLICH, B., and JANZ, G.J., *Trans. Faraday Soc.*, 65, 3361 (1969)

39. POLLACK, M.I., and SUNDHEIM, B.R., *J. Phys. Chem.*, 78, 1957 (1974)

40. PAINTER, J.L., and SUNDHEIM, B.R., New York University, New York, private communication,
        1978

41. RUDICH, S.W., and LIND, J.E., Jr., *J. Chem. Phys.*, 50, 3035 (1969)

42. NEWMAN, D.S., MORGAN, D.P., and TILLACK, R.T., *J. Chem. Eng. Data*, 21, 279 (1976)

43. EASTEAL, A.J., and ANGELL, C.A., *J. Phys. Chem.*, 74, 3987 (1970)

44. BLOOM, H., and REINSBOROUGH, V.C., *Aust. J. Chem.*, 21, 1525 (1968)

45. NEWMAN, D.S., TILLACK, R.T., MORGAN, D.P., and WAI-CHING WAN, *Proc. Int. Symp. Molten Salts*,
        The Electrochemical Soc., Princeton, 168-181 (1976)

46. MEISEL, T., HALMOS, Z., and PUNGOR, E., *Periodica Polytechnica Chem. Eng.*, 17, 89 (1973),
        and private communication

47. HALMOS, Z., MEISEL, T., SEYBOLD, K., and ERDEY, L., *Talanta*, 17, 1191 (1970)

48. LEONESI, D., CINGOLANI, A., and BERCHIESI, G., *Z. Naturforsch.*, 31a, 1609 (1976)

49. HAZLEWOOD, F.J., RHODES, E., and UBBELOHDE, A.R., *Trans. Faraday Soc.*, 62, 3101 (1966)

50. DURUZ, J.J., MICHELS, H.J., and UBBELOHDE, A.R., *Proc. Roy. Soc.* (London), A322, 281 (1971)

51. EKWUNIFE, M.E., NWACHUKWU, M.U., RINEHART, F.P., and SIME, S.J., *JCS Faraday Trans. I*, 71
        1432 (1975)

52. ADEOSUN, S.O., SIME, W.J., and SIME, S.J., *JCS Faraday Trans. I*, 72, 2470 (1976)

53. GRINDLEY, T., and LIND, J.E., Jr., *J. Chem. Phys.*, 56, 3602 (1972)

54. MORRISON, G., and LIND, J.E., Jr., *J. Chem. Phys.*, 49, 5310 (1968)

55. WALDEN, P., ULICH, H., and BIRR, E.J., *Z. physik. Chem.*, 130, 495 (1927)

56. WALDEN, P., and BIRR, E.J., *Z. physik. Chem.*, A160, 45 (1932)

57. OVENDEN, P.J., University of Southampton, private communication, 1965

58. SEWARD, R., *J. Amer. Chem. Soc.*, 73, 515 (1951)

59. BARREIRA, M.L., and LAMPREIA, M.I., *Rev. Port. Quim.*, 15, 65 (1973)

60. BARTON, A.F.M., and SPEEDY, R.J., *JCS Faraday Trans. I*, 70, 506 (1974)

61. CLEAVER, B., and SPENCER, P.N., *High Temp.-High Pressures*, 7, 539 (1975)

62. REINSBOROUGH, V.C., and VALLEAU, J.P., *Aust. J. Chem.*, 21, 2905 (1968)

63. LIND, J.E., Jr., in: *"Advances in Molten Salt Chemistry"*, ed. by Braunstein, J., Mamantov,
        G., and Smith, G.P., Vol. 2, 1-26, Plenum Press, New York, 1973

64. WAISMAN, E., and LEBOWITZ, J.L., *J. Chem. Phys.*, 52, 4307 (1970)

65. GOLDMANN, G., and TÖDHEIDE, K., *Z. Naturforsch.*, 31a, 769 (1976)

66. KUSS, E., and TASLIMI, M., *Chem. Ing. Techn.*, 42, 1073 (1970)

67. BANNARD, J.E., and BARTON, A.F.M., *JCS Faraday Trans. I*, 74, 153 (1978)

68. OWENS, B.B., *J. Chem. Phys.*, 44, 3918 (1966)

69. CLEAVER, B., and ZANI, P., *High Temp.-High Pressures*, 10, 437 (1978)

70. SCHAMM, R., and TÖDHEIDE, K., *High Temp.-High Pressures*, 8, 59 (1976)

71. REISS, H., in: *"Advances in Chemical Physics"*, ed. by Prigogine, I., Vol. 9, 1-84, Interscience Publ., London, New York, 1965

72. WYCKOFF, R.W.G., *"Crystal Structures"*, Interscience Publ., New York, 1971

73. KENNARD, O., WATSON, D., ALLEN, F., MOTHERWELL, W., TOWN, W., and ROGERS, J., *Chem. Brit.*, 11, 213 (1975)

74. IMMIRZI, A., and PERINI, B., *Acta Cryst.*, A33, 216 (1977)

75. COKER, T.G., AMBROSE, J., and JANZ, G.J., *J. Amer. Chem. Soc.*, 92, 5293 (1970)

76. VACATELLO, M., and CORRADINI, P., *Gazz. Chim. Ital.*, 103, 1027 (1973)

77. GALLOT, B., and SKOULIOS, A., *Kolloid-Z. Polym.*, 209, 164 (1966)

78. SKODA, W., *Kolloid-Z. Polym.*, 234, 1128 (1969)

79. BENTON, D.P., HOWE, P.G., FARNAND, R., and PUDDINGTON, I.E., *Can. J. Chem.*, 33, 1798 (1955)

80. VOLD, M.J., MACOMBER, M., and VOLD, R.D., *J. Amer. Chem. Soc.*, 63, 168 (1941)

81. HATIBARUA, J., and PARRY, G.S., *Acta Cryst.*, B28, 3099 (1972)

82. GALLOT, B., and SKOULIOS, A., *Kolloid-Z. Polym.*, 210, 143 (1966)

83. FISCHER, W., and LEMKE, A., *Z. physik. Chem.*, A151, 56 (1930)

84. GALLOT, B., and SKOULIOS, A., *Kolloid-Z. Polym.*, 213, 143 (1966)

85. TAIT, P.G., *"Scientific Papers"*, Vol. 2, 334, Cambridge University Press, 1898

86. BRIDGMAN, P.W., *Proc. Amer. Acad. Arts Sci.*, 64, 51 (1929)

87. BRIDGMAN, P.W., *Proc. Amer. Acad. Arts Sci.*, 76, 71 (1948)

88. BRIDGMAN, P.W., *Proc. Amer. Acad. Arts Sci.*, 76, 89 (1948)

89. LIU, C.Y., ISHIZAKI, K., PAAUWE, J., and SPAIN, L., *High Temp.-High Pressures*, 5, 359 (1973)

90. BABB, St.-E., Jr., in: *"High-Pressure Measurement"*, ed. by Giardini, A.A., and Lloyd, E.C., 115-123, Butterworths, Washington, 1963

Chapter 1.2

THERMAL PROPERTIES

M. Sanesi, A. Cingolani, P. L. Tonelli and P. Franzosini

TEMPERATURES AND ENTHALPIES OF PHASE TRANSITIONS

SALTS WITH ORGANIC ANION

Alkali n.alkanoates

In dealing with the phase transitions of alkali n.alkanoates one of the most important features to be pointed out is that in each salt family characterized by the same cation the passage from the crystalline solid to a conventional isotropic liquid occurs in a single step only in the case of homologues with lower values of the number of carbon atoms, n_C, whereas at least two steps are involved in the case of higher n_C's. In particular, it can be said that: (i) in sodium, potassium, rubidium and cesium salts, starting from n_C = 3 (or 4), 4, 5, and 6, respectively, the crystal lattice first transforms into an anisotropic liquid phase (liquid crystal), which at a more or less significantly higher temperature turns into a clear isotropic melt; (ii) additional plastic phases, intermediate between the crystalline solid and the anisotropic liquid, may occur in the long-chain homologues of the same salt families; (iii) in the lithium salts series conventional melting takes place up to n_C = 11; at higher n_C values plastic and perhaps (starting from n_C = 18) also mesomorphic liquid phases are formed.

The complexity of the above phase relationships is reflected in a complicated and to a some extent confused terminology regarding both the phase changes and the designation of single phases; moreover wall effects may modify the observations when the salts are confined in very small enclosures (microscope cells).

Let us first consider the field of existence of an anisotropic liquid. Its lower and upper limits were formerly indicated by Vorländer (1910) as 1st melting point, transition crystalline solid → liquid crystal ("1. Schmelzpunkt, Übergang kr.-fest → kr.-fl.") and 2nd melting point ("2. Schmelzpunkt"), respectively. Several Russian authors appear to consider the lower limit as a poorly defined transition and the upper one as the true melting point. British authors, on the contrary, more satisfactorily define the upper limit, i.e., the transition from the anisotropic to the isotropic (clear) liquid, as a "clearing" and the corresponding temperature value as a "clearing temperature" (see, e.g., Duruz, Michels & Ubbelohde, 1971). Accepting the latter terminology, it seems consistent to take the lower limit as the fusion temperature, independently of the nature of the phase (either crystalline solid or plastic) which on heating is transformed into the anisotropic liquid at this temperature, although a few authors prefer to consider as a fusion the transition (occurring on heating) from a crystalline solid to any other phase.

The isotropic liquid, which looks dark when viewed through crossed polaroids, is sometimes also called the "nigre" phase. With other terms borrowed from the soap manufacturing practice, the anisotropic liquid is often indicated as the "neat" phase (inasmuch as it is substantially continuous with the homonymous phase occurring in any binary formed with soap and water), and, in a similar way, the various plastic and crystalline phases are distinguished as "waxy", "subwaxy", "superwaxy", etc., and "curd", "supercurd", etc.

Alternatively to the above terminology, based on mere phenomenological observations, a different nomenclature drawn from the results of structural investigations may be encountered, which employs terms as "labile lamellar", "ribbon", "disc", "crystalline lamellar", etc., structures (see, e.g., Gallot & Skoulios, 1966 d).

Finally, other authors list the transitions they observed without specifying (or only partially specifying) the nature of the phases involved.

From this equivocal and somewhat arbitrary lexicon the fact derives that, although terms such as "anisotropic liquid", "liquid crystal", "neat phase", "labile lamellar structure" can undoubtedly be considered synonyms, in other cases, and in particular when the plastic region is concerned, to establish a correct correspondence among the phases detected, e.g., in a given long-chain Na alkanoate, by different authors, is often rather puzzling.

Moreover, this difficulty is remarkably enhanced by the fact that the actual possibility of detecting a given phase transition largely differs depending on the experimental method employed. Thus, a transition involving a very small volume change might escape dilatometric investiga-

tion, whereas it might be easily detected through microscopic observation if sufficient optical contrast exists between the two phases concerned.

The above remarks can be illustrated by means of two examples. Thus, concerning Na pentanoate,
Vorländer (1910) gave a 1st and a 2nd melting point of ≈483 and ≈608 K respectively (capillary
method), while Sokolov by the so-called "visual-polythermal method" reported fusion at 630 K
(Sokolov, 1954 a) and transitions occurring at 489, 482 and 453 K (Sokolov, 1956), and Michels
& Ubbelohde (1972) recorded by differential scanning calorimetry clearing and fusion at 626
and 502 K respectively. It is apparent that correct correspondences can be immediately stated
as follows:

Vorländer's 2nd melting point ≡ Sokolov's fusion   ≡ Michels & Ubbelohde's clearing

Vorländer's 1st melting point ≡ Sokolov's highest transition ≡ Michels & Ubbelohde's fusion

As a second example, we may refer to the case of sodium octadecanoate which appears as much more
complicated, even if one neglects for the moment the polymorphs existing at temperatures lower
than ≈360 K, to be discussed later. Although a reasonable convergence of results seems to allow
for the occurrence of at least five different mesomorphic phases (four plastic and one liquid,
i.e., subwaxy, waxy, superwaxy, subneat and neat), a more detailed analysis of data obtained
through different experimental methods by different authors reveals discrepancies not only for
what concerns the temperature ranges where each of the mentioned phases can exist, but also on
the total number of existing phases. Hereafter a schematical comparison is given for twelve
series of data: each transition temperature is supplied with the authors' comments and/or with
the designation, if any, of the concerned phases, as reported in the original papers. From this
comparison the existence (from ≈360 K) of up to ten transitions might be inferred: for better
evidence of the correspondences among different series a (somewhat arbitrary) numbering from 1
to 10 has been given, although no single series includes more than seven data.

	McBain, Vold & Frick (1940): polarizing microscopy		Vold, M.J., Macomber & Vold (1941): dilatometry		Vold, M.J. (1941): hot-wire method	
	T/K	Obs. phenomena	T/K	Transition	T/K	Transition
1	≈363	"gradual brighten= ing of field"	363	nstr (a)	–	–
2	381-391	"gradual appear= ance of brighter golden material"	390	nstr	–	–
3	396	appearance of "bright, fine grained, pebbled structure"	405	nstr	405	formation of waxy soap
4	439	"structure coarser and brighter, and the color still (...) golden"	440	nstr	440	waxy → superwaxy
5	–	–	–	–	–	–
6	476	"some softening", and appearance of "some red, green and blue color"	471	nstr	476	superwaxy → subneat
7	–	–	–	–	–	–
8	529	"marked softening (...) variety and depth of color in= creased (...) fo= cal conics"	530	nstr	530	subneat → neat
9	–	–	–	–	–	–
10	558	"field (...) completely dark"	561	nstr	–	–

a: non-specified transition

	Stross & Abrams (1951): DTA		Benton, Howe & Puddington (1955): photometric method		Skoulios & Luzzati (1961): X-ray diffraction	
	T/K	Transition	*T*/K	Transition	*T*/K	Transition
1	359-369	nstr	-	-	-	-
2	385-391	nstr	-	-	-	-
3	402-403	nstr	405	subwaxy → waxy	406	subwaxy → waxy (ribbon structures)
4	-	-	438	waxy → superwaxy	448	waxy → superwaxy (ribbon structures)
5	-	-	-	-	-	-
6	471	nstr	461	superwaxy → subneat	483	superwaxy → subneat (ribbon structures)
7	-	-	493	minor transition	-	-
8	516-524	nstr (gel → jelly)	531	subneat → neat	529	subneat → neat, smectic (ribbon structure → lamellar structure)
9	544-546	nstr	-	-	-	-
10	553	nstr (birefringence → no birefringence)	556	neat → isotropic	563	neat → isotropic

	Trzebowski(1965): DTA		Lawson & Flautt (1965): NMR		Baum, Demus & Sackmann (1970): polarizing microscopy	
	T/K	Transition	*T*/K	Transition	*T*/K	Transition
1	366	curd → supercurd	358	crystalline → crystalline	-	-
2	391	supercurd → subwaxy	387	crystalline → subwaxy	389	solid → subwaxy
3	409	subwaxy → waxy	404	subwaxy → waxy	408	subwaxy → waxy
4	≃439	nstr	430-432	waxy → superwaxy	438	waxy → superwaxy
5	≃469	nstr	-	-	-	-
6	≃475	nstr	451-454	superwaxy → subneat	481	superwaxy → subneat
7	-	-	-	-	-	-
8	515	formation of neat	-	-	528	subneat → neat
9	-	-	-	-	-	-
10	-	-	-	-	556	neat → isotropic

	Ripmeester & Dunell (1971 b): DSC		Wasik (1976): DSC		Wasik (1976): GLC	
	T/K	Transition	*T*/K	Transition	*T*/K	Transition
1	355÷363	nstr	363-368	curd → curd	368	curd → curd
2	384÷389	nstr	388-392	curd → subwaxy	390	curd → subwaxy

3	404÷406	nstr	403–409	subwaxy → waxy	408	subwaxy → waxy	
4	–	–	–	– *(a)*	–	– *(a)*	
5	–	–	–	–	–	–	
6	–	–	478–480	superwaxy → subneat	478	superwaxy → subneat	
7	–	–	–	–	–	–	
8	–	–	529–533	subneat → neat	530	subneat → neat	
9	–	–	–	–	–	–	
10	–	–	558	neat → isotropic	558	neat → isotropic	

a: the temperature of the transition waxy → superwaxy could not be detected by the author

The phase transition temperatures and enthalpies were measured by different authors using a wide variety of procedures. A short account of these techniques is given here, although, of course, they are generally valid for all the other salt groups considered in the present chapter.

Several more or less advanced techniques, which may be included under the common designation of *thermal analysis* were used to collect the largest amount of significant data. The classical thermal analysis (temperature *vs* time curves, usually detected on slowly cooling a sample), and some modifications of it also allowing a direct observation of the sample (e.g., the "synthetic method" by Vold, R.D., 1939, and the "visual-polythermal method" preferred by a number of Russian authors), only provide information on phase transition temperatures.

For the simultaneous determination of the pertinent enthalpy changes, instrumental methods, such as low temperature *adiabatic calorimetry*, *differential thermal analysis* (DTA), Calvet *microcalorimetry* and *differential scanning calorimetry* (DSC) have been employed, among which DTA and DSC are here thought to be worthy of some comment.

It is well known that: (i) in both methods a small sample and an inert reference material are heated or cooled at the same programmed rate; (ii) in DTA, the differential temperature, ΔT, is continuously recorded *vs* T, so that, should a phase change involving a heat absorption or emission take place in the sample, ΔT will become different from zero and a "peak" will appear in a ΔT *vs* T trace; (iii) in DSC, proper energy amounts are continuously supplied in order to maintain the sample at the same temperature as that programmed for the reference material, so that the power change required when a phase transition occurs is also recorded as a peak in a $\partial \Delta q/\partial t$ *vs* t plot (t ∝ T). The correct way of singling out a transition temperature from a DSC (or DTA) peak is that shown in Fig. 1. In the literature, however, transition temperature values corresponding to peak maxima are sometimes found: whenever possible, this circumstance has been indicated in the tables. Moreover, it must be noted that higher-order transitions do not give rise to peaks but only to steps (or other minor peculiarities) in the recorded traces, and are not unfrequently neglected by the authors in data tabulation.

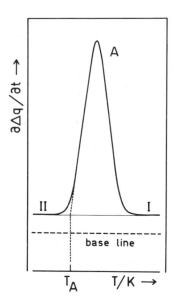

Fig. 1. DSC trace referring to transition A occurring at T_A/K from phase II to phase I, without change in specific heat capacity.

Occasionally, *cryometry* has also been employed to evaluate the enthalpy of fusion, $\Delta H_{F,1}$, of a few salts taken as solvent (component 1) on the basis of the usual equation:

$$\Delta H_{F,1} = R\, T_{F,1}^2 / K_1$$

where the unified cryoscopic constant, K_1/K, is defined as

$$K_1 = \{ \lim_{x_2 \to 0} (\Delta T / x_2) \} / \nu$$

and: R = gas constant;

$T_{F,1}$ = freezing point of component 1;

x_2 = stoichiometric mole fraction of the solute;

ΔT = $T_{F,1} - T_F$ (freezing point depression; T_F= freezing point of the solution);

ν = number of cryoscopycally active, i.e., "foreign to the solvent" species.

Among procedures which may be classified as optical, the Kofler's *hot-stage polarizing micro-scopy* found a wide application.

Worthy of mention is an independently developed microscopic technique described as the *hot-wire method* by Vold, M.J. (1941). The essential part of the apparatus is schematized in Fig. 2; the Transite base a, resting on the movable stage of a polarizing microscope, is fitted with a Chromel wire b stretched over the observation hole c. The sample is prepared in the form of a thin film between two cover glasses: this sandwich, resting on dd, is kept in close contact with the wire, which can be heated up to, and held at any desired temperature by means of a controlled current. Within the range between the temperature of the wire and that at dd (room temperature), an approximately constant gradient is established through the sample: thus each phase transition gives rise to a sharp boundary parallel to the wire. After proper calibration, the measurement of the distance of any boundary from the wire allows statement of the temperature of the transition pertinent to that boundary. The possibility of recognizing a transition is of course conditioned by the existence of sufficient optical contrast between the involved phases.

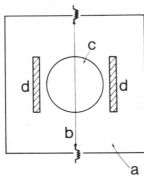

Fig. 2

Finally, photometric recording of light transmission as a function of temperature during a heating (or cooling) run allowed, e.g., Benton, Howe & Puddington to detect phase transitions in alkali metal octadecanoates.

Techniques developed for different kinds of structural investigation, such as *X-ray diffraction* and *nuclear magnetic resonance* (NMR), when carried out over a suitable temperature range, could give useful information about transition temperatures.

In the case of NMR, however, both instrumental and structural factors not infrequently limit the actual ability of this method to single out different phases: thus Barr & Dunell found that the recorded spectra did not allow one to distinguish on one hand among the subwaxy, waxy, superwaxy and subneat phases, and on the other hand between the neat and isotropic liquid phases in sodium octadecanoate.

The following other experimental procedures, which were more or less extensively employed, can still be mentioned: *dilatometry, dielectric measurements* and *gas-liquid-chromatography* (GLC). Concerning the latter, it was recently stated by Wasik that GLC allows one to correlate T-dependent structural changes occurring in a "solvent" with the corresponding variations in the interaction energy between the solvent itself and a "probe" solute molecule: thus he succeeded in singling out phase transition temperatures in sodium octadecanoate, through plots of log specific retention volumes *vs* $1/T$ for the elution of different solutes.

A short discussion of a few more points may be useful for a better understanding of the origin of further discrepancies apparent among the experimental data by different authors reported in the tables.

First of all it must be said that salt specimens might happen not to be fully crystalline: this would in particular affect the results of DTA or DSC investigations, inasmuch as traces recorded during the first heating run might exhibit glass transitions ([+]) and exothermic peaks due

---

([+]) A few details on glass-forming salts will be given at the end of this chapter.

to transformation from amorphous to crystalline solid. Glass formation may also occur as a consequence of an unusually pronounced tendency to undercooling of the melt, observed with a number of salts among which lithium methanoate and ethanoate may be mentioned: this would of course
affect traces recorded during runs subsequent to the first one.

Apart from the case of glass formation, the (reversible) transitions which a given salt undergoes are often characterized by different degrees of reversibility: thus, e.g., in the DSC traces
recorded during successive heating runs it may happen that the positions (with respect to the
temperature scale) and the areas of the peaks reassume the original values only after rest of
the sample at a suitably low temperature for a sufficient time. An example of such hysteresis
phenomena is offered by sodium butanoate as shown in Fig. 3.

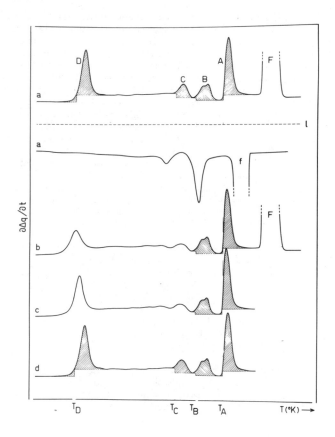

Fig. 3. DSC traces taken in the same operational conditions on the same
sample of sodium butanoate. Peaks A-D, F and f correspond to solid state
transitions, fusion and freezing, respectively. Trace a: recorded on heating; trace a': recorded on cooling; traces b, c and d: recorded on successive re-heating after 20 min, 1 h and 14 h rest at a lower temperature,
respectively. Concerning the areas of peaks A and B, a close coincidence
was found in all heating runs. For peaks C and D, on the contrary, coincidence in the pertinent areas occurred only in traces a and d: this proves
that a prolonged rest at a lower temperature is needed to overcome the hysteresis affecting the corresponding transitions.
(Ferloni & Franzosini; reproduced by kind permission of *Gazz. Chim. Ital.*)

An important role for what concerns the reliability of transitions detected at "intermediate"
temperatures (i.e., between room temperature and a not generally specifiable higher temperature at which all possible traces of residual water are completely removed) is played by the
way a given sample has been handled before measurements. This is particularly significant with

long-chain alkanoates: the case of sodium octadecanoate is hereafter discussed in some detail, although without attempting to give a complete review of the abundant literature dealing with this much debated subject.

According to Vold, R.D. (1945) this salt, when (as usually) obtained by precipitation from aqueous ethanol, is a semi-hydrated α form ($C_{17}H_{35}COONa.\frac{1}{2}H_2O$). He then suggested using the term β form (previously used loosely to designate soap forms resulting from several different procedures) for the (nearly) anhydrous one in which α is transformed on moderate heating. Wasik (1976) underlined the fact that this transition is reversible only in the presence of water vapour. A second order transition leads β to a new form, called λ by Vold, the field of existence of which seemed limited between about 344 and 363 K, the latter being the temperature of formation of supercurd soap (in Vold's terminology). A fourth form, γ, is reversibly originated when supercurd is cooled. Neglecting other designations occasionally adopted by a few other authors (see, e.g., Buerger, Smith, Ryer & Spike), from the above remarks the following outline was derived for the behaviour of sodium octadecanoate (temperatures given by Vold, 1945; the temperature of formation of supercurd is coincident with that of the lowest nstr by Vold, M.J., Macomber & Vold, 1941, reported in Table 1):

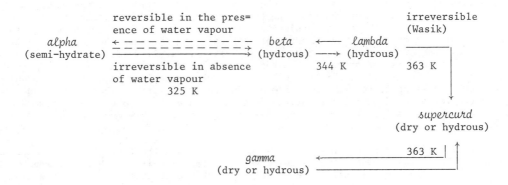

X-ray investigations, however, allowed e.g. Vold, R.D., Grandine & Schott (1952) to characterize at the intermediate temperatures a still larger number of "modifications", each to be intended as "a material having clearly different properties but without commitment as to whether it is or not a discrete phase". Among these modifications the following are of interest here: α, prepared by crystallization from aqueous ethanol; β, prepared by heating α up to 327-329 K, or by drying it over P_2O_5; γ, prepared by heating α up to 377-385 K; and σ, prepared either from α by heating up to 391-408 K or from the melt by cooling down to 395 K and then quenching in dry ice. Mention is no longer made of the λ form, whereas it seems that (above 363 K) a difference of only a few degrees in the heating treatment is sufficient to give rise, on successive cooling, to two different modifications, γ and σ.

Trzebowski (1965) recorded DTA traces on three different sodium octadecanoate specimens designated and handled as follows: α, air dried at room temperature; β, dried at 328-333 K; and σ, obtained according to Vold, Grandine & Schott's procedure. The significant temperatures he put into evidence were the following:

α	β	σ	Transitions or obs. phenomena
326	–	–	α → β
≈368	–	–	water evaporation
372	370.7		removal of hydration water (?)
378	375	366	curd → supercurd
395	395	391	supercurd → subwaxy
409	409	409	subwaxy → waxy

In Table 1 only transition temperatures taken on σ (presumably, the unique anhydrous modification) have been reported.

Ripmeester & Dunell (1971 b) submitted to DSC analysis four salt samples (A: dried under vacuum 24 h at room temperature; B: dried under vacuum 14 h at 383-393 K, and scanned 24 h later; C: dried under vacuum 28 h at the same temperature, and scanned 24 h later; and D: fused under vacuum, and scanned about a week later), obtaining the transition temperatures tabulated hereafter:

```
A:              321         355         389         405
B:              -           -           389         403
C:              no significant data drawn by the authors from the pertinent DSC trace
D:              -           363         384         406
```

In the above authors'opinion, their sample D ought to correspond to Trzebowski's σ, whereas "other correspondences are doubtful".

It was now thought proper to summarize (see Table 1) the above figures as follows [+]:

```
321         355÷363       384÷389     403÷406
```

i.e., taking the temperature limits pertinent to each transition, included that at 321 K inasmuch as stating a correspondence between sample A (vacuum dried, possibly anhydrous) and Trzebowski's α (air dried, surely hydrous) could be incorrect.

Worthy of mention here is the "monotrop Umwandlung" observed at 324.7 K by Thiessen & Stauff (1936) on a specially prepared single crystal of "pure neutral sodium stearate".

Coming back to the β→λ transition, a few more remarks may be added. In 1932-1933 Thiessen & Ehrlich pointed out that changes of some characteristic properties of long-chain sodium alkanoates occurred at temperatures close to the melting points of the parent fatty acids and proposed the term "genotypical" to designate such effect. In a subsequent paper Thiessen & von Klenck (1935) determined for sodium hexadecanoate and octadecanoate the genotypical transition temperatures and pertinent heat effects as reported in Table 1. Vold, R.D (1945) identified as the genotypical his β→λ transition in the latter salt and still in 1976 Wasik seemed inclined to accept this point of view. Wirth & Wellman, however, since 1956 came to the conclusion, on the ground of dielectric constant measurements on sodium palmitate, that in this salt the presumed detection of a genotypical transition was simply due to the presence of hexadecanoic acid impurities in the samples: these results give rise to considerable doubt also of the real existence of the genotypical effect in other salts.

It may be finally pointed out that troubles in investigating the thermal behaviour of alkali alkanoates also arise from the hygroscopicity they often exhibit and which can reach deliquescence, e.g., in some Cs salts. The extent to which the absorption of moisture can affect the results of DSC analysis is shown, as an example, in Fig. 4, referring to cesium pentanoate.

At the present state of knowledge it seems a hard job to attempt to draw from the thermal data taken on alkali n.alkanoates general correlations as functions of either the nature of the cation (for a given anion) or the anion chain length (for a given cation). For the latter case, however, it was thought of interest to illustrate at least the trends of clearing and fusion in each salt family by plotting selected pertinent temperatures vs n_c (see Figures 5, 7-10).

Concerning lithium salts two remarks can be made. (i) The propanoate may undergo both a stable and a metastable fusion (see the thermal cycle of Fig. 6): the formation of a metastable solid from the melt is proved by the coincidence in absolute value of the heat effects pertinent to peaks f and F". (ii) As already pointed out, in Li salts the formation of mesomorphic liquid phases is generally excluded: in the case of octadecanoate, however, Cox & McGlynn (1957) reported they could observe with the hot-stage polarizing microscope the softening (at 498 K) of the salt to "a jelly-like, brightly birefringent, liquid-crystalline phase", followed (4 K higher) by melting to "a mobile isotropic liquid".

In the case of sodium salts, the results by several authors with different experimental techniques agree on the occurrence of a mesomorphic liquid phase starting from butanoate: very recently, however, Roth, Meisel, Seybold & Halmos (1976) on the basis of simultaneous DTA and electrical conductivity measurements claimed a liquid crystalline phase to occur already in propanoate between 545 and 561 K [++].

[+] In tabulating, the notation $T_1 \div T_2$ (indicating, for the results of a given paper, the lower and upper limits of transition temperatures detected on samples of different history) was adopted whenever, on one hand, not fully coincident sets of data were reported and, on the other hand, no sufficiently valid reasons could lead to the preference of a particular set.

[++] For the sake of completeness, it can be added that, as a provisional result, also Ubbelohde, Michels & Duruz (1970) mentioned for this salt "a fugitive mesophase extending only 1° above the melting point", the existence of which was not confirmed in subsequent papers by the same authors.

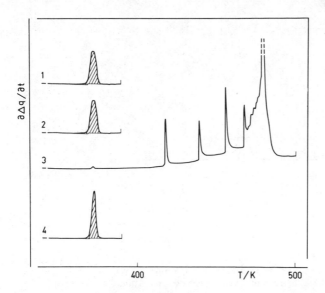

Fig. 4. DSC traces taken in the same operational conditions on the same Cs pentanoate sample in the temperature range around the transition occurring at ≃370 K (see Table 1). Traces 1 and 2: recorded in subsequent heating runs on the dry salt sealed in an Al pan; trace 3: recorded after having drilled a pinhole in the pan and leaving the sample exposed to atmospheric moisture until nearly one mole water per mole salt was absorbed; trace 4: recorded after complete removal of the absorbed water (proved through check of the sample weight) by heating up to 500 K. A strict reproducibility exists among the shaded areas (i.e., also among the pertinent enthalpy changes) in traces 1, 2 and 4; in trace 3 the peak corresponding to the transition is almost effaced, whereas a series of sharp new peaks occurs at $410<T/K<470$ followed by a large endothermic effect at 470–490 K to be referred to water removal. (Sanesi, Ferloni, Zangen & Franzosini, reproduced by kind permission of *Z. Naturforsch.*)

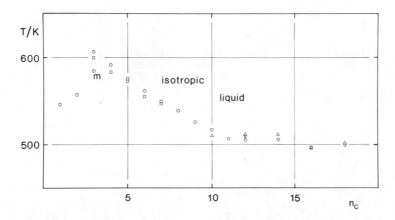

Fig. 5. Fusion temperatures of Li n.alkanoates (from methanoate to octa-decanoate). Circles: Ferloni, Sanesi & Franzosini, 1975 (n_C= 1–4); Sanesi, Ferloni & Franzosini (n_C= 5–7); Ferloni, Zangen & Franzosini, 1977 a (n_C= 8–12). Squares: Michels & Ubbelohde, 1972 (n_C= 3–7). Triangles: Gallot & Skoulios, 1966 b (n_C= 10, 12, 14, 16, 18). Inverted triangles: Baum, Demus & Sackmann (n_C= 12, 14, 16, 18). For point m (metastable fusion), see text and Fig. 6.

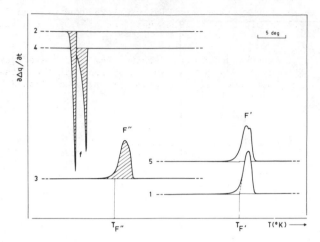

Fig. 6. DSC traces taken in the same operational conditions on the same Li propanoate sample in the melting and freezing region. Trace 1 (recorded on heating from room temperature): the sample after having undergone a transition at T_A= 533 K (see Table 1) melts at $T_{F'}$= 607 K. Trace 2 (recorded on cooling): the melt freezes at $T_f > T_A$ and cooling is stopped at 550 K. Trace 3 (recorded on heating from 550 K): melting occurs at $T_{F''}$= 584 K. Trace 4 (recorded on cooling): the melt freezes at $T_f > T_A$ and cooling is stopped at 400 K. Trace 5 (recorded on heating from 400 K, after 100 min rest at this temperature to overcome the salt hysteresis in undergoing transition at 533 K): melting again occurs at 607 K. A strict coincidence exists on one side among the areas of peaks f and F'', and on the other side between those (remarkably smaller) of peaks F'. (Ferloni, Sanesi & Franzosini, 1975, reproduced by kind permission of Z. *Naturforsch*.)

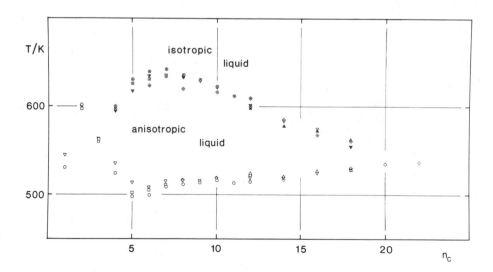

Fig. 7. Clearing (dot-centered symbols) and fusion (empty symbols) temperatures of Na n.alkanoates (from methanoate to docosanoate). Circles: Braghetti, Berchiesi & Franzosini (n_C= 1); Ferloni, Sanesi & Franzosini, 1975 (n_C= 2, 3); Ferloni & Franzosini (n_C= 4); Sanesi, Ferloni & Franzosini, 1977 (n_C= 5-7); Ferloni, Zangen & Franzosini, 1977 a (n_C= 8-12). Squares: Michels & Ubbelohde, 1972 (n_C= 2-7). Triangles: Skoulios & Luzzati, 1961 (n_C= 12, 14, 16, 18). Inverted triangles: Baum, Demus & Sackmann (n_C= 1-10, 12, 14, 16, 18). Rhombs: Vold, M.J., Macomber & Vold (even homologues from n_C= 6 to n_C= 22).

The field of existence of the mesomorphic liquid is substantially enlarged on passing from the sodium to the potassium series, where butanoate is once more the first homologue exhibiting this feature.

With rubidium salts the occurrence of the mesomorphic liquid starts from pentanoate, while the width of its field of existence begins to slightly decrease in comparison with the potassium family. This decrease becomes much more remarkable in the cesium series, where liquid crystals are formed only from hexanoate on.

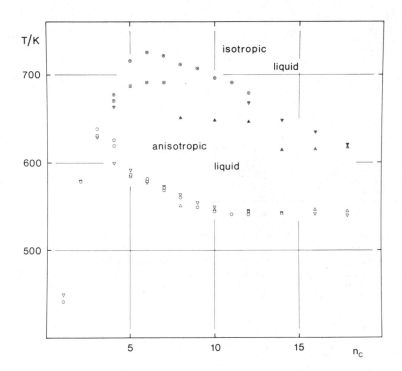

Fig. 8. Clearing (dot-centered symbols) and fusion (empty symbols) temperatures of K n.alkanoates (from methanoate to octadecanoate). Circles: Braghetti, Berchiesi & Franzosini (n_C= 1); Ferloni, Sanesi & Franzosini, 1975 (n_C= 2, 3); Ferloni & Franzosini (n_C= 4); Sanesi, Ferloni & Franzosini (n_C= 5, 6, 7); Ferloni, Spinolo, Zangen & Franzosini (n_C= 8-12). Squares: Michels & Ubbelohde, 1972 (n_C= 3-7). Triangles: Gallot & Skoulios, 1966 c (even homologues from n_C=8 to n_C=18). Inverted triangles: Baum, Demus & Sackmann (n_C= 1-10, 12, 14, 16, 18).

Data available from the literature up to early 1978 are listed in Table 1. The figures dating back to the XIX century, however, were not taken into account when apparently obsolete. A few more results could not be tabulated for having been presented by the authors in graphical form only: this is in particular the case of the paper by Pacor & Spier on thermal analysis and calorimetry of some fatty acid sodium soaps, and (in part) of the dilatometric investigation by Skoda.

Although the list is thought to be reasonably comprehensive, gaps do surely exist at least for what concerns some papers for which on one hand neither reprints nor photocopies could be obtained, and on the other hand the Chemical Abstracts information was too scanty.

A special case is represented by Sokolov's *Tezisy Dokl. X Nauchn. Konf. S.M.I. (Summaries of Papers Presented at the 10th Scientific Conference of the Smolensk Medical Institute)*, very frequently quoted by the authors of the Smolensk research group as their primary source of information on a large number of phase transitions in alkali alkanoates. For the sake of simplicity, the pertinent data are hereafter reported under the reference Sokolov (1956), in spite of the fact that this publication could not be directly consulted.

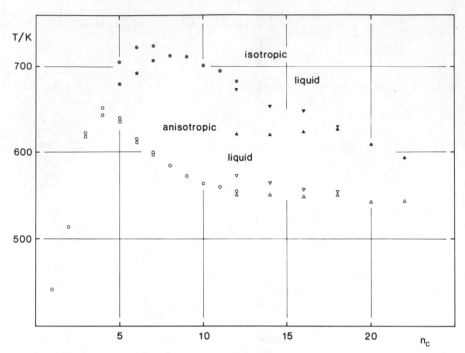

Fig. 9. Clearing (dot-centered symbols) and fusion (empty symbols) temperatures of Rb n.alkanoates (from methanoate to docosanoate). Circles: Braghetti & Berchiesi (n_C= 1, 2); Ferloni, Sanesi & Franzosini, 1975 (n_C= 3, 4);
Sanesi, Ferloni & Franzosini (n_C= 5-7); Ferloni, Zangen & Franzosini, 1977 b
(n_C= 8-12). Squares: Michels & Ubbelohde, 1972 (n_C= 3-7). Triangles: Gallot
& Skoulios, 1966 a (even homologues from n_C= 12 to n_C= 22). Inverted triangles: Baum, Demus & Sackmann (n_C= 12, 14, 16, 18).

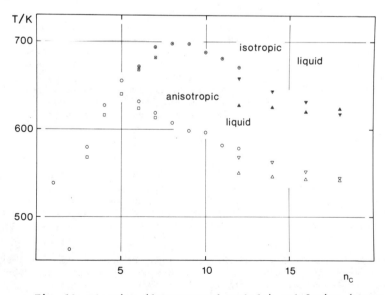

Fig. 10. Clearing (dot-centered symbols) and fusion (empty symbols) temperatures of Cs n.alkanoates (from methanoate to octadecanoate). Circles: Braghetti & Berchiesi (n_C= 1, 2); Ferloni, Sanesi & Franzosini, 1975 (n_C= 3, 4); Sanesi, Ferloni, Zangen & Franzosini (n_C= 5-12). Squares: Michels & Ubbelohde,
1972 (n_C= 3-7). Triangles: Gallot & Skoulios, 1966 d (n_C= 12, 14, 16, 18). Inverted triangles: Baum, Demus & Sackmann (n_C= 12, 14, 16, 18).

For each salt the transition temperatures (K) and enthalpies (kJ mol^{-1}) are those taken at at-mospheric pressure, and are tabulated in chronological order with the indication of the perti-nent experimental technique (when known) abbreviated as follows:

ad-cal, *adiabatic calorimetry*

cal, *calorimetry*

capill, *classical capillary m.p. method* [+]

cond, *electrical conductivity*

cryom, *cryometry*

dens, *density*

diel, *dielectric constant*

dilat, *dilatometry*

DSC, *differential scanning calorimetry*

DTA, *differential thermal analysis*

GLC, *gas-liquid chromatography*

h-w, *hot-wire method*

micr, *microscopic observation*

microcal, *Calvet microcalorimetry*

NMR, *nuclear magnetic resonance*

phot, *photometric method*

TA, *thermal analysis*

vis, *visual observation*

vis-pol, *visual-polythermal method*

X-ray, *X-ray diffractometry*

Other abbreviations (listed hereafter) were borrowed from the terminology used by French authors (e.g., Gallot & Skoulios, 1966) to designate a number of structurally defined phases:

BO, "structure à rubans *bidimensionnelle oblique*"

BR, "structure à rubans *bidimensionnelle rectangulaire*"

D, "structure à *disques*"

F, *fondu*

I, "structure à *îlots*"

LC, "structure *lamellaire cristalline*"

LL, "structure *lamellaire labile*"

LSC, "structure *lamellaire semi-cristalline*"

The notation *nstr* has been introduced to indicate any *non-specified transition*, i.e., any tran-sition for which neither phenomenological nor structural information for the phases in equilib-rium was given.

Finally, figures which in the original papers were either given in brackets for any reason or indicated as provisional are here reported in brackets, while those indicated as approximate are marked with the symbol ≃.

[+] This method, however, has been expressly mentioned only when employed in a given case al-ternatively to a different procedure.

TABLE 1. Temperatures and enthalpies of phase transitions of alkali n.alkanoates

a) Li salts

Anion	Transition	T/K	$\dfrac{\Delta H}{\text{kJ mol}^{-1}}$	Method	Year	Ref.	Notes
Methanoate (Formate) CHO_2	fusion	552–553			1922	1	
	nstr	505		vis-pol	1956	2	
	nstr	388					
	nstr	360					
	fusion	546		vis-pol	1958	3	a
	fusion	546			1968	4	
	fusion	546	16.2	DSC	1975	5	
	nstr	496±2	1.8				
Ethanoate (Acetate) $C_2H_3O_2$	fusion	559			1922	1	
	fusion	545		TA	1930	6	
	fusion	564		vis-pol	1956	7	
	nstr	530–536					
	fusion	557		vis-pol	1961	8	
	fusion	563			1964	9	
	fusion	564			1968	4	
	fusion	562		DTA	1970	10	
	fusion	553		DTA	1970	11	
	fusion	553			1971	12	
	fusion	561		vis-pol	1974	13	
		557		DTA			
	nstr	405		vis-pol	1974	14	
	fusion	557±2	11.9	DSC	1975	5	
Propanoate (Propionate) $C_3H_5O_2$	nstr	538		vis-pol	1956	2	
	fusion	602		vis-pol	1958	15	
	fusion	600±1		DSC	1972	16	
	fusion	606.8	15.9	DSC	1975	5	
	fusion (m)	584	17.8				b
	nstr	533±2	3.3				
Butanoate (Butyrate) $C_4H_7O_2$	nstr	371		vis-pol	1956	2	
	fusion	602		vis-pol	1958	17	
	fusion	603		vis-pol	1961	8	

a: extrapolated
b: metastable

TABLE 1 (Continued); a) Li salts (Continued)

	fusion	583±1		DSC	1972	16
	fusion	591.7	20.8	DSC	1975	5
Pentanoate (Valerate) $C_5H_9O_2$	fusion	585		vis-pol	1961	8
	fusion	573±1		DSC	1972	16
	fusion nstr	576.0±0.2 201±2	21.2±0.2 1.4±0.1	DSC	1977	18
Hexanoate (Caproate) $C_6H_{11}O_2$	fusion	570		vis-pol	1961	8
	fusion	555±1		DSC	1972	16
	fusion	562	23.8±0.3	DSC	1977	18
Heptanoate (Enanthate) $C_7H_{13}O_2$	fusion	547±1		DSC	1972	16
	fusion nstr	550 317.9±0.8	25.4±0.3 5.69±0.08	DSC	1977	18
Octanoate (Caprylate) $C_8H_{15}O_2$	fusion nstr	539.2±0.3 217.5±0.8	25.8±0.3 1.09±0.04	DSC	1977	19
Nonanoate (Pelargonate) $C_9H_{17}O_2$	fusion nstr	526.1±0.3 346.2±0.1	27.7±0.3 10.1±0.1	DSC	1977	19
Decanoate (Caprate) $C_{10}H_{19}O_2$	fusion $LC_1 \rightarrow LC_2$	511 312		X-ray	1966	20
	fusion nstr	517.8±0.5 308.5±0.3	29.5±0.3 3.47±0.04	DSC	1977	19
Hendecanoate (Undecanoate) $C_{11}H_{21}O_2$	fusion nstr	507.2±0.3 360.3±0.1	28.7±0.5 11.5±0.2	DSC	1977	19
Dodecanoate (Laurate) $C_{12}H_{23}O_2$	fusion	502.4–503.0			1916	21
	fusion $LC_2 \rightarrow BR$ $LC_1 \rightarrow LC_2$	512 502 340		X-ray	1966	20
	waxy → isotropic solid → waxy	510 500		micr	1970	22
	fusion nstr nstr	506 503 345.8±0.6	18.9±0.3 11.7±1.3 6.07±0.21	DSC	1977	19
Tetradecanoate (Myristate) $C_{14}H_{27}O_2$	fusion	496.0–497.4			1916	21
	fusion	512		X-ray	1966	20

TABLE 1 (Continued); a) Li salts (Continued)

	$BR_1 \rightarrow BR_2$	504		*X-ray*	*1966*	*20*	
	$LC_2 \rightarrow BR_1$	483					
	$LC_1 \rightarrow LC_2$	360					
	waxy → isotropic	506		micr	1970	22	
	solid → waxy	481					
Hexadecanoate	form III → melt	497–498		vis	1943	23	*a*
(Palmitate)	form II → form III	460					
$C_{16}H_{31}O_2$	form I → form II	375					
	fusion	496	24.10	DTA	1945	24	
		496		micr			
	nstr	464	16.32	DTA			
		469		micr			
	inter-crystal	374	14.23	DTA			
		375		micr			
	fusion	497–498			1916	21	
	fusion	496		X-ray	1966	20	
	$BR_1 \rightarrow BR_2$	484					
	$LC_2 \rightarrow BR_1$	463					
	$LC_1 \rightarrow LC_2$	375					
	waxy I → isotropic	496		micr	1970	22	
	waxy II → waxy I	488					
	solid → waxy II	470					
	form III → melt	495.3±0.6	23.4±0.2	DSC	1976	25	*b*
	form II → form III	481.3±2.0					*c*
	form I → form II	379.0±0.2	21.3				*c*
Octadecanoate	fusion	409			1938	26	
(Stearate)	nstr	351					
$C_{18}H_{35}O_2$	nstr	450	12.0	DTA	1949	27	
	nstr	364	13.1				
	nstr	322					
	nstr	497±3		DTA	1949	28	
	nstr	458±3					
	nstr	386±3					
	waxy → isotropic	494		DTA	1950	29	
	crystalline I → waxy	458					
	crystalline II → cryst. I	387					
	fusion	502		phot	1955	30	
	fusion	502		dens	1955	31	
	nstr	449					
	nstr	388					
	clearing	502		DTA, micr	1957	32	*d*
	fusion	498					
	nstr	473					
	nstr	390					

a: form III, "resembling the waxy forms of Na salts"
b: form III, "viscous, double refracting fluid mesophase"
c: forms I and II, crystalline solid phases
d: temperatures taken in correspondence with the peak maxima

TABLE 1 (Continued); a) Li salts (Continued)

fusion	493.7–494.7				1916	21
fusion	502		X-ray		1962	33
nstr	488					
nstr	463					
fusion	≈498		DTA		1965	34
nstr	≈491					
nstr	467					
nstr	390.7					
nstr	460±5		NMR		1965	35
nstr	387					
fusion	502		X-ray		1966	20
$BR_1 \rightarrow BR_2$	488					
$LC_2 \rightarrow BR_1$	463					
$LC_1 \rightarrow LC_2$	395					
waxy I → isotropic	499		micr		1970	22
waxy II → waxy I	488					
solid → waxy II	464					
nstr	464÷468		DSC		1971	36
nstr	383÷385					

b) Na salts

Anion	Transition	T/K	$\dfrac{\Delta H}{\text{kJ mol}^{-1}}$	Method	Year	Ref.	Notes
Methanoate (Formate) CHO_2	fusion	526			1903	37	
	fusion	528±1			1921	38	
	fusion	524–526			1932	39	
	fusion	531			1939	40	
	fusion	528	17±4		1951	41	a
	fusion	531		vis-pol	1954	42	
	nstr	515		vis-pol	1956	2	
	fusion	531			1966	43	
	fusion	530.7	17.0±0.4	cryom	1968	44	
	fusion	530.7	17.2	DSC	1969	45	
	nstr	502	1.2				
	fusion	545		micr	1970	22	
	fusion	528 }	17.6±2.1	DTA	1974	46	b
	nstr	–					
Ethanoate (Acetate) $C_2H_3O_2$	fusion	592			1857	47	
	fusion	595–597			1908	48	

a: enthalpy value estimated
b: the reported enthalpy value was taken on cooling and refers to crystallization + nstr

TABLE 1 (Continued); b) Na salts (Continued)

fusion	593.2		TA	1915	49	
fusion	603			1922	1	
fusion	593–594			1932	39	
fusion	601.5		TA	1939	50	
fusion	604		vis-pol	1954	42	
nstr	21.1		ad-cal	1955	51	
fusion	599		vis-pol	1956	52	
nstr	527					
fusion	610		vis-pol	1956	7	
nstr	596					
nstr	511–513		vis-pol	1956	2	
nstr	403					
nstr	391					
nstr	331					
nstr	599		vis-pol	1958	53	
fusion	600		vis-pol	1958	54	
nstr	583–584					
fusion	601		vis-pol	1960	55	
fusion	608		vis-pol	1964	56	
fusion	602.4±0.2	(28±8)		1966	43	*a*
fusion	601.3±0.2	18.4±0.4	cryom	1968	57	
fusion	595		vis-pol	1969	58	
fusion	600		micr	1970	22	
fusion	605		DTA	1970	11	
fusion	598			1971	12	
fusion	597±1		DSC	1972	16	
fusion	604	17.6±2.5	DTA	1973	59	
fusion	608		DTA	1974	60	*b*
fusion	605		vis-pol	1974	61	
	610		cond			
fusion	601	15±2	DTA	1974	62	*c*
fusion	601.3	17.9	DSC	1975	5	
nstr	527±15	(0.2)				
nstr	465±3	(0.4)				
nstr	414±10	(0.3)				
fusion	605		DTA	1976	63	
nstr	337					

a: enthalpy value estimated
b: temperature taken in correspondence with the peak maximum
c: the reported enthalpy value was taken on cooling

TABLE 1 (Continued); b) Na salts (Continued)

	fusion	607±2		DTA	1977	64	
Propanoate (Propionate) $C_3H_5O_2$	fusion	571		vis-pol	1954	42	
	nstr	560		vis-pol	1956	2	
	nstr	490					
	nstr	468					
	nstr	350					
	fusion	560±1			1969	65	
	fusion	563		micr	1970	22	
	fusion	566	(8.8)	DSC	1970	66	
	fusion	561±1		DTA	1971	67	
	nstr	498±1					
	nstr	475±1					
	fusion	560±1		DSC	1972	16	
	fusion	563	12.6±2.1	DTA	1974	46	a
	fusion	562.4	13.4	DSC	1975	5	
	nstr	494 }	7.4				
	nstr	470.2					
	clearing	561		DTA	1976	63	
	fusion	545					
	nstr	478					
Butanoate (Butyrate) $C_4H_7O_2$	2nd fusion	≈523		micr	1910	68	b, c
		≈583		capill			
	1st fusion	≈493		micr			d
		≈483		capill			
	fusion	559		TA	1935	69	
	fusion	603		vis-pol	1954	42	b
	nstr	589		vis-pol	1956	2	
	nstr	525					
	nstr	505					
	nstr	390					
	neat → isotropic	594		micr	1970	22	
	solid → neat	535					
	clearing	605	(1.5)	DSC	1970	66	
	fusion	529	(6.2)				
	nstr	519					
	nstr	451					
	clearing	597±1		DTA	1971	67	
	fusion	525±1					
	nstr	510±1					
	nstr	449±1					

a: the reported enthalpy value was taken on cooling
b: actually clearing
c: no explanation for the large discrepancy in the original paper
d: actually fusion

TABLE 1 (Continued); b) Na salts (Continued)

	clearing	597		DSC	1972	16	
	fusion	524					
	clearing	600.4±0.2	2.22±0.04	DSC	1975	70	
	fusion	524.5±0.5	10.4±0.1				
	nstr	508.4±0.5	1.8±0.1				
	nstr	498.3±0.3	0.75±0.04				
	nstr	489.8±0.2	0.54±0.04				
	nstr	450.4±0.5	1.72±0.04				
	clearing	585		DTA	1976	63	
	fusion	517					
	nstr	479					
	nstr	393					
Pentanoate (Valerate) $C_5H_9O_2$	2nd fusion	≈608			1910	68	a
	1st fusion	≈483					b
	fusion	630		vis-pol	1954	42	a
	nstr	489		vis-pol	1956	2	
	nstr	482					
	nstr	453					
	neat → isotropic	617		micr	1970	22	
	solid → neat	514					
	clearing	628±1		DTA	1971	67	
	fusion	501±1					
	nstr	479±1					
	clearing	626±1		DSC	1972	16	
	fusion	502±1					
	fusion	629		vis-pol	1972	71	a
	clearing	631±4	2.0±0.1	DSC	1977	18	
	fusion	498±2	10.9±0.1				
Hexanoate (Caproate) $C_6H_{11}O_2$	2nd fusion	≈623			1910	68	a
	1st fusion	≈498					b
	fusion	498–508			1930	72	
	neat → isotropic	623		dilat, h-w	1941	73	
	nstr	499		dilat, h-w			
	nstr	480		dilat			
	nstr	445		dilat			
	fusion	638		vis-pol	1954	42	a
	nstr	615		vis-pol	1956	2	
	nstr	499					
	nstr	476					
	neat → isotropic	634		micr	1970	22	
	superwaxy → neat	508					
	solid → superwaxy	483					
	clearing	631±1		DSC	1972	16	
	fusion	505±1					

a: actually clearing
b: actually fusion

TABLE 1 (Continued); b) Na salts (Continued)

	clearing	605		DTA	1976	63	
	fusion	491					
	nstr	481					
	clearing	639.0±0.5	1.7±0.1	DSC	1977	18	
	fusion	499.6±0.6	10.8±0.2				
	nstr	473±2	0.9±0.1				
	nstr	386±2	1.5±0.1				
Heptanoate (Enanthate) $C_7H_{13}O_2$	2nd fusion	≈623			1910	68	a
	1st fusion	≈513					b
	neat → isotropic	636		micr	1970	22	
	superwaxy → neat	515					
	solid → superwaxy	471					
	clearing	634±1		DSC	1972	16	
	fusion	511±1					
	clearing	642±2	1.5±0.1	DSC	1977	18	
	fusion	509±2	9.0±0.2				
Octanoate (Caprylate) $C_8H_{15}O_2$	2nd fusion	≈628			1910	68	a
	1st fusion	≈498					b
	nstr	620		dilat	1941	73	a
	nstr	516					b
	nstr	461					
	nstr	387					
	nstr	348					
	neat → isotropic	633		micr	1970	22	
	superwaxy → neat	516					
	solid → superwaxy	462					
	clearing	595		DTA	1976	63	
	fusion	495					
	nstr	409					
	nstr	390					
	clearing	635.5±0.6	1.34±0.04	DSC	1977	19	
	fusion	512	8.5±0.1				
	nstr	385	(6.3)				
Nonanoate (Pelargonate) $C_9H_{17}O_2$	2nd fusion	≈515			1910	68	a
	1st fusion	≈491					b
	neat → isotropic	628		micr	1970	22	
	superwaxy → neat	516					
	solid → superwaxy	458					
	clearing	630.6±0.8	1.30±0.04	DSC	1977	19	
	fusion	514±2	8.5±0.1				
	nstr	412±3	(10.0)				
Decanoate (Caprate) $C_{10}H_{19}O_2$	2nd fusion	≈591			1910	68	a
	1st fusion	≈493					b

a: actually clearing
b: actually fusion

TABLE 1 (Continued); b) Na salts (Continued)

	nstr	616		dilat	1941	73	*a*
	nstr	520					*b*
	nstr	456					
	nstr	412					
	nstr	358–373					
	neat → isotropic	621		micr	1970	22	
	superwaxy → neat	518					
	waxy → superwaxy	454					
	solid → waxy	413					
	clearing	622.7±0.6	1.17±0.04	DSC	1977	19	
	fusion	516.8±0.6	8.4±0.2				
	nstr	410±2	(8.4)				
Hendecanoate (Undecanoate) $C_{11}H_{21}O_2$	clearing	611.8±0.8	1.09±0.04	DSC	1977	19	
	fusion	513.6±0.9	7.8±0.2				
	nstr	417.9±0.7	(10.5)				
Dodecanoate (Laurate) $C_{12}H_{23}O_2$	fusion	528–533			1899	74	
	2nd fusion	≈583			1910	68	*a*
	1st fusion	≈493					*b*
	nstr	583		vis	1938	75	*a, c*
	nstr	499					*d*
	neat → isotropic	609		vis	1941	76	
	subneat → neat	517		h–w			
	superwaxy → subneat	493		h–w			
	waxy → superwaxy	455		h–w			
	nstr	609		dilat, h–w	1941	73	*a*
	nstr	528		dilat			
	nstr	493		dilat, h–w			
	nstr	455		dilat			
	nstr	414		dilat			
	nstr	373		dilat			
	neat → isotropic	597	3.4	cal	1941	77	
	subneat → neat	–					*e*
	superwaxy → subneat	493	8.2				
	waxy → superwaxy	460	2.5				
	subwaxy → waxy	403	13.8				
	curd → subwaxy	371	4.8				
	fusion	600–606		DTA	1952	78	*a*
	neat → isotropic	598		X-ray	1961	79	
	subneat → neat	525					
	"new phase" → subneat	488					
	superwaxy → "new phase"	473					
	waxy → superwaxy	456					
	subwaxy → waxy	415					
	superwaxy → subneat	>473		NMR	1965	80	
	waxy → superwaxy	463–465					
	subwaxy → waxy	420–422					

a: actually clearing
b: actually fusion
c: "upper temperature limit of existence of liquid crystalline soap phase"
d: "upper temperature limit of existence of crystalline soap"
e: absent according to the authors

TABLE 1 (Continued); b) Na salts (Continued)

	crystalline → subwaxy	377		*NMR*	*1965*	*80*
	genotypical	315				
	neat → isotropic	602		micr	1970	22
	subneat → neat	519				
	superwaxy → subneat	491				
	waxy → superwaxy	452				
	solid → waxy	414				
	neat → isotropic	592		DTA	1973	81	*a*
	subneat → neat	514					*a*
	disc phase → subneat	497					*a*
	superwaxy → disc phase	465.7					*a*
	waxy II → superwaxy	456					*a, b*
	waxy I → waxy II	448.7					*b, c*
	subwaxy II → waxy I	414					*a*
	subwaxy I → subwaxy II	394					*a*
	crystalline → subwaxy I	372					*a*
	clearing	593		DTA	1976	63
	fusion	513				
	nstr	502				
	nstr	413				
	nstr	373				
	clearing	600.7±0.8	0.88±0.04	DSC	1977	19
	fusion	515±2	7.2±0.3			
	nstr	411±2	(7.9)			
Tetradecanoate (Myristate) $C_{14}H_{27}O_2$	fusion	≈523			1899	74
	2nd fusion	≈603			1910	68	*d*
	1st fusion	≈513					*e*
	nstr	307		various	1936	82	*f*
	nstr	583, 589		micr, vis	1941	73
	nstr	518		dilat, h-w		
	nstr	490		dilat, h-w		
	nstr	449		h-w		
	nstr	414		dilat, h-w		
	nstr	380		dilat, h-w		
	neat → isotropic	570	2.05	cal	1941	77
	subneat → neat	506	5.59			
	superwaxy → subneat	488	7.99			
	waxy → superwaxy	–					*g*
	subwaxy → waxy	406	14.25			
	curd → subwaxy	379	14.75			
	curd → curd	353	0.90			
	neat → isotropic	588		micr	1941	83
		582		h-w		
		(585)		dilat		
	subneat → neat	519		micr		
		524		h-w		
		530		dilat		

a: soap melted under vacuum
b: soap dried at 120 °C (24 h)
c: second order transition
d: actually clearing
e: actually fusion
f: "monotrop Umwandlung"
g: absent according to the authors

TABLE 1 (Continued); b) Na salts (Continued)

	superwaxy → subneat	491		micr	*1941*	*83*	
		480		h–w			
		477		dilat			
	waxy → superwaxy	450		micr			
		448		h–w			
		449		dilat			
	subwaxy → waxy	414		micr			
		407		h–w			
		408		dilat			
	curd → subwaxy	371		micr			
		386		dilat			
	nstr	352		dilat			
	neat → isotropic	578		X-ray	1961	79	
	subneat → neat	521					
	superwaxy → subneat	483					
	waxy → superwaxy	455					
	subwaxy → waxy	415					
	superwaxy → subneat	475–477		NMR	1965	80	
	waxy → superwaxy	455					
	–	(435)					
	subwaxy → waxy	414					
	crystalline → subwaxy	381					
	crystalline → crystalline	353					
	fusion	581		dilat	1969	84	*a*
	neat → isotropic	584		micr	1970	22	
	subneat → neat	519					
	superwaxy → subneat	488					
	waxy → superwaxy	444					
	subwaxy → waxy	411					
	solid → subwaxy	386					
	clearing	589		DTA	1976	63	
	fusion	520					
	nstr	483					
	nstr	411					
	nstr	384					
Hexadecanoate (Palmitate) $C_{16}H_{31}O_2$	fusion	543			1899	74	
	2nd fusion	≈538, ≈589		micr	1910	68	*a, b*
	1st fusion	≈493, ≈488					*c*
	nstr	563		vis	1930	85	*a, d*
	nstr	407					*e*
	genotypical	337	0.96	TA, cal	1935	86	
	nstr	315.9		various	1936	82	*f*
	neat → nigre	570±3		dilat, micr	1939	87	
	subneat → neat	528±2					
	waxy → subneat	468±1					
	curd → waxy	398					*g*

a: actually clearing
b: no explanation for the large discrepancy in the original paper
c: actually fusion
d: "upper temperature of existence of liquid crystalline soap phase"
e: "upper temperature of existence of crystalline soap"
f: "monotrop Umwandlung"
g: mean value between 409±1 and 390±1 K

TABLE 1 *(Continued); b) Na salts (Continued)*

neat → isotropic	565		dilat	1939	88	
subneat → neat	526					
waxy → subneat	481					
subwaxy → waxy	408					
curd → subwaxy	390					
neat → isotropic	563		X-ray	1940	89	
subneat → neat	530					
waxy → subneat	478					
subwaxy → waxy	411					
curd II → subwaxy	390					
curd I → curd II	340					
neat → isotropic	568		vis	1941	76	
subneat → neat	526		h-w			
superwaxy → subneat	481		h-w			
waxy → superwaxy	445		h-w			
formation of waxy	408, 390		h-w, dilat			
nstr	568		dilat, h-w	1941	73	*a*
nstr	526		dilat, h-w			
nstr	481		dilat, h-w			
nstr	445		h-w			
nstr	408		dilat, h-w			
nstr	390		dilat, h-w			
neat → isotropic	565	2.1	cal	1941	77	
subneat → neat	510	6.4				
superwaxy → subneat	482	7.7				
waxy → superwaxy	-					*b*
subwaxy → waxy	408	16.2				
curd → subwaxy	387	20.1				
superwaxy → subneat	479		diel	1956	90	
waxy → superwaxy	435					
subwaxy → waxy	410					
crystalline → subwaxy	391					
neat → isotropic	573		X-ray	1961	79	
subneat → neat	527					
superwaxy → subneat	484					
waxy → superwaxy	449					
subwaxy → waxy	413					
superwaxy → subneat	>473		NMR	1965	80	
waxy → superwaxy	445-448					
subwaxy → waxy	415-417					
crystalline → subwaxy	386					
crystalline → crystalline	355					
formation of waxy	409.7		DTA	1965	34	
formation of subwaxy	389.7					
fusion	570		dilat	1969	84	*a*
neat → isotropic	575		micr	1970	22	
subneat → neat	524					
superwaxy → subneat	485					
waxy → superwaxy	441					
subwaxy → waxy	409					
solid → subwaxy	390					

a: actually clearing
b: absent according to the authors

TABLE 1 (Continued); b) Na salts (Continued)

Octadecanoate (Stearate) $C_{18}H_{35}O_2$	fusion	≈533				1899	74
	2nd fusion	≈543, ≈578		micr		1910	68 *a, b*
	1st fusion	≈498, ≈493					*c*
	genotypical	344	1.02	TA, cal		1935	86
	nstr	324.7		various		1936	82 *d*
	neat → isotropic	561		vis		1941	76
	subneat → neat	530		h–w			
	superwaxy → subneat	476		h–w			
	waxy → superwaxy	440		h–w			
	formation of waxy	405		h–w			
	formation of subwaxy	390		dilat			
	nstr	561		dilat, h–w		1941	73 *a*
	nstr	530		dilat			
		531–533		h–w			
	nstr	471		dilat			
		478–479		h–w			
	nstr	440		dilat			
		439–442		h–w			
	nstr	405		dilat			
		403		h–w			
	nstr	390		dilat			
	nstr	363		dilat			
	neat → isotropic	553	2.24	cal		1941	77
	subneat → neat	511	6.53				
	superwaxy → subneat	481	6.69				
	waxy → superwaxy	–					*e*
	subwaxy → waxy	407	16.86				
	curd → subwaxy	387	21.67				
	curd → curd	362	3.95				
	clearing	555		DTA, vis		1950	91
	nstr	553	0.59	DTA		1951	92 *a*
	nstr	544–546	0.08				
	nstr	516–524	8.03				
	nstr	471	4.90				
	nstr	402–403					
	nstr	385–391 }	29.29				
	nstr	359–369					
	fusion	565–571		DTA		1952	78 *a*
	fusion	581		vis-pol		1954	93 *a*
	neat → isotropic	556		phot		1955	30
	subneat → neat	531					
	minor transition	493					
	superwaxy → subneat	461					
	waxy → superwaxy	438					
	subwaxy → waxy	405					
	waxy → superwaxy	438		NMR		1956	94
	subwaxy → waxy	403					
	supercurd → subwaxy	387					

a: actually clearing
b: no explanation for the large discrepancy in the original paper
c: actually fusion
d: "monotrop Umwandlung"
e: absent according to the authors

TABLE 1 (Continued); b) Na salts (Continued)

	supercurd → subwaxy	386–387	NMR	1960	95	
	neat → isotropic	563	X-ray	1961	79	
	subneat → neat	529				
	superwaxy → subneat	483				
	waxy → superwaxy	448				
	subwaxy → waxy	406				
	formation of neat	519±3	NMR	1964	96	
	supercurd → subwaxy	387±1				
	superwaxy → subneat	451–454	NMR	1965	80	
	waxy → superwaxy	430–432				
	subwaxy → waxy	404				
	crystalline → subwaxy	387				
	crystalline → crystalline	358				
	formation of neat	515	DTA	1965	34	a
	nstr	≈475				
	nstr	≈469				
	nstr	≈439				
	subwaxy → waxy	409				
	supercurd → subwaxy	391				
	curd → supercurd	366				
	neat → isotropic	556	micr	1970	22	
	subneat → neat	528				
	superwaxy → subneat	481				
	waxy → superwaxy	438				
	subwaxy → waxy	408				
	solid → subwaxy	389				
	nstr	403÷406	DSC	1971	36	
	nstr	384÷389				
	nstr	355÷363				
	nstr	321				
	neat → isotropic	558	DSC	1976	97	
		558	GLC			
	subneat → neat	529–533	DSC			
		530	GLC			
	superwaxy → subneat	478–480	DSC			
		478	GLC			
	waxy → superwaxy	–				b
	subwaxy → waxy	403–409	DSC			
		408	GLC			
	curd → subwaxy	388–392	DSC			
		390	GLC			
	curd → curd	363–368	DSC			
		368	GLC			
Eicosanoate	nstr	535	dilat, h–w	1941	73	
(Arachidate)	nstr	473	dilat, h–w			
$C_{20}H_{39}O_2$	nstr	436	dilat, h–w			
	nstr	404	dilat, h–w			
	nstr	383	dilat			
	formation of waxy	408	DTA	1965	34	a
	formation of subwaxy	393				

a: data taken on the σ modification
b: absent according to the author

TABLE 1 (Continued); b) Na salts (Continued)

Docosanoate (Behenate) $C_{22}H_{43}O_2$	Transition	T/K		Method	Year	Ref.	Notes
Docosanoate	nstr	537		dilat, h–w	1941	73	
(Behenate)	nstr	475					
$C_{22}H_{43}O_2$	nstr	431					
	nstr	392					
	nstr	365					
	formation of waxy	407		DTA	1965	34	*a*
	formation of subwaxy	394					

c) K salts

Anion	Transition	T/K	$\dfrac{\Delta H}{kJ\ mol^{-1}}$	Method	Year	Ref.	Notes
Methanoate (Formate) CHO_2	fusion	430			1903	37	
	fusion	440.7±0.5			1921	38	
	nstr	430		vis-pol	1956	2	
	nstr	408					
	nstr	333					
	fusion	440		vis-pol	1958	98	
	fusion	440		vis-pol	1958	99	
	fusion	441		vis-pol	1961	100	
	fusion	440			1966	43	
	fusion	441.9	11.9±0.1	cryom	1968	44	
	fusion	441.9	11.8	DSC	1969	45	
	nstr	418	0.8				
	fusion	449		micr	1970	22	
	fusion	442	11.5±0.1	microcal	1971	101	*b*
Ethanoate (Acetate) $C_2H_3O_2$	fusion	565			1857	47	
	fusion	568.2			1915	49	
	fusion	565		TA	1930	6	
	fusion	570			1935	69	
	fusion	565		TA	1938	102	
	fusion	575		vis-pol	1956	52	
	nstr	428		vis-pol	1956	2	
	nstr	331					
	fusion	583		vis-pol	1957	103	
	nstr	565–566					
	fusion	583.7		vis-pol	1958	53	
	nstr	569					

a: data taken on the σ modification
b: enthalpy change cumulative of fusion + premelting, the latter estimated as ≈4.4 kJ mol^{-1}

TABLE 1 (Continued); c) K salts (Continued)

	fusion	579		vis-pol	1958	104	
	fusion	574		vis-pol	1958	105	
	fusion	575			1961	100	
	fusion nstr	577±1 423	9.7	dilat, DTA	1966	106	
	fusion nstr nstr nstr	577.9±0.2 (503) (433) (353)	19±8		1966	43	*a*
	fusion	578.7±0.3	15.2±0.3	cryom	1968	107	
	fusion	579–581		vis-pol	1969	58	
	fusion	575		DTA	1970	11	
	fusion	580		micr	1970	22	
	fusion	565±1		DSC	1972	16	
	monoclinic → orthorhombic monoclinic → monoclinic	428 ≃348		X-ray	1972	108	*b*
	fusion	581	22.6±2.5	DTA	1973	59	
	fusion	579		vis-pol	1974	61	
	fusion	575		vis-pol	1975	109	
	fusion nstr	578.7 422.2	15.2 0.4	DSC	1975	5	
	fusion	581		vis-pol	1975	110	
	fusion nstr	571–573 413–423		DTA	1976	111	
	fusion	584±2		DTA	1977	64	
Propanoate (Propionate) $C_3H_5O_2$	nstr nstr	603 341		vis-pol	1956	2	
	fusion	638		vis-pol	1958	105	
	fusion	639		vis-pol	1961	100	
	fusion	628		micr	1970	22	
	fusion	631±1		DSC	1972	16	
	fusion nstr nstr	638.3 352.5 258±2	20.1 1.7 0.3	DSC	1975	5	
Butanoate (Butyrate) $C_4H_7O_2$	2nd fusion	≃618 ≃628		micr capill	1910	68	

a: enthalpy value estimated
b: with eightfold decrease in cell volume

TABLE 1 (Continued); c) K salts (Continued)

	1st fusion	≈568		micr	1910	68	
		≈588		capill			
	nstr	618		vis-pol	1956	2	a
	nstr	553–558					
	nstr	463					
	fusion	677		vis-pol	1958	112	b
	neat → isotropic	≈663		micr	1970	22	
	solid → neat	600					
	clearing	670±1		DSC	1972	16	
	fusion	620±1					
	clearing	677.3±0.5	4.98±0.08	DSC	1975	70	
	fusion	626.1±0.7	10.84±0.08				
	nstr	562.2±0.6 } 4.44±0.08					
	nstr	540.8±1.1					
	nstr	467.2±0.5 } 1.17±0.08					
	nstr	461.4±1.0					
	nstr	143±2 } 0.7		DSC	1975	5	
	nstr	123±2					
Pentanoate (Valerate) $C_5H_9O_2$	nstr	580		vis-pol	1956	2	a
	fusion	717		vis-pol	1961	8	b
	neat → isotropic	>673		micr	1970	22	
	solid → neat	592					
	clearing	687±1		DSC	1972	16	
	fusion	585±1					
	clearing	716±2	4.8±0.2	DSC	1977	18	
	fusion	586.6±0.7	15.5				
	nstr	399.5±0.9	0.46±0.08				
Hexanoate (Caproate) $C_6H_{11}O_2$	nstr	575		vis-pol	1956	2	a
	fusion	717.7		vis-pol	1959	113	b
	neat → isotropic	>673		micr	1970	22	
	solid → neat	578					
	clearing	691±1		DSC	1972	16	
	fusion	580±1					
	clearing	725.8±0.8	4.2	DSC	1977	18	
	fusion	581.7±0.5	16.9±0.3				
Heptanoate (Enanthate) $C_7H_{13}O_2$	2nd fusion	≈673		micr	1910	68	
	1st fusion	≈498					
	fusion	725		vis-pol	1961	100	b
	fusion	723		vis-pol	1961	8	b

a: actually fusion
b: actually clearing

TABLE 1 (Continued); c) K salts (Continued)

	neat → isotropic	>673		micr	1970	22	
	solid → neat	574					
	clearing	691±1		DSC	1972	16	
	fusion	569±1					
	clearing	722±3	3.43±0.08	DSC	1977	18	
	fusion	571.3±0.9	17.7±0.2				
	nstr	345.4±0.6	5.5±0.2				
	nstr	332.0±0.8	4.8±0.2				
Octanoate (Caprylate) $C_8H_{15}O_2$	nstr	328±3		X-ray	1952	114	
	fusion	717		vis-pol	1961	100	*a*
	nstr	558±5		NMR	1963	115	*b*
	nstr	327±2					
	LL → F	652		X-ray	1966	116	
	LC_2 → LL	551					
	LC_1 → LC_2	331					
	neat → isotropic	>673		micr	1970	22	
	solid → neat	564					
	clearing	712±2	2.64±0.08	DSC	1977	117	
	fusion	560.6±0.8	18.3±0.2				
	nstr	326.6±0.1	9.5±0.2				
Nonanoate (Pelargonate) $C_9H_{17}O_2$	fusion	694		vis-pol	1961	100	*a*
	neat → isotropic	>673		micr	1970	22	
	solid → neat	555					
	clearing	707.4±0.8	2.5±0.1	DSC	1977	117	
	fusion	549.1±0.8	14.8±0.4				
	nstr	390.5±0.4	1.3±0.1				
	nstr	367.5±0.5	11.3±0.1				
Decanoate (Caprate) $C_{10}H_{19}O_2$	nstr	349±3		X-ray	1952	114	
	mesomorphic transition	519		NMR	1961	118	
	crystal transition	347					
	LL → F	649		X-ray	1966	116	
	LC_2 → LL	547					
	LC_1 → LC_2	350					
	neat → isotropic	>673		micr	1970	22	
	solid → neat	550					
	clearing	696	2.09±0.08	DSC	1977	117	
	fusion	544	12.2±0.3				
	nstr	382.9±0.6	1.00±0.04				
	nstr	348.7±0.5	8.20±0.08				
Hendecanoate (Undecanoate) $C_{11}H_{21}O_2$	clearing	691.4±0.8	1.84±0.04	DSC	1977	117	
	fusion	541±2	12.0±0.2				
	nstr	400±2	2.8±0.2				
	nstr	355.0±0.6	14.9±0.3				

a: actually clearing
b: actually fusion

TABLE 1 (Continued); c) K salts (Continued)

Dodecanoate (Laurate) $C_{12}H_{23}O_2$	clearing	649		micr	1926	119	
	fusion	537					
	clearing	649.2		vis	1933	120	*a*
	fusion	537.2					
	nstr	327±3		X-ray	1952	114	
	mesomorphic transition	485		NMR	1961	118	
	crystal transition	326					
	LL → F	647		X-ray	1966	116	
	LC_3 → LL	545					
	LC_2 → LC_3	408					
	LC_1 → LC_2	328					
	neat → isotropic	668		micr	1970	22	
	solid → neat	546					
	clearing	679.2±0.5	1.84±0.04	DSC	1977	117	
	fusion	540.8±0.5	11.8±0.3				
	nstr	404±2	3.0±0.3				
	nstr	327.2±0.4	11.7±0.1				
Tetradecanoate (Myristate) $C_{14}H_{27}O_2$	nstr	334±3		X-ray	1952	114	
	mesomorphic transition	477		NMR	1961	118	
	crystal transition	333					
	LL → F	615		X-ray	1966	116	
	D_4 → LL	543					
	D_3 → D_4	517					
	D_2 → D_3	503					
	D_1 → D_2	491					
	LC_3 → D_1	468					
	LC_2 → LC_3	415					
	LC_1 → LC_2	330					
	neat → isotropic	648		micr	1970	22	
	waxy → neat	544					
Hexadecanoate (Palmitate) $C_{16}H_{31}O_2$	formation of isotr. liq.	648		DTA, micr	1945	24	
	nstr	532	10.46	DTA			
		538		micr			
	nstr	447	6.28	DTA			
		446		micr			
	nstr	425	4.18	DTA			
		425		micr			
	nstr	404	3.77	DTA			
	inter-crystal	333	20.50	DTA			
		332		micr			
	nstr	341±3		X-ray	1952	114	
	2nd order transition	336	6.71	cal	1954	121	
	mesomorphic transition	463		NMR	1961	118	
	crystal transition	339					
	LL → F	616		X-ray	1966	116	
	BO_4 → LL	547					

a: designated as "the first temperature at which the homogeneous isotropic liquid becomes tur=
bid (...) on cooling"

TABLE 1 (Continued); c) K salts (Continued)

	$BO_3 \rightarrow BO_4$	533	*X-ray*	*1966*	*116*
	$BO_2 \rightarrow BO_3$	514			
	$BO_1 \rightarrow BO_2$	499			
	$? \rightarrow BO_1$	474			*a*
	$LC_3 \rightarrow ?$	462			
	$LC_2 \rightarrow LC_3$	419			
	$LC_1 \rightarrow LC_2$	335			
	neat → isotropic	635	micr	1970	22
	waxy → neat	542			
	solid → waxy	468			
Octadecanoate (Stearate) $C_{18}H_{35}O_2$	clearing	623	TA	1951	122
	nstr	351±3	X-ray	1952	114
	clearing	626	phot	1955	30
	nstr	618			
	minor transition	583			
	nstr	540			
	nstr	515			
	nstr	433–438			
	mesomorphic transition	444	NMR	1960	123
	crystal transition	335			
	clearing	618	X-ray	1961	124
	nstr	545			
	nstr	511			
	nstr	498			
	nstr	483			
	nstr	458			
	nstr	443			*b*
	nstr	538	DTA	1965	34
	nstr	≈458			
	nstr	447			
	nstr	≈436			
	nstr	421			
	nstr	407			
	nstr	≈379			
	nstr	350			
	nstr	341			
	$LL \rightarrow F$	618	X-ray	1966	116
	$BO_5 \rightarrow LL$	545			
	$BO_4 \rightarrow BO_5$	511			
	$BO_3 \rightarrow BO_4$	498			
	$BO_2 \rightarrow BO_3$	483			
	$BO_1 \rightarrow BO_2$	458			
	$LC_3 \rightarrow BO_1$	443			
	$LC_2 \rightarrow LC_3$	412			
	$LC_1 \rightarrow LC_2$	342			
	neat → isotropic	621	micr	1970	22
	waxy → neat	540			
	solid → waxy	443			

a: according to the authors, between 462 and 474 K the system instability and complexity made it difficult to state the structure surely

b: according to the authors, at 443 K the salt changes from the crystalline to the mesomorphic state; moreover (see Ref. 33), between 443 and 458 K two different structures may exist dependently on the history of the sample

TABLE 1 (Continued); c) K salts (Continued)

	nstr	422÷426		DSC	1971	36
	nstr	403÷405				
	nstr	375÷377				
	nstr	368÷369				
	nstr	336÷348				

d) Rb salts

Anion	Transition	T/K	$\dfrac{\Delta H}{\text{kJ mol}^{-1}}$	Method	Year	Ref.	Notes
Methanoate (Formate) CHO_2	fusion	443			1922	1	
	fusion	443	11.9	DSC	1969	125	
	nstr	368.1	0.25	DSC	1975	5	
Ethanoate (Acetate) $C_2H_3O_2$	fusion	519			1922	1	
	fusion	509		vis-pol	1958	54	
	nstr	489–493					
	fusion	511–513		vis-pol	1964	126	
	fusion	510		vis-pol	1965	127	
	fusion	509			1968	4	
	fusion	514	11.0	DSC	1969	125	
	nstr	498	2.2				
	fusion	514		DTA	1970	11	
	fusion	515	14.6±2.5	DTA	1973	59	
	nstr	479		X-ray	1974	14	
	fusion	510		vis-pol	1974	61	
Propanoate (Propionate) $C_3H_5O_2$	fusion	618±1		DSC	1972	16	
	fusion	623.1	14.6	DSC	1975	5	
	nstr	564.3	3.0				
	nstr	317±2	1.5				
Butanoate (Butyrate) $C_4H_7O_2$	fusion	643±1		DSC	1972	16	
	fusion	652	15.7	DSC	1975	5	
	nstr	466.0	2.3				
	nstr	346±2	1.0				
	nstr	191	2.4				
Pentanoate (Valerate) $C_5H_9O_2$	clearing	679±1		DSC	1972	16	
	fusion	635±1					
	clearing	703±2	5.2±0.3	DSC	1977	18	
	fusion	640.1±0.5	10.9±0.1				
	nstr	502.7±0.4	5.0±0.1				
	nstr	258±2	3.6±0.1				

TABLE 1 (Continued); d) Rb salts (Continued)

Hexanoate (Caproate) $C_6H_{11}O_2$	clearing fusion	693±1 611±1		DSC	1972	16
	clearing fusion nstr nstr nstr	723±2 615.4±0.7 526.0±0.4 515.9±0.9 241±2	4.6±0.2 10.3±0.1 } 2.7±0.1 3.1±0.3	DSC	1977	18
Heptanoate (Enanthate) $C_7H_{13}O_2$	clearing fusion	708±1 597±1		DSC	1972	16
	clearing fusion nstr	724.4±0.8 600 320.1±0.8	4.2 11.3±0.3 8.9±0.1	DSC	1977	18
Octanoate (Caprylate) $C_8H_{15}O_2$	clearing fusion nstr nstr	713±2 585±3 303.3 283.5±0.4	3.7±0.2 11.4±0.3 (8.4) (2.5)	DSC	1977	128
Nonanoate (Pelargonate) $C_9H_{17}O_2$	clearing fusion nstr	712±2 572.8±0.7 339	3.5±0.3 10.3±0.1 (16.7)	DSC	1977	128
Decanoate (Caprate) $C_{10}H_{19}O_2$	clearing fusion nstr nstr nstr	701.9±0.4 564±3 545±2 323 318	2.7±0.3 10.1±0.3 2.5 } (15.9)	DSC	1977	128
Hendecanoate (Undecanoate) $C_{11}H_{21}O_2$	clearing fusion nstr nstr	695±3 560±3 529.1±0.2 349.1	2.6±0.1 10.4±0.3 2.9 (21.3)	DSC	1977	128
Dodecanoate (Laurate) $C_{12}H_{23}O_2$	LL → F LSC → LL LC$_2$ → LSC LC$_1$ → LC$_2$	622 551 506 333		X-ray	1966	129
	neat → isotropic solid → neat	≈673 573		micr	1970	22
	clearing fusion nstr nstr nstr	683±3 555 515.3±0.5 360±2 334	2.0 9.0±0.2 3.8 (2.1) (18.4)	DSC	1977	128
Tetradecanoate (Myristate) $C_{14}H_{27}O_2$	LL → F LSC → LL LC$_2$ → LSC LC$_1$ → LC$_2$	621 551 497 336		X-ray	1966	129
	neat → isotropic solid → neat	≈653 564		micr	1970	22

TABLE 1 (Continued); d) Rb salts (Continued)

Hexadecanoate (Palmitate) $C_{16}H_{31}O_2$	formation of isotr. liq.	653		DTA, micr	1945	24
	nstr	530	9.58	DTA		
		529		micr		
	nstr	483		DTA		
	nstr	454		DTA		
	nstr	399	11.30	DTA		
		395		micr		
	inter-crystal	340	32.22	DTA		
		339		micr		
	LL → F	624		X-ray	1966	129
	D_2 → LL	549				
	D_1 → D_2	509				
	LSC → D_1	490				
	LC_2 → LSC	462				
	LC_1 → LC_2	351				
	neat → isotropic	648		micr	1970	22
	solid/waxy → neat	557				
Octadecanoate (Stearate) $C_{18}H_{35}O_2$	clearing	630		phot	1955	30
	nstr	541				
	nstr	495				
	nstr	433–438				
	nstr	416		X-ray, NMR	1962	130
	nstr	350				
	LL → F	627		X-ray	1966	129
	D_2 → LL	551				
	D_1 → D_2	515				
	LSC → D_1	493				
	LC_2 → LSC	450				
	LC_1 → LC_2	357				
	neat → isotropic	629		micr	1970	22
	solid/waxy → neat	554				
	nstr	445÷464		DSC	1971	36
	nstr	413÷415				
	nstr	408÷410				
	nstr	401÷402				
	nstr	351÷355				
	nstr	331÷332				
Eicosanoate (Arachidate) $C_{20}H_{39}O_2$	LL → F	609		X-ray	1966	129
	D_2 → LL	543				
	D_1 → D_2	505				
	LSC → D_1	481				
	LC_2 → LSC	442				
	LC_1 → LC_2	362				
Docosanoate (Behenate) $C_{22}H_{43}O_2$	LL → F	594		X-ray	1966	129
	BR_2 → LL	544				
	BR_1 → BR_2	528				
	BO_4 → BR_1	513				
	BO_3 → BO_4	498				
	BO_2 → BO_3	485				
	BO_1 → BO_2	464				
	LC_2 → BO_1	444				
	LC_1 → LC_2	365				

TABLE 1 (Continued); e) Cs salts

e) Cs salts

Anion	Transition	T/K	$\dfrac{\Delta H}{\text{kJ mol}^{-1}}$	Method	Year	Ref.	Notes
Methanoate (Formate) CHO_2	fusion	538			1922	1	
	fusion	537			1953	131	
	fusion	539	6.8	DSC	1969	125	
	nstr	312	4.5				
Ethanoate (Acetate) $C_2H_3O_2$	fusion	467			1922	1	
	fusion	453		vis-pol	1960	132	
	nstr	447					
	fusion	458		vis-pol	1964	9	
	fusion	453		vis-pol	1964	126	
	fusion	459		vis-pol	1965	133	
	fusion	455		vis-pol	1965	134	
	fusion	467			1968	4	
	fusion	463	12.0	DSC	1969	125	
	fusion	470			1970	11	
	fusion	465	9.6±1.7	DTA	1973	59	
	fusion	454		vis-pol	1974	135	
	fusion	460		vis-pol	1974	136	
	fusion	467.0			1976	137	
	fusion	467±2		DTA	1977	64	
Propanoate (Propionate) $C_3H_5O_2$	fusion	568±1		DSC	1972	16	
	fusion	580	11.7	DSC	1975	5	
	nstr	419±2	1.9				
	nstr	314.2	1.3				
Butanoate (Butyrate) $C_4H_7O_2$	fusion	616±1		DSC	1972	16	
	fusion	628±2	13.8	DSC	1975	5	
	nstr	343.9	1.5				
	nstr	263±3	1.3				
Pentanoate (Valerate) $C_5H_9O_2$	fusion	640±1		DSC	1972	16	
	fusion	655.5±0.9	16.6±0.3	DSC	1977	138	
	nstr	369.4±0.6	3.26±0.04				

TABLE 1 (Continued); e) Cs salts (Continued)

Hexanoate (Caproate) $C_6H_{11}O_2$	clearing fusion	668±1 624±1		DSC	1972	16
	clearing fusion nstr	672 632 385.2±0.4	4.6±0.2 10.9±0.8 1.88±0.08	DSC	1977	138
Heptanoate (Enanthate) $C_7H_{13}O_2$	clearing fusion	682±1 613±1		DSC	1972	16
	clearing fusion nstr	694±2 618.4±0.7 383.3±0.5	4.8±0.2 10.3±0.1 1.72±0.04	DSC	1977	138
Octanoate (Caprylate) $C_8H_{15}O_2$	clearing fusion nstr	698.0±0.8 607.7±0.3 373.4±0.3	4.0±0.2 10.9±0.3 1.59±0.04	DSC	1977	138
Nonanoate (Pelargonate) $C_9H_{17}O_2$	clearing fusion nstr	697.8±0.7 598.6±0.4 374.5±0.2	3.6±0.3 10.6±0.3 1.55±0.08	DSC	1977	138
Decanoate (Caprate) $C_{10}H_{19}O_2$	clearing fusion nstr	688±2 586.9±0.6 373.9±0.2	3.0±0.1 10.0±0.4 1.76±0.04	DSC	1977	138
Hendecanoate (Undecanoate) $C_{11}H_{21}O_2$	clearing fusion nstr	681 582.4±0.5 378.1±0.2	2.4±0.3 11.0±0.4 1.80±0.08	DSC	1977	138
Dodecanoate (Laurate) $C_{12}H_{23}O_2$	LL → F LC$_3$ → LL LC$_2$ → LC$_3$ LC$_1$ → LC$_2$	628 551 325 313		X-ray	1966	139
	neat → isotropic solid → neat	658 568		micr	1970	22
	clearing fusion nstr	671 569 377.1±0.4	2.0±0.1 12.2±0.3 1.8±0.1	DSC	1977	138
Tetradecanoate (Myristate) $C_{14}H_{27}O_2$	LL → F LC$_3$ → LL LC$_2$ → LC$_3$ LC$_1$ → LC$_2$	626 547 342 322		X-ray	1966	139
	neat → isotropic solid → neat	643 563		micr	1970	22
Hexadecanoate (Palmitate) $C_{16}H_{31}O_2$	formation of isotr. liq. nstr nstr nstr inter-crystal	648 642 (543) 498 368 367 335 330	15.06 27.20	DTA micr micr DTA, micr DTA micr DTA micr	1945	24

TABLE 1 (Continued); e) Cs salts (Continued)

	LL → F	621	X-ray	1966	139
	LC$_3$ → LL	544			
	LC$_2$ → LC$_3$	368			
	LC$_1$ → LC$_2$	336			
	neat → isotropic	631	micr	1970	22
	solid → neat	552			
Octadecanoate	clearing	624	phot	1955	30
(Stearate)	nstr	543			
C$_{18}$H$_{35}$O$_2$	nstr	433–438			
	nstr	373	NMR, X-ray	1962	130
	nstr	342			
	LL → F	624	X-ray	1966	139
	I → LL	543			
	LC$_3$ → I	479			
	LC$_2$ → LC$_3$	377			
	LC$_1$ → LC$_2$	343			
	neat → isotropic	618	micr	1970	22
	solid/waxy → neat	546			
	nstr	455÷457	DSC	1971	36
	nstr	401÷405			
	nstr	387÷388			
	nstr	373÷375			
	nstr	344÷345			
	nstr	332÷336			

1. Sidgwick & Gentle
2. Sokolov (1956)
3. Tsindrik
4. Ubbelohde
5. Ferloni, Sanesi & Franzosini (1975)
6. Davidson & McAllister
7. Diogenov (1956 a)
8. Sokolov, Tsindrik & Dmitrevskaya
9. Diogenov & Sarapulova (1964 a)
10. Bartholomew (1970)
11. Halmos, Meisel, Seybold & Erdey
12. Van Uitert, Bonner & Grodkiewicz
13. Sarapulova, Kashcheev & Diogenov
14. Diogenov, Erlykov & Gimel'shtein
15. Tsindrik & Sokolov (1958 a)
16. Michels & Ubbelohde
17. Tsindrik & Sokolov (1958 b)
18. Sanesi, Ferloni & Franzosini
19. Ferloni, Zangen & Franzosini (1977 a)
20. Gallot & Skoulios (1966 b)
21. Jacobson & Holmes
22. Baum, Demus & Sackmann
23. Vold, M.J. (1943)
24. Vold, R.D., & Vold (1945)
25. Vold, M.J., Funakoshi & Vold
26. Lawrence
27. Hattiangdi, Vold & Vold
28. Vold, M.J., & Vold (1949)
29. Vold, M.J., & Vold (1950)
30. Benton, Howe & Puddington
31. Benton, Howe, Farnand & Puddington
32. Cox & McGlynn
33. Gallot & Skoulios (1962)
34. Trzebowski
35. Dunell & Janzen
36. Ripmeester & Dunell (1971 b)
37. Groschuff
38. Kendall & Adler
39. Symons & Buswell
40. Takagi
41. Schwab, Paparos & Tsipouris
42. Sokolov (1954 a)
43. Hazlewood, Rhodes & Ubbelohde
44. Leonesi, Piantoni, Berchiesi & Franzosini
45. Braghetti, Berchiesi & Franzosini
46. Storonkin, Vasil'kova & Potemin (1974 a)

47. Schaffgotsch
48. Green
49. Baskov
50. Lehrman & Skell
51. Strelkov
52. Bergman & Evdokimova
53. Diogenov & Erlykov
54. Gimel'shtein & Diogenov
55. Nesterova & Bergman
56. Diogenov & Sarapulova (1964 b)
57. Piantoni, Leonesi, Braghetti & Franzosini
58. Pavlov & Golubkova
59. Potemin, Tarasov & Panin
60. Judd, Plunkett & Pope
61. Nadirov & Bakeev (1974 a)
62. Storonkin, Vasil'kova & Potemin (1974 b)
63. Roth, Meisel, Seybold & Halmos
64. Storonkin, Vasil'kova & Tarasov
65. Duruz, Michels & Ubbelohde (1969)
66. Ubbelohde, Michels & Duruz
67. Duruz, Michels & Ubbelohde (1971)
68. Vorländer
69. Bakunin & Vitale
70. Ferloni & Franzosini
71. Sokolov & Khaitina (1972)
72. Godbole & Joshi
73. Vold, M.J., Macomber & Vold
74. Krafft
75. McBain, Brock, Vold & Vold
76. Vold, M.J. (1941)
77. Vold, R.D. (1941)
78. Ravich & Nechitaylo
79. Skoulios & Luzzati
80. Lawson & Flautt
81. Madelmont & Perron
82. Thiessen & Stauff
83. Vold, R.D., Reivere & McBain
84. Skoda
85. McBain, Lazarus & Pitter
86. Thiessen & v. Klenck
87. Vold, R.D., Rosevear & Ferguson
88. Vold, R.D., & Vold
89. Chesley
90. Wirth & Wellman
91. Stross & Abrams (1950)
92. Stross & Abrams (1951)

93. Sokolov (1954 b)
94. Grant, Hedgecock & Dunell
95. Grant & Dunell (1960 b)
96. Barr & Dunell
97. Wasik
98. Dmitrevskaya
99. Sokolov & Pochtakova (1958 a)
100. Sokolov & Minich
101. Berchiesi & Laffitte
102. Lehrman & Leifer
103. Diogenov, Nurminskii & Gimel'shtein
104. Golubeva, Bergman & Grigor'eva
105. Sokolov & Pochtakova (1958 b)
106. Bouaziz & Basset
107. Braghetti, Leonesi & Franzosini
108. Hatibarua & Parry
109. Diogenov & Chumakova
110. Diogenov & Morgen (1975)
111. Poppl
112. Sokolov & Pochtakova (1958 c)
113. Pochtakova (1959)
114. Lomer
115. Janzen & Dunell
116. Gallot & Skoulios (1966 c)
117. Ferloni, Spinolo, Zangen & Franzosini
118. Grant & Dunell (1961)
119. McBain & Field (1926)
120. McBain & Field (1933)
121. Shmidt
122. Penther, Abrams & Stross
123. Grant & Dunell (1960 a)
124. Gallot & Skoulios (1961)
125. Braghetti & Berchiesi
126. Diogenov & Sarapulova (1964 c)
127. Diogenov & Gimel'shtein
128. Ferloni, Zangen & Franzosini (1977 b)
129. Gallot & Skoulios (1966 a)
130. Shaw & Dunell
131. Rice & Kraus
132. Nurminskii & Diogenov
133. Diogenov, Borzova & Sarapulova
134. Diogenov, Bruk & Nurminskii
135. Nadirov & Bakeev (1974 b)
136. Diogenov & Morgen (1974)
137. Leonesi, Cingolani & Berchiesi
138. Sanesi, Ferloni, Zangen & Franzosini
139. Gallot & Skoulios (1966 d)

Alkali branched alkanoates

These salts have received much less attention than the linear ones: in particular information is so far available only for a few lithium, sodium and potassium branched homologues with $4 \leq n_C \leq 6$.

Branching, as expected, strongly affects the fusion mechanism: as an example, Na and K *iso*butyrates, *iso*valerates and *iso*caproates exhibit in comparison with the corresponding linear isomers clearing temperatures heavily abated, although the mesomorphic liquid field might result either enlarged (in K *iso*salts), or reduced (in Na *iso*salts) even down to obliteration.

TABLE 2. Temperatures and enthalpies of phase transitions of alkali branched alkanoates

Cation	Anion	Transition	T/K	$\dfrac{\Delta H}{\text{kJ mol}^{-1}}$	Method	Year	Ref.	Notes
Li	2-Methylpropanoate (*iso*butyrate) $C_4H_7O_2$	fusion nstr	502.7±0.9 435.7±0.6	8.6±0.2 7.4±0.3	DSC	1978	1	
	3-Methylbutanoate (*iso*valerate) $C_5H_9O_2$	fusion nstr	534±2 365.4±0.8	11.7 1.8±0.1	DSC	1978	1	
	2,2-Dimethyl= propanoate (Trimethylacetate) $C_5H_9O_2$	fusion nstr nstr	≈650 502.6±0.7 446±2	1.9±0.1 5.9±0.3	vis DSC DSC	1978	2	*a*
	4-Methylpentanoate (*iso*caproate) $C_6H_{11}O_2$	fusion	500.0±0.9	9.5±0.3	DSC	1978	1	
	2-Ethylbutanoate (Diethylacetate) $C_6H_{11}O_2$	fusion nstr	655.5±0.1 591.6±0.2	17.0±0.3 1.00±0.04	DSC	1978	2	
	3,3-Dimethyl= butanoate (*tert*-butylacetate) $C_6H_{11}O_2$	fusion nstr nstr	603.3±0.9 536.8±0.9 508.1±0.5	8.7±0.2 2.09±0.08 8.1±0.1	DSC	1978	2	
Na	2-Methylpropanoate (*iso*butyrate) $C_4H_7O_2$	2nd fusion 1st fusion	≈593 ≈523		micr	1910	3	*b* *c*
		fusion	533		vis-pol	1954	4	
		nstr nstr nstr	493 364 340		vis-pol	1956	5	
		fusion	535		vis-pol	1960	6	

a: this salt exhibits a sharp tendency to sublimation above 570 K
b: this ought to be a clearing point, not confirmed by subsequent authors
c: actually fusion

TABLE 2 (Continued)

		fusion	527±1		DSC	1971	7	
		solid → solid	493±1					
		solid → solid	468±1					
		fusion	526.9±0.7	8.5±0.3	DSC	1978	1	
	3-Methylbutanoate (*iso*valerate) $C_5H_9O_2$	2nd fusion	≈558		capill	1910	3	*a*
		1st fusion	≈441					*b*
		fusion	535		vis-pol	1954	4	*a*
		nstr	451		vis-pol	1956	5	*b*
		nstr	425					
		fusion	533		vis-pol	1963	8	*a*
		clearing	556		DSC	1970	9	
		fusion	460					
		solid → solid	351					
		clearing	553±1	0.75	DSC	1971	7	
		fusion	461±1	4.18				
		clearing	559	1.26±0.08	DSC	1978	1	
		fusion	461.5±0.6	7.0±0.1				
	2,2-Dimethyl= propanoate (Trimethylacetate) $C_5H_9O_2$	fusion	683.7±0.7	8.9±0.2	DSC	1978	2	
		nstr	630.6±0.7	9.9±0.2				
	4-Methylpentanoate (*iso*caproate) $C_6H_{11}O_2$	clearing	626±2	1.59±0.08	DSC	1978	1	
		fusion	534±2	10.5±0.8				
	2-Ethylbutanoate (Diethylacetate) $C_6H_{11}O_2$	fusion	651±3	6.3±0.8	DSC	1978	2	
		nstr	436±3	3.5±0.1				
		nstr	419.4±0.3	2.3±0.2				
	3,3-Dimethyl= butanoate (*tert*-butylacetate) $C_6H_{11}O_2$	fusion	652±3	16.4±0.3	DSC	1978	2	
K	2-Methylpropanoate (*iso*butyrate) $C_4H_7O_2$	2nd fusion	≈608		capill	1910	3	*a*
		1st fusion	≈513					*b*
		nstr	621		vis-pol	1956	5	
		nstr	546					
		nstr	481					
		fusion	629		vis-pol	1960	10	*a*
		fusion	633		vis-pol	1960	6	*a*
		fusion	638		vis-pol	1961	11	*a*

a: actually clearing
b: actually fusion

TABLE 2 (Continued)

	clearing	625.6±0.8	4.52±0.08	DSC	1978	1	
	fusion	553.9±0.5	8.8±0.1				
	nstr	424±3	8.2±0.1				
3-Methylbutanoate (*iso*valerate) $C_5H_9O_2$	nstr	618		vis-pol	1956	5	
	nstr	493					
	nstr	473					
	nstr	327					
	fusion	669		vis-pol	1961	11	*a*
	clearing	679±2	4.23±0.08	DSC	1978	1	
	fusion	531±3	7.6±0.2				
2,2-Dimethyl= propanoate (Trimethylacetate) $C_5H_9O_2$	clearing	613.5±0.8 }	11.4±0.2	DSC	1978	2	
	fusion	612.6±1.0					
	nstr	598.5±0.9	(0.4)				
4-Methylpentanoate (*iso*caproate) $C_6H_{11}O_2$	clearing	713±2	3.9±0.2	DSC	1978	1	
	fusion	516±2	9.6				
2-Ethylbutanoate (Diethylacetate) $C_6H_{11}O_2$	clearing	541.5±0.8	1.7±0.1	DSC	1978	2	
	fusion	424.3±0.2	6.15±0.08				
3,3-Dimethyl= butanoate (*tert*-butylacetate) $C_6H_{11}O_2$	clearing	654±2	3.4±0.2	DSC	1978	2	
	fusion	572.1±0.9	4.4±0.2				
	nstr	517±2	4.81±0.08				

a: actually clearing

1. Ferloni, Sanesi, Tonelli & Franzosini

2. Sanesi, Ferloni, Spinolo & Tonelli

3. Vorländer

4. Sokolov (1954 a)

5. Sokolov (1956)

6. Sokolov & Pochtakova (1960)

7. Duruz, Michels & Ubbelohde (1971)

8. Pochtakova (1963)

9. Ubbelohde, Michels & Duruz

10. Dmitrevskaya & Sokolov

11. Sokolov & Minich

Ammonium alkanoates

It is well known that ammonium salts of fatty acids, when heated, easily undergo (besides sublimation) decomposition reactions of different kinds such as

$$2 \ R-COONH_4 \rightarrow R-COOH.R-COONH_4 + NH_3 \uparrow$$

$$R-COOH.R-COONH_4 \rightarrow 2 \ R-COOH + NH_3 \uparrow$$

or

$$R-COONH_4 \rightarrow R-CONH_2 + H_2O$$

The consequent difficulties in studying their phase transitions may explain the somewhat conflicting data reported by different authors.

In particular, Kench & Malkin (1939) on one hand claimed that sintering over a wide range of temperatures often prevented a trustworthy melting point determination of neutral salts, and on the other hand determined for some acid salts the fusion temperatures, here shown as filled circles in Fig. 11.

Derivatographic measurements, however, recently allowed Roth, Halmos & Meisel (1974) to obtain for "clean, anhydrous and neutral salts" the T_F values represented with open circles.

Concerning the higher homologues a reasonable agreement exists between the latter T_F data and the temperatures pertinent to the crystal $\rightarrow$ anisotropic-liquid-transition detected by Tamamushi & Matsumoto (1974) and shown as triangles in the same Figure.

The Japanese authors also put into evidence (for $12 \leq n_C \leq 18$) clearing points at about 390 K (dot-centered triangles): in this connection it is interesting to note that keeping the salts in an atmosphere of ammonia Zuffanti determined on a series of lower homologues melting points all lying in the same region (crosses).

In Table 3, neglecting the existing discrepancies, each transition is designated as in the original papers.

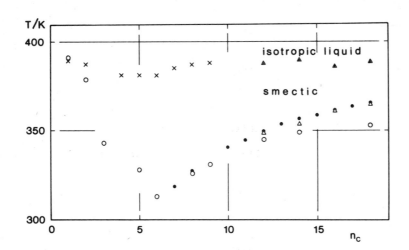

Fig. 11

TABLE 3. Phase transitions of ammonium alkanoates

Anion	Transition	T/K	$\dfrac{\Delta H}{\text{kJ mol}^{-1}}$	Method	Year	Ref.	Notes
Methanoate (Formate)	fusion	387–389			1902	1	
CHO_2	fusion	390			1913	2	

TABLE 3 (Continued)

	fusion	390.5±0.2		1921	3
	fusion	389±1		1941	4
	fusion	391	DTA, micr	1974	5
Ethanoate (Acetate) $C_2H_3O_2$	fusion	385.7–387		1902	1
	fusion	386		1913	2
	fusion	385–386		1930	6
	fusion	387±1		1941	4
	fusion	390		1944	7
	fusion	379	DTA, micr	1974	5
Propanoate (Propionate) $C_3H_5O_2$	fusion	380±1		1941	4
	fusion	343	DTA, micr	1974	5
Butanoate (Butyrate) $C_4H_7O_2$	fusion	381±1		1941	4
2-Methylpropanoate (*iso*butyrate) $C_4H_7O_2$	fusion	391±1		1941	4
Pentanoate (Valerate) $C_5H_9O_2$	fusion	381±1		1941	4
	fusion	328	DTA, micr	1974	5
3-Methylbutanoate (*iso*valerate) $C_5H_9O_2$	fusion	364±1		1941	4
Hexanoate (Caproate) $C_6H_{11}O_2$	fusion	381±1		1941	4
	fusion	313	DTA, micr	1974	5
4-Methylpentanoate (*iso*caproate) $C_6H_{11}O_2$	fusion	375±1		1941	4
Heptanoate (Enanthate) $C_7H_{13}O_2$	fusion	385±1		1941	4
Octanoate (Caprylate) $C_8H_{15}O_2$	fusion	387±1		1941	4
	fusion	326	DTA, micr	1974	5

TABLE 3 (Continued)

Nonanoate (Pelargonate)	fusion	388±1		1941	4
$C_9H_{17}O_2$	fusion	331	DTA, micr	1974	5
Dodecanoate (Laurate)	fusion	≃403		1939	8
$C_{12}H_{23}O_2$	fusion	345	DTA, micr	1974	5
	anisotropic → isotropic liq.	387.7		1974	9
	crystal → anisotropic liq.	348.7			
Tetradecanoate (Myristate)	fusion	349	DTA, micr	1974	5
$C_{14}H_{27}O_2$	smectic → isotropic liquid	389.7±0.3		1974	10
	crystal → smectic	353.9±0.2			
Hexadecanoate (Palmitate)	smectic → isotropic liquid	386±1		1974	10
$C_{16}H_{31}O_2$	crystal → smectic	361±1			
Octadecanoate (Stearate)	fusion	380		1938	11
	nstr	323			
$C_{18}H_{35}O_2$	fusion	353	DTA, micr	1974	5
	smectic → isotropic liquid	388.7±0.5		1974	10
	crystal → smectic	364.7±1			

1. Reik	7. Davidson, Sisler & Stoenner
2. Escales & Koepke	8. Kench & Malkin
3. Kendall & Adler	9. Tamamushi & Matsumoto
4. Zuffanti	10. Tamamushi
5. Roth, Halmos & Meisel	11. Lawrence
6. Davidson & McAllister	

Thallium(I) alkanoates

In the present salt family, as in most alkali alkanoates, the solid turns into the isotropic liquid directly in lower homologues and through the intermediate formation of liquid crystals in higher ones.

Concerning the normal alkanoates, among which the highest melting is propanoate (as in the potassium series) and liquid crystals occur from pentanoate on (as in the rubidium series), the above phase relationships are known in sufficient detail from formate up to octadecanoate (see Fig. 12).

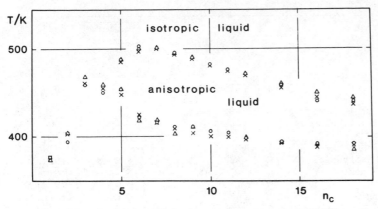

Fig. 12. Clearing and fusion temperatures of Tl(I) n.alkanoates (from meth-anoate to octadecanoate). Circles: Baum, Demus & Sackmann; triangles: Meisel, Seybold, Halmos, Roth & Melykuti; crosses: Walter.

TABLE 4. Temperatures and enthalpies of phase transitions of thallium(I) alkanoates

Anion	Transition	T/K	$\dfrac{\Delta H}{\text{kJ mol}^{-1}}$	Method	Year	Ref.	Notes
Methanoate (Formate) CHO_2	fusion	377		micr	1926	1	
	fusion	374			1929	2	
	fusion	374	5.65	DSC	1969	3	
	nstr	371	6.74				
	fusion	377		micr	1970	4	
	fusion	374	10.9	DTA	1976	5	
Ethanoate (Acetate) $C_2H_3O_2$	fusion	403		micr	1926	1	
	fusion	404			1929	2	
	fusion	400			1930	6	
	fusion	404			1968	7	
	fusion	394		micr	1970	4	
	fusion	404	17.6	DTA	1976	5	
Propanoate (Propionate) $C_3H_5O_2$	fusion	461		micr	1926	1	
	fusion	459		micr	1970	4	
	fusion	468	9.2	DTA	1976	5	
	solid → solid	365	0.4				
Butanoate (Butyrate) $C_4H_7O_2$	fusion	455		micr	1926	1	
	fusion	450		micr	1970	4	
	fusion	459	6.7	DTA	1976	5	

TABLE 4 (Continued)

2-Methylpropanoate (*iso*butyrate) $C_4H_7O_2$	2nd fusion	397		micr	1926	1	*a, b*
Pentanoate (Valerate) $C_5H_9O_2$	fusion	418–420			1924	8	
	liq.crystal → isotropic liq.	485		micr	1926	1	*c*
	solid I → liquid crystal	448					*a*
	solid II → solid I	351					
	neat → isotropic	488		micr	1970	4	
	solid → neat	454					
	clearing	488.7		micr	1971	9	
	fusion	454.2					
	mesophase → isotropic liq.	487	3.1	DTA	1976	5	*c*
	solid → mesophase	455	5.4				*a*
	solid → solid	354.7	2.3				
3-Methylbutanoate (*iso*valerate) $C_5H_9O_2$	1st fusion	447		micr	1926	1	*c*
	2nd fusion	429.7					*a*
Hexanoate (Caproate) $C_6H_{11}O_2$	1st fusion	496–497		micr	1926	1	*c*
	2nd fusion	425					*a*
	neat → isotropic	503		micr	1970	4	
	solid → neat	422					
	clearing	502.7		micr	1971	9	
	fusion	422.2					
	mesophase → isotropic liq.	500	3.3	DTA	1976	5	*c*
	solid → mesophase	418	4.5				*a*
	solid → solid	412	0.8				
	solid → solid	395	0.2				
2-Ethylbutanoate (Diethylacetate) $C_6H_{11}O_2$	fusion	385		micr	1926	1	
Heptanoate (Enanthate) $C_7H_{13}O_2$	1st fusion	500		micr	1926	1	*c*
	2nd fusion	416					*a*
	neat → isotropic	501		micr	1970	4	
	solid → neat	416					
	clearing	500.7		micr	1971	9	
	fusion	416.0					
	mesophase → isotropic liq.	501	3.1	DTA	1976	5	*c*
	solid → mesophase	419	6.3				*a*
	solid → solid	299	2.8				
Octanoate (Caprylate) $C_8H_{15}O_2$	1st fusion	493		micr	1926	1	*c*
	2nd fusion	408–409					*a*

a: actually fusion   *b*: first fusion "nicht bestimmbar"   *c*: actually clearing

TABLE 4 (Continued)

	neat → isotropic	495		micr	1970	4	
	solid → neat	411					
	clearing	495.7		micr	1971	9	
	fusion	408.9					
	mesophase → isotropic liq.	494	2.7	DTA	1976	5	*a*
	solid → mesophase	403	4.7				*b*
2-Propylpentanoate (Dipropylacetate) $C_8H_{15}O_2$	fusion	395–396		micr	1926	1	
Nonanoate (Pelargonate) $C_9H_{17}O_2$	1st fusion	488		micr	1926	1	*a*
	2nd fusion	403					*b*
	neat → isotropic	490		micr	1970	4	
	solid → neat	411					
	clearing	490.2		micr	1971	9	
	fusion	411.8					
	mesophase → isotropic liq.	490	2.6	DTA	1976	5	*a*
	solid → mesophase	410	5.0				*b*
	solid → solid	330	7.5				
	solid → solid	315	2.6				
	solid → solid	300	1.7				
Decanoate (Caprate) $C_{10}H_{19}O_2$	1st fusion	480		micr	1926	1	*a*
	2nd fusion	400					*b*
	neat → isotropic	482		micr	1970	4	
	solid → neat	406					
	clearing	482.7		micr	1971	9	
	fusion	404.2					
Hendecanoate (Undecanoate) $C_{11}H_{21}O_2$	1st fusion	473–474		micr	1926	1	*a*
	2nd fusion	399					*b*
	neat → isotropic	476		micr	1970	4	
	solid → neat	404					
	clearing	476.7		micr	1971	9	
	fusion	403.6					
Dodecanoate (Laurate) $C_{12}H_{23}O_2$	fusion	398–399			1925	10	
	1st fusion	470		micr	1926	1	*a*
	2nd fusion	396					*b*
	neat → isotropic	471		micr	1970	4	
	solid → neat	398					
	clearing	470.2		micr	1971	9	
	fusion	397.9					

a: actually clearing
b: actually fusion

TABLE 4 (Continued)

	mesophase → isotropic liq.	471	1.9	DTA	1976	5	*a*
	solid → mesophase	398	5.9				*b*
	solid → solid	354	2.4				
	solid → solid	312	3.8				
Tetradecanoate (Myristate) $C_{14}H_{27}O_2$	fusion	393–396			1925	10	
	1st fusion	454.7		micr	1926	1	*a*
	2nd fusion	392.7					*b*
	neat → isotropic	458		micr	1970	4	
	solid → neat	394					
	clearing	458.2		micr	1971	9	
	fusion	393.4					
	mesophase → isotropic liq.	460	1.6	DTA	1976	5	*a*
	solid → mesophase	393	5.4				*b*
	solid → solid	371	3.1				
	solid → solid	313	11.7				
Hexadecanoate (Palmitate) $C_{16}H_{31}O_2$	clearing	443			1925	11	
	fusion	388					
	fusion	388–390			1925	12	
	1st fusion	445		micr	1926	1	*a*
	2nd fusion	387–389					*b*
	neat → isotropic	440		micr	1970	4	
	solid → neat	392					
	clearing	448.7		micr	1971	9	
	fusion	391.4					
	mesophase → isotropic liq.	450	1.4	DTA	1976	5	*a*
	solid → mesophase	390	8.8				*b*
	solid → solid	327	10.9				
Octadecanoate (Stearate) $C_{18}H_{35}O_2$	fusion	390			1922	13	
	fusion	392			1925	11	
	fusion	392			1925	12	
	1st fusion	436		micr	1926	1	*a*
	2nd fusion	391					*b*
	clearing	436			1933	14	
	fusion	391					
	fusion	436			1938	15	*a*
	nstr	401					*b*
	neat → isotropic	440		micr	1970	4	
	solid → neat	392					
	clearing	440.2		micr	1971	9	
	fusion	392.2					

a: actually clearing
b: actually fusion

TABLE 4 (Continued)

	mesophase → isotropic liq.	444	1.4	DTA	1976	5	a
	solid → mesophase	385	5.4				b
	solid → solid	380	4.6				
	solid → solid	324	9.6				
Hexacosanoate	1st fusion	398		micr	1926	1	a
(Cerotate)	2nd fusion	386-387					b
$C_{26}H_{51}O_2$							

a: actually clearing
b: actually fusion

1. Walter	9. Pelzl & Sackmann
2. Sugden	10. Holde & Takehara
3. Braghetti, Berchiesi & Franzosini	11. Christie & Menzies
4. Baum, Demus & Sackmann	12. Holde & Selim
5. Meisel, Seybold, Halmos, Roth & Melykuti	13. Meigen & Neuberger
6. Davidson & McAllister	14. Herrmann
7. Ubbelohde	15. Lawrence
8. Menzies & Wilkins	

Alkali-earth alkanoates

The thermal behaviour of alkali-earth alkanoates was never systematically investigated: information is available for some ten magnesium salts and for an even more restricted number of calcium, strontium and barium salts.

Concerning the long-chain homologues it may be mentioned that:

(i) Vold, R.D., Grandine & Vold (1948), who submitted calcium octadecanoate to investigation with a variety of experimental techniques, could draw for this salt (on the basis of the results collected both in heating and cooling runs) a quite complicated "transformation scheme": however, only transformations which the anhydrous material undergoes on heating are listed in the table.

(ii) Hattiangdi, Vold & Vold (1949), in a study performed on hexadecanoates and octadecanoates of alkali-earth (and other di- and trivalent) metals, pointed out *inter alia* the fact that "fluidity and transparency alone are not sufficient evidence that true liquid has formed with these substances; even in cases where examination of the soaps in the usual glass capillary between crossed polaroids leads to the observation that the material is isotropic, orientation of the actually mesomorphic phase may be readily confused with true melting. This is the case, for example, with calcium stearate, which appears to be melted in a capillary at 150°C but undergoes a further transition on heating at 190°".

Among short-chain homologues, the peculiar behaviour offered by Ba butanoate has been interpreted by Ferloni, Sanesi & Franzosini (1976) on the assumption of monotropy, complicated by enantiotropy both of the metastable and of the stable forms of this salt, and by possible formation of glass-like phases from the melt.

TABLE 5. Temperatures and enthalpies of phase transitions of alkali-earth alkanoates

Cation	Anion	Transition	T/K	$\dfrac{\Delta H}{kJ\ mol^{-1}}$	Method	Year	Ref.	Notes
Mg	Methanoate	nstr	413		vis-pol, DTA	1974	1	
	Ethanoate	fusion	596			1912	2	a
		fusion	630			1928	3	
		nstr	445–449		vis-pol,	1974	1	
		nstr	425		DTA			
		fusion	603–608		DTA	1975	4	a
	Propanoate	fusion	559			1928	3	
		fusion	577		vis-pol,	1974	1	
		nstr	519		DTA			
		nstr	490					
		nstr	473					
		nstr	458					
		fusion	577		DSC	1976	5	
	Butanoate	fusion	548			1928	3	
		fusion	575		vis-pol, DTA	1974	1	
		fusion	567		DSC	1976	5	
	Pentanoate	fusion	531			1928	3	
		fusion	537		vis-pol, DTA	1974	1	b
	Dodecanoate	fusion	423.6			1962	6	
	Tetradecanoate	fusion	404.8			1962	6	
	Hexadecanoate	nstr	388	28.9	DTA	1949	7	c
		nstr	363	39.3				
		fusion	394–395			1962	6	
	Octadecanoate	nstr	367	55.3	DTA	1949	7	d
		nstr	330					
		nstr	463		X-ray	1962	8	
		nstr	383					

a: with decomposition
b: extrapolated
c: visual melting point determined by Hattiangdi, 402.7 K
d: visual melting point determined by Hattiangdi, 382 K

TABLE 5 (Continued)

		fusion	405			1962	6	
	3-Methylbutanoate	fusion	497			1928	3	
Ca	Ethanoate	fusion	713		DTA	1974	9	*a*
	Propanoate	fusion	666	15.7	DSC	1976	5	
		nstr	541±2	1.5				
		fusion	≈673		micr	1977	10	
	Butanoate	fusion	604.3	11.9	DSC	1976	5	
	Nonanoate	fusion	489			1905	11	
	Hexadecanoate	nstr	420		DTA	1949	7	*b*
		nstr	379	28.7				
		nstr	352					
		nstr	323					
	Octadecanoate	fusion	623		DTA, micr	1948	12	*a*
		nstr	468	2.1				
		nstr	423	(< 1)				
		nstr	396	46.0				
		nstr	359	18.8				
		nstr	336	2.5				
		nstr	460÷463		DTA	1949	7	*c*
		nstr	419÷425					
		nstr	379÷380	31÷38				
		nstr	362					
		nstr	331					
		nstr	≈475		X-ray	1960	13	
		nstr	≈388					
Sr	Ethanoate	fusion	594±1	17±2	DTA	1966	14	*d*
	Propanoate	fusion	725±4	10.5	DSC	1976	5	
		fusion	≈723		micr	1977	10	
	Butanoate	fusion	681.0	12.3	DSC	1976	5	
	Hexadecanoate	nstr	408	14.7	DTA	1949	7	
		nstr	388	48.6				
		nstr	348	35.3				
	Octadecanoate	nstr	406	11.6	DTA	1949	7	
		nstr	378	77.7				

a: with decomposition
b: visual melting point determined by Hattiangdi, 419 K
c: visual melting point determined by Hattiangdi, 439 K
d: temperature taken in correspondence with the peak maximum

TABLE 5 (Continued)

		nstr	358					
		nstr	(330)					
Ba	Propanoate	fusion	647.8	14.0	DSC	1976	5	
		fusion	≈623		micr	1977	10	
		nstr	513					
	Butanoate	fusion	586	7.3	DSC	1976	5	
		nstr	500±2	3.3				
		nstr (m)	433.8	11.3				
	Heptanoate	fusion	511–512			1877	15	a
	Hexadecanoate	nstr	403	67.3	DTA	1949	7	
		nstr	338					
	Octadecanoate	nstr	391÷409	76÷102	DTA	1949	7	
		nstr	331÷347					

a: with decomposition

1. Pochtakova (1974)
2. Späth
3. Ivanoff
4. Balarew & Stoilova
5. Ferloni, Sanesi & Franzosini (1976)
6. Jacobson & Holmes
7. Hattiangdi, Vold & Vold
8. Spegt & Skoulios (1962)
9. Judd, Plunkett & Pope
10. Gobert-Ranchoux & Charbonnier
11. Harries & Thieme
12. Vold, R.D., Grandine & Vold
13. Spegt & Skoulios (1960)
14. McAdie
15. Mehlis

Zn, Cd, Hg and Pb alkanoates

To enlarge the picture on the thermal behaviour of metal alkanoates, the known information concerning zinc, cadmium, mercury and lead n.alkanoates is summarized in Table 6, whereas random data existing for salts with other cations have been neglected.

TABLE 6. Temperatures and enthalpies of phase transitions of Zn, Cd, Hg and Pb n.alkanoates

Cation	Anion	Transition	T/K	$\dfrac{\Delta H}{\text{kJ mol}^{-1}}$	Method	Year	Ref.	Notes
Zn	Ethanoate	fusion	514–515			1879	1	
		fusion	517			1914	2	
		fusion	515.6			1939	3	
		fusion	531	≈18.8	DTA	1966	4	a
		fusion	511–513		DTA	1967	5	
		fusion	509 510		vis–pol cond	1974	6	
		fusion	523–533			1975	7	b
	Hexanoate	fusion	415			1929	8	
	Octanoate	fusion	407.7–408.7			1869	9	
		fusion	408–409			1874	10	
	Nonanoate	fusion	404–405			1924	11	
	Decanoate	fusion	403.2			1948	12	
	Dodecanoate	fusion	401			1930	13	
		fusion	403.7			1948	12	
	Tetradecanoate	fusion	401.2			1948	12	
	Hexadecanoate	fusion	402			1930	13	
		nstr nstr	394 358	14.0 91.0	DTA	1949	14	
	Octadecanoate	fusion	≈403			1930	13	
		fusion	391			1938	15	
		fusion	402.2			1948	12	
		nstr nstr	366 331	64.0	DTA	1949	14	
Cd	Ethanoate	fusion	527–529			1912	16	b
		fusion	532–533		DTA	1967	5	

a: temperature taken in correspondence with the peak maximum
b: with decomposition

TABLE 6 (Continued); Cd n.alkanoates (Continued)

		fusion	530–531		vis-pol	1969	17	
		fusion	≃523		DTA	1975	7	
	Hexadecanoate	nstr	404		DTA	1949	14	
		nstr	389					
		nstr	359	58.7				
		nstr	331					
	Octadecanoate	nstr	423		DTA	1949	14	
		nstr	401					
		nstr	390					
		nstr	380					
		nstr	355	45.4				
		nstr	322					
Hg(I)	Propanoate	fusion	498			1887	18	*a*
	Octadecanoate	fusion	390.7			1938	15	
Hg(II)	Ethanoate	fusion	453		DTA	1967	5	
	Propanoate	fusion	383			1887	18	
	Heptanoate	fusion	379.7			1907	19	
	Dodecanoate	fusion	≃373			1930	13	
	Hexadecanoate	nstr	395		DTA	1949	14	
		nstr	358	(140.4)				
		nstr	315	5.9				
	Octadecanoate	fusion	385.4			1930	13	
		nstr	402		DTA	1949	14	
		nstr	362	75.1				
		nstr	309	4.9				
Pb(II)	Ethanoate	fusion	477			1914	2	
		fusion	477			1930	20	
		fusion	477			1938	21	
		fusion	463–473			1939	22	
	Propanoate	fusion	463–473			1939	22	
	Butanoate	fusion	433			1939	22	
	Hexanoate	fusion	346–347			1912	23	

a: with decomposition

TABLE 6 (Continued); Pb(II) n.alkanoates (Continued)

	fusion	368			1939	22
	V$_2$(cubic isomorphous) → liquid	350.2	1.3±0.1	DTA	1976	24
	G(smectic) → V$_2$	338.5	15.8±0.5			
	crystal → G	335.9	6.02±0.46			
Heptanoate	fusion	351			1877	25
	fusion	363.7–364.7			1912	23
	fusion	358			1939	22
Octanoate	fusion	356.7–357.7			1869	9
	fusion	356.7–357.7			1912	23
	fusion	373			1939	22
	fusion	355.2–356.0			1939	26
	fusion	357–358			1949	27
	V$_2$(cubic isomorphous) → liquid	381.2	1.2	DTA	1976	24
	G(smectic) → V$_2$	356.2	7.9±0.4			
	crystal → G	352.5	30±1			
Nonanoate	fusion	367–368			1912	23
	fusion	368–373			1939	22
	fusion	371.2–371.7			1939	26
	fusion	367–368			1949	27
Decanoate	fusion	373			1912	23
	fusion	369.7–370.2			1939	26
	fusion	372–373			1949	27
	V$_2$(cubic isomorphous) → liquid	385.2	1.0	DTA	1976	24
	G(smectic) → V$_2$	367.4	20.0			
	crystal → G	354.9	32.7			
Hendecanoate	fusion	363–365			1939	22
Dodecanoate	fusion	376–377			1912	23
	fusion	379			1930	13
	fusion	371			1939	22
	fusion	377.0–377.4			1939	26
	fusion	377.2–377.7			1949	27
	fusion	377.8–378.0			1962	28

TABLE 6 (Continued); Pb(II) n.alkanoates (Continued)

	V_2(cubic isomorphous) $\rightarrow$ liquid	381.2	0.9	DTA	1976	24
	G(smectic) $\rightarrow V_2$	377.2	29.8			
	crystal $\rightarrow$ G	365.2	48.3			
Tetradecanoate	fusion	380			1912	23
	fusion	382.8-383.4			1939	26
	fusion	381-382			1949	27
	fusion	381.8-382.0			1962	28
	G(smectic) $\rightarrow$ liquid	382.7	41.6	DTA	1976	24
	crystal $\rightarrow$ G	372.7	59.1			
Hexadecanoate	fusion	385			1912	23
	fusion	386			1930	13
	fusion	386.2-386.8			1939	26
	fusion	385-386			1949	27
	fusion	385.4-385.6			1962	28
	G_1(smectic) $\rightarrow$ liquid	384.2	46.4	DTA	1976	24
	G_2(smectic) $\rightarrow G_1$	380.6				
	crystal $\rightarrow G_2$	373.4	55±3			
Octadecanoate	fusion	398			1912	23
	fusion	386-387			1914	29
	fusion	388			1930	13
	fusion	379			1938	15
	fusion	393			1939	22
	fusion	388.2-388.7			1939	26
	nstr	(406)		DTA	1949	14
	nstr	374	39.5			
	nstr	342	71.7			
	fusion	388.8-389.0			1962	28
	G(smectic) $\rightarrow$ liquid	387.2	56.9	DTA	1976	24
	crystal $\rightarrow$ G	381.2	62.6			
Hexacosanoate	fusion	385.2-386.7			1939	26

1. Franchimont
2. Petersen
3. Lehrman & Skell
4. McAdie
5. Bernard & Busnot
6. Nadirov & Bakeev (1974 a)
7. Balarew & Stoilova
8. Haworth
9. Zincke
10. Van Renesse
11. Robinson
12. Martin & Pink
13. Whitmore & Lauro
14. Hattiangdi, Vold & Vold

15. Lawrence   23. Neave

16. Späth   24. Adeosun & Sime

17. Pavlov & Golubkova   25. Mehlis

18. Renard   26. Piper, Fleiger, Smith & Kerstein

19. Bornwater   27. Naves

20. Davidson & McAllister   28. Jacobson & Holmes

21. Lehrman & Leifer   29. Wagner, Muesmann & Lampart

22. Kenner & Morton

Li, Na, K, Rb, Cs, NH₄ and Tl(I) oleates

Among the salts of unsaturated fatty acids, information is here limited to a number of oleates.

TABLE 7. Temperatures and enthalpies of phase transitions of alkali, ammonium and thallium(I) 9.10-octadecenoates*(cis)* (oleates)

Cation	Transition	T/K	$\dfrac{\Delta H}{\text{kJ mol}^{-1}}$	Method	Year	Ref.	Notes
Li	nstr	418		NMR	1971	1	
	nstr	295					
Na	fusion	505–508			1899	2	
	fusion	514			1939	3	
	superwaxy → subneat	474		dilat			
	waxy → superwaxy	452		dilat			
	subwaxy → waxy	395		dilat			
	curd → subwaxy	341		dilat			
	subneat → isotropic	529		dilat, micr	1941	4	
		510	8.0	cal			
	superwaxy → subneat	491		dilat, micr			
		475	3.6	cal			
	waxy → superwaxy	452		dilat, micr			
		460	3.7	cal			
	subwaxy → waxy	391		dilat, micr			
		388	12.1	cal			
	curd → subwaxy	338		dilat, micr			
		339	1.3	cal			
	curd → curd	311		dilat, micr			
		313	6.7	cal			
	nstr	337		NMR	1971	1	
	nstr	260					

TABLE 7 (Continued)

K	clearing	597.6			1933	5	*a*
	fusion	531.7					
	nstr	407	NMR		1971	1	
	nstr	280					
Rb	nstr	348	NMR		1971	1	
	nstr	280					
Cs	nstr	297	NMR		1971	1	
NH$_4$	smectic → isotropic liquid	342.6±1			1974	6	
	crystal → smectic	314.8±1					
Tl(I)	fusion	355–356			1925	7	
	fusion	351–355			1925	8	
	1st fusion	404–405			1926	9	*b*
	2nd fusion	354					*c*

a: designated as "the first temperature at which the homogeneous isotropic liquid becomes tur-
bid (...) on cooling"
b: actually clearing
c: actually fusion

1. Ripmeester & Dunell (1971 a)

2. Krafft

3. Vold, R.D. (1939)

4. Vold, R.D. (1941)

5. McBain & Stewart

6. Tamamushi

7. Holde & Selim

8. Christie & Menzies

9. Walter

The above Tables 1-7 summarize information on those families of salts with organic anion (alkanoates and oleates) for which temperatures and enthalpies of phase transitions have been
more extensively investigated.

Salts belonging to other families, no matter how important either for theoretical or technical
purposes, were not (or are impossible to be) submitted to a systematic study from the thermal
point of view. Listing the random data at disposal would therefore exceed the aim of the present report.

SALTS WITH ORGANIC CATION

The number of salts with organic cation, on which information can be found in the literature, is rather overwhelming. The largest part of them, however, has not been studied with a physico-chemical interest, having been prepared essentially for identification purposes in organic chemical work.

Since a crude listing of fusion temperatures would not be significant for the present report, attention is to be given here only to those materials the thermal behaviour of which has been investigated in some detail.

Table 8 summarizes the relevant data available for the alkyl substituted ammonium salts type

$$R_{4-n}NH_nX \qquad\qquad\qquad\qquad\qquad\qquad\qquad\qquad\qquad (I)$$

where n = 0, 1, 2, 3 and the R groups may be either equal or different. For the notations adopted in the table reference is made to the preceding sections.

The existence of polymorphs in many alkylammonium halides and picrates was already put into evidence long ago by Wagner (1907) and Ries (1915), respectively, through crystallographic investigations.

In a few homogeneous salt groups interesting features could be observed, and discussed in the light of the phase structures and of the mechanism of phase transitions. Thus Levkov et al., on the basis of DSC measurements supplemented by IR spectra on tetraalkylammonium iodides (R_4NI, where R = C_nH_{2n-1} with n = 2÷7), could develop considerations on the trend of the transition enthalpies and entropies for both the odd- and even-n salts.

Tsau & Gilson (1968), found that in the series of monosubstituted alkylammonium chlorides and bromides from C_1 to C_{16} the melting temperature shows "an initial increase to a maximum for the n.pentylammonium salts, followed by a gradual decrease". Evidence of pronounced hysteresis phenomena, similar to those mentioned in a previous section for a number of salts with organic anion, is also apparent from Tsau & Gilson's results: in some materials DSC traces "differed, both in transition temperature and enthalpy, if the same sample was scanned more than once", and did not reassume the original characteristics even after about two months rest at room temperature.

A dependence of the thermal behaviour (as detected by DSC) on the thermal history of the samples was found by Coker, Ambrose & Janz (1970) in a large group of quaternary ammonium compounds. In this connection, they stated that "such irreproducibility is not due to decompositions or impurities, but rather may be ascribed to the thermal properties of the quaternary alkylammonium cation, i.e., the freezing in of disorder" (see also: Coker, Wunderlich & Janz; Southard, Milner & Hendricks). As for the observed trends, the above authors reported that, for a given anion, the sum $\Sigma \, \Delta H_{tr}$ generally increases with increasing alkyl chain length in the cation, while for a given cation it progressively decreases (together with the fusion temperature) when I, Br and Cl are successively taken into account as the anion.

Methyl substituted ammonium chlorides, bromides and iodides were submitted to systematic investigation by Stammler (1967). Although his DTA results (having been given only in graphical form) could not be included in Table 8, it can be mentioned here that the first two members of the chloride series (n = 3, 2 in (I)) have two polymorphic phase transitions while the remaining two members (n = 1, 0) exhibit only one such transition.

Finally, it is worthwhile to mention the fact that for some salts, which were thought to decompose before melting, the existence of a definite melting point has been proved by more recent research. This is the case, e.g., of the monoalkylammonium chlorides (with even n from 8 to 18), which were reported as thermally unstable by Ralston et al. (1941, 1942); Tsau & Gilson (1968), on the contrary, were able to determine by DSC both their fusion temperatures and enthalpies.

TABLE 8. Temperatures and enthalpies of phase transitions of alkyl-substituted ammonium salts

Cation	Anion	Transition	T/K	$\dfrac{\Delta H}{kJ\ mol^{-1}}$	Method	Year	Ref.	Notes
CH_3NH_3	Br	fusion	537		DSC	1968	1	a
		nstr	389	1.3				
		nstr	285	4.4				
	Cl	form $\gamma \rightarrow$ form α	264.5	2.82±0.04	ad-cal	1946	2	
		form $\beta \rightarrow$ form γ	220.4	1.78±0.02				
		fusion	506		DSC	1968	1	a
		nstr	264	0.7				b
		nstr	223	2.7				b
		nstr	264	3.6				
		nstr	223	0.6				
$(CH_3)_4N$	Cl	"λ" transition	184.85		ad-cal	1962	3	
		1st order trans.	75.76					
		1st order trans.	528±1		DTA	1971	4	
		1st order trans.	434					
		nstr	418					
	ClO_4	nstr	613			1966	5	
	NO_3	nstr	287.5		DTA	1974	6	
			≈285		NMR			
$C_2H_5NH_3$	Br	fusion	434	8.4	DSC	1968	1	
		nstr	363	11.3				
	Cl	fusion	381	9.8	DSC	1968	1	
$(C_2H_5)_4N$	Br	nstr	468	1.3	DTA, NMR	1974	7	
		tetragonal $\rightarrow$ cubic	448	16.7				
		nstr	415	0.05				
	I	nstr	465	21.3±0.4	DSC	1971	8	
$C_3H_7NH_3$	Br	fusion	456	12.5	DSC	1968	1	
		nstr	138	1.9				
	Cl	fusion	439	5.9	DSC	1968	1	
		nstr	408	4.2				
		nstr	188	1.2				
		nstr	415		NMR	1973	9	
		nstr	409					
		nstr	200					
		nstr	148					

a: sublimation occurs
b: transition "observed in the first, but not subsequent, scans"

TABLE 8 (Continued)

$(C_3H_7)_4N$	I	nstr	419±0.5	32.6±3 %	DSC	1970	10	
		nstr	418	13.8±1.3	DSC	1971	8	
		nstr	224	2.9±0.1				
$(C_3H_7)(C_4H_9)_3N$	I	fusion	468±0.5	13.8±3 %	DSC	1970	10	
$(C_3H_7)(C_8H_{17})_3N$	Br	fusion	351±0.5	44.4±3 %	DSC	1970	10	
$C_4H_9NH_3$	Br	fusion	479		DSC	1968	1	*a*
		nstr	257	5.2				
		nstr	198	0.9				
	Cl	fusion	487		DSC	1968	1	*a*
		nstr	253	2.5				
		nstr	237	0.8				
		nstr	225		NMR	1973	9	
		nstr	190					
$(C_4H_9)_4N$	Br	fusion	395±0.5	15.5±3 %	DSC	1970	10	
		nstr	367–372	14.2±3 %				
	Cl	fusion	314±0.5	20.5±3 %	DSC	1970	10	
	I	nstr	390.9		micr	1965	11	
		fusion	419±0.5	9.6±3 %	DSC	1970	10	
		nstr	392±0.5	28.0±3 %				
		fusion	418	9.2	DSC	1971	8	
		nstr	392	28.0±0.4				
	NO_3	fusion	392±0.5	14.6±3 %	DSC	1970	10	
		nstr	366±0.5	0.2±3 %				
$(C_4H_9)(i.C_5H_{11})_3N$	Br	fusion	395±0.5	15.9±3 %	DSC	1970	10	
		nstr	342±0.5	4.0±3 %				
C_5H_6N	Cl	fusion	419.2	9.67±0.17	cryom	1967	12	*b*
$C_5H_{11}NH_3$	Br	fusion	501		DSC	1968	1	*a*
		nstr	275	5.9				
	Cl	nstr	246.5±0.5		ad-cal	1933	13	
		nstr	221.5±0.5					*c*
			222.0±0.5					*d*
		fusion	502		DSC	1968	1	*a*
		nstr	259	6.2				

a: sublimation occurs
b: C_5H_6N = pyridinium
c: quenched material
d: annealed material

TABLE 8 (Continued)

		nstr	240		NMR	1973	9	
		nstr	230					
$(C_5H_{11})_4N$	Br	fusion	376±0.5	41.4±3 %	DSC	1970	10	
	Cl	fusion	295±0.5	1.3±3 %	DSC	1970	10	
		nstr	281	2.8±3 %				
	CNS	nstr	313		micr	1965	11	
		fusion	322.7±1.5	19.3±3 %	DSC	1969	14	
		nstr	315.2±1.5	22.6±3 %				
		fusion	322.7±0.5	19.7±3 %	DSC	1970	10	
		nstr	315±0.5	22.6±3 %				
	I	fusion	412±0.5	39.3±3 %	DSC	1970	10	
		nstr	405±0.5	17.2±3 %				
		fusion	410	37.7±0.4	DSC	1971	8	
		nstr	403	13.8±1.7				
	NO_3	fusion	387±0.5	28.5±3 %	DSC	1970	10	
		nstr	366±0.5	12.6±3 %				
$(i.C_5H_{11})_4N$	I	fusion	422±0.5	15.9±3 %	DSC	1970	10	
		nstr	352±0.5	5.9±3 %				
		nstr	345±0.5	28.9±3 %				
$C_6H_{13}NH_3$	Br	fusion	502		DSC	1968	1	a
		nstr	316					
		nstr	282	0.8				
		nstr	254	3.3				
	Cl	fusion	493		DSC	1968	1	a
		nstr	272	3.4				
		nstr	265	0.9				
		nstr	214 }	0.7				
		nstr	208					
		nstr	266		NMR	1973	9	
		nstr	255					
		nstr	205					
		nstr	190					
$(C_6H_{13})_4N$	Br	fusion	377±0.5	15.9±3 %	DSC	1970	10	
		nstr	315	12.1±3 %				
		nstr	305	6.7±3 %				
	ClO_4	nstr	368.2		micr	1965	11	
		nstr	358.1					
		fusion	383±0.5	18.4±3 %	DSC	1970	10	
		nstr	369±0.5	2.5±3 %				

a: sublimation occurs

TABLE 8 (Continued)

		nstr	358±0.5	5.9±3 %			
		nstr	335±0.5	23.0±3 %			
		fusion	379.18±0.01	16.35±0.1 %	ad-cal	1973	15
		nstr	367.51±0.01	2.658±0.1 %			
		nstr	355.91±0.01	5.839±0.1 %			
		nstr	333.57±0.01	22.99±0.1 %			
	I	fusion	378	17.2±0.4	DSC	1971	8
		nstr	352	5.86±0.08			
		nstr	344	24.3±1.3			
	NO_3	fusion	345±0.5	17.6±3 %	DSC	1970	10
		nstr	323±0.5	22.2±3 %			
$(C_6H_{13})_3(C_7H_{15})N$	Br	nstr	321		micr	1965	11
		fusion	381±0.5	7.5±3 %	DSC	1970	10
		nstr	356±0.5	7.5±3 %			
		nstr	335±0.5	29.3±3 %			
	ClO_4	fusion	376±0.5	24.3±3 %	DSC	1970	10
		nstr	362±0.5	16.3±3 %			
	I	fusion	371±0.5	20.5±3 %	DSC	1970	10
		nstr	365±0.5	21.3±3 %			
	NO_3	fusion	345±0.5	33.5±3 %	DSC	1970	10
$(C_6H_{13})_2(C_7H_{15})_2N$	ClO_4	fusion	378±0.5	25.9±3 %	DSC	1970	10
		nstr	365±0.5	14.2±3 %			
	I	fusion	373±0.5	26.8±3 %	DSC	1970	10
		nstr	362±0.5	13.0±3 %			
$C_7H_{15}NH_3$	Br	fusion	490	5.8	DSC	1968	1
		nstr	326	2.0			
		nstr	301	1.6			
		nstr	260	11.7			
	Cl	fusion	480	6.6	DSC	1968	1
		nstr	303	1.9			
		nstr	260	3.2			
		nstr	310		NMR	1973	9
		nstr	280				
		nstr	260				
		nstr	230				
$(C_7H_{15})_4N$	Br	fusion	369±0.5	36.0±3 %	DSC	1970	10
		nstr	343±0.5	5.4±3 %			
	Cl	fusion	264±0.5		DSC	1970	10

TABLE   8 (Continued)

		nstr	215±0.5				*a*
	ClO_4	fusion	399±0.5	31.8±3 %	DSC	1970	10
		nstr	388±0.5	4.2±3 %			
		nstr	365±0.5	4.6±3 %			
		nstr	356±0.5	4.6±3 %			
	I	fusion	396±0.5	37.2±3 %	DSC	1970	10
		nstr	391±0.5	3.1±3 %			
		nstr	356±0.5	9.6±3 %			
		fusion	396	36.4±0.4	DSC	1971	8
		nstr	392	2.5±0.2			
		nstr	358	9.2±0.4			
	NO_3	fusion	370.5	35.8	DSC	1976	16
		nstr	335.2	4.6			
$C_8H_{17}NH_3$	Br	fusion	483	6.0	DSC	1968	1
		nstr	312	12.6			*b*
		nstr	326	2.1			
	Cl	fusion	477	6.4	DSC	1968	1
		nstr	313 ⎤				
		nstr	308 ⎬	4.6			
		nstr	305 ⎦				
		nstr	276	1.5			
		nstr	310		NMR	1973	9
		nstr	290				
		nstr	210				
		nstr	195				
$C_9H_{19}NH_3$	Br	fusion	477	6.4	DSC	1968	1
		nstr	306	7.7			*b*
		nstr	333 ⎤				
		nstr	327 ⎬	2.2			
		nstr	324 ⎦				
		nstr	255	5.2			
	Cl	fusion	469	7.5	DSC	1968	1
		nstr	325	3.1			
		nstr	270	0.2			
		nstr	259	1.3			
		nstr	324		NMR	1973	9
		nstr	318				
		nstr	190				
		nstr	175				
$C_{10}H_{21}NH_3$	Br	fusion	471	6.5	DSC	1968	1
		nstr	324	16.6			*b*
		nstr	337	2.2			
		nstr	295	7.0			

a: exothermic
b: transition "observed in the first, but not subsequent, scans"

TABLE 8 (Continued)

	Cl	fusion	465	7.2	DSC	1968	1	
		nstr	323	12.6				a
		nstr	305	1.3				a
		nstr	321	4.1				
		nstr	303	2.5				
		nstr	320		NMR	1973	9	
		nstr	305					
		nstr	278					
$(C_{10}H_{21})_2NH_2$	Br	fusion	519.8	19.2	DSC	1976	16	
		nstr	318.2	39.4				
$C_{11}H_{23}NH_3$	Br	fusion	467	7.2	DSC	1968	1	
		nstr	320	25.0				a
		nstr	341	4.5				
		nstr	284	7.1				
	Cl	fusion	462	7.4	DSC	1968	1	
		nstr	339					
		nstr	331	5.3				
		nstr	304	3.2				
$C_{12}H_{25}NH_3$	Br	nstr	345		X-ray	1953	17	
		nstr	334					
		nstr	330					
		fusion	465	7.0	DSC	1968	1	
		nstr	334	19.6				a
		nstr	347	6.1				
		nstr	310	12.1				
	Cl	monoclinic → mono- or triclinic	332		X-ray	1953	17	
		fusion	458	7.7	DSC	1968	1	
		nstr	333	30.9				a
		nstr	347	2.8				
		nstr	334	2.6				
		nstr	324	5.0				
$(C_{12}H_{25})_2NH_2$	Br	fusion	516.8	17.0	DSC	1976	16	
		nstr	317.5	34.0				
$C_{13}H_{27}NH_3$	Br	fusion	463	7.7	DSC	1968	1	
		nstr	331	16.0				a
		nstr	353	6.1				
		nstr	295	9.8				
	Cl	fusion	453	9.1	DSC	1968	1	
		nstr	320	10.3				a
		nstr	359	3.9				
		nstr	343	3.5				
		nstr	318	} 2.3				
		nstr	315					

a: transition "observed in the first, but not subsequent, scans"

TABLE 8 (Continued)

$C_{14}H_{29}NH_3$	Br	fusion	453	7.5	DSC	1968	1	
		nstr	355	} 26.7				a
		nstr	343					a
		nstr	360	8.1				
		nstr	319	12.8				
	Cl	fusion	447	8.4	DSC	1968	1	
		nstr	343	30.2				a
		nstr	365	3.9				
		nstr	343	2.4				
		nstr	334	6.4				
$C_{15}H_{31}NH_3$	Br	fusion	451	7.2	DSC	1968	1	
		nstr	333	22.6				a
		nstr	362	8.0				
		nstr	313	10.4				
	Cl	fusion	443	9.0	DSC	1968	1	
		nstr	337	28.7				a
		nstr	374	5.6				
		nstr	348	4.3				
		nstr	327	} 4.9				
		nstr	323					
$C_{16}H_{33}NH_3$	Br	fusion	447	7.8	DSC	1968	1	
		nstr	353	12.1				a
		nstr	322	13.5				a
		nstr	360	8.7				
		nstr	319	9.7				
	Cl	fusion	436	8.9	DSC	1968	1	
		nstr	345	21.4				a
		nstr	372	5.7				
		nstr	343	} 9.5				
		nstr	335					

a: transition "observed in the first, but not subsequent, scans"

1. Tsau & Gilson (1968)

2. Aston & Ziemer

3. Chang & Westrum

4. Albert, Gutowsky & Ripmeester

5. Stammler, Bruenner, Schmidt & Orcutt;
   taken from Ref. 6

6. Jurga, Depireux & Pajak

7. Goc, Pajak & Szafranska

8. Levkov, Kohr & Mackay

9. Tsau & Gilson (1973)

10. Coker, Ambrose & Janz

11. Gordon

12. Bloom & Reinsborough

13. Southard, Milner & Hendricks

14. Coker, Wunderlich & Janz

15. Andrews & Gordon

16. Lippman & Rudman

17. Gordon, Stenhagen & Vand

HEAT CAPACITIES

Although in principle heat capacity measurements can afford a great deal of fundamental thermodynamic information, this possibility has not been so far largely exploited for what in particular concerns organic salts.

SALTS WITH ORGANIC ANION

Attention has been given to the following.

Na n.alkanoates

(i) Methanoate - An adiabatic calorimetric investigation has been performed between 5 and 350
K by Westrum, Chang & Levitin (1960). For the molar heat capacity at constant pressure, the
enthalpy increment ($H° - H_0°$) and the standard entropy at 298 K the values 82.68±0.04 J $K^{-1}mol^{-1}$,
15.76 kJ mol^{-1} and 103.8 J $K^{-1}mol^{-1}$, respectively, have been found (for the latter quantity the
figure 135.6 had been previously given by Sirotkin). Further measurements have been carried
out with the (DSC) base-line displacement method by Ferloni, Sanesi & Franzosini (1975) from
340 K up to the melt region.

(ii) Ethanoate - The C_p vs T curve, determined by adiabatic calorimetry between 13 and 291 K
(Strelkov, 1955), exhibits a small sharp peak at 21.1 K: the standard entropy at 298 K was
found to be 123.1 J $K^{-1}mol^{-1}$. DSC data at higher temperatures were supplied by Ferloni, Sanesi
& Franzosini (1975).

(iii) Propanoate and butanoate - Only results of DSC investigations are available, respectively due to Ferloni, Sanesi & Franzosini and to Ferloni & Franzosini (1975).

The dependence of the molar heat capacities on temperature for the above salts is shown in Fig.
13. Good overlapping is apparent of the calorimetric and DSC data on methanoate.

Miscellaneous ethanedioates

(i) Concerning the sodium salt, mean specific heat capacities were supplied by Cherbov & Chernyak (1937) as follows:
between 273 and 290 K: 0.9686 J $K^{-1}g^{-1}$
between 290 and 323 K: 1.0577
between 290 and 373 K: 1.0523

(ii) Nogteva, Luk'yanova, Naumov & Paukov (1974) investigated the heat capacity of $(C_2O_4)Cs_2$
between 13 and 300 K, and gave a standard entropy value (at 298 K) equal to 238.2±0.4
J $K^{-1}mol^{-1}$.

(iii) Kapustinskii, Strelkov, Ganenko, Alapina, Stakhanova & Selivanova (1960) on the basis of
heat capacity measurements carried out in the temperature range 66-300 K deduced a value of
146.0±6.7 J $K^{-1}mol^{-1}$ for the standard entropy of lead ethanedioate at 298 K.

Other salts

(i) Calorimetric measurements were performed by Shmidt (1954) on potassium hexadecanoate at
310-357 K in order to obtain information on the heat of conversion of this salt near 366 K
where a second order transition occurs (see Table 1).

(ii) For sodium butanedioate (succinate) an adiabatic calorimeter was employed by Strelkov,
Ganenko & Alapina (1955) to determine the molar heat capacities between 70 and 300 K: for the
standard entropy at 298 K they calculated 211.7±8.4 J $K^{-1}mol^{-1}$.

(iii) A quantitative thermographic method was employed by Barskii, Chechik, Vendel'shtein,
Andreev & Kon'kov (1969) to study the heat capacity dependence on temperature of potassium
benzoate and terephthalate in the temperature ranges 373-673 and 373-773, respectively.

(iv) Finally, Adeosun & Sime recently (1976) presented a study on lead n.alkanoates with even

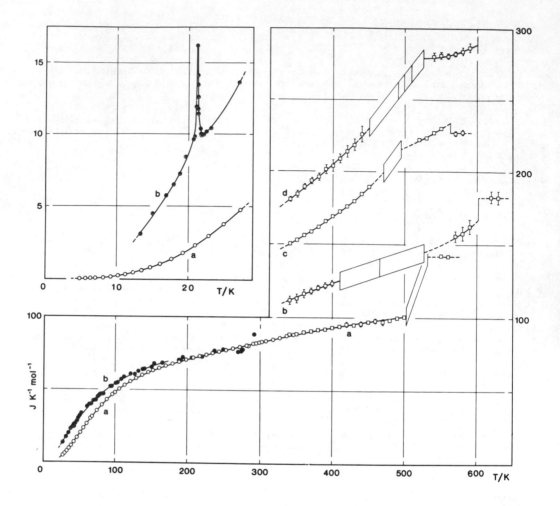

Fig. 13. Molar heat capacities of Na methanoate (a), ethanoate (b), propanoate (c) and butanoate (d). Empty circles: Westrum, Chang & Levitin; dot-centered circles: Strelkov; squares: Ferloni et al. (the "windows" indicate the temperature regions where the occurrence of phase transitions close to each other did not allow satis-factory C_p measurements by DSC). In the upper left-hand corner the initial portions of Westrum's and Strelkov's curves are shown in enlarged scale.

n_c from 6 to 18, where *inter alia* curves C_p *vs* T ($T > 300$ K) are shown for the solids and nu-merical values of the (mean) heat capacity are reported for the mesomorphic and liquid phases.

SALTS WITH ORGANIC CATION

The results available for alkyl-substituted ammonium salts are reviewed hereafter.

(i) Aston & Ziemer (1946) proved that for crystalline methylammonium chloride three polymorphs (α, stable above 264.5 K; β, stable below 220.4 K; and γ, stable between 220.4 and 264.5 K but obtainable in a metastable state down to the hydrogen region) can exist. Employing an adiabatic calorimeter able to work from room temperature to 11 K, each of them was submitted to C_p measure-ment in the proper range: the standard entropy at 298 K was evaluated as 138.6±0.2 J K^{-1} mol^{-1}.

(ii) The heat capacities of tetramethylammonium bromide and chloride, and of tetramethylam-monium iodide were studied respectively by Chang & Westrum (1962, down to 5 K) and by Coulter, Pitzer & Latimer (1940, down to 14 K). A first order and a λ-type transition were observed in the chloride at 75.76 and at 184.85 K (see Table 9), whereas no anomaly occurs in the C_p vs T curves of the bromide and iodide. The standard entropy values at 298 K were: for the bromide 200.8, for the chloride 190.7 and for the iodide 207.9±0.8 J $K^{-1}mol^{-1}$.

(iii) A differential scanning calorimetric technique was adopted by Burns & Verrall (1974) to draw heat capacity curves in the temperature range 273-373 K for several salts among which tetramethyl-, tetraethyl- and tetrabutylammonium bromide.

(iv) Southard, Milner & Hendricks (1933) carried out C_p measurements on "quenched" and "anneal-ed" samples of pentylammonium chloride between 20 and 280 K, and stated the standard entropy value at 298 K to be 266.7 J $K^{-1}mol^{-1}$.

(v) Finally, Andrews & Gordon (1973) determined the heat capacities of tetrahexylammonium per-chlorate at 300-390 K, a region where this salt undergoes four phase transitions: the para-meters pertinent to the latter are reported in Table 9.

Adiabatic low-temperature heat capacities of Na methanoate, ethanoate and butanedioate, as well as of tetramethylammonium bromide, chloride and iodide are given in detail in Table 9.

TABLE 9. Low-temperature (T/K) molar heat capacities (C_p/J $K^{-1}mol^{-1}$) of some organic salts

Na methanoate	T	4.84	5.65	6.39	7.03	8.19
(Westrum, Chang &	C_p	0.025	0.046	0.059	0.071	0.109
Levitin, 1960)	T	9.21	10.22	11.27	12.39	13.55
	C_p	0.167	0.251	0.347	0.460	0.611
	T	14.77	16.12	17.65	19.27	20.96
	C_p	0.808	1.059	1.393	1.820	2.326
	T	22.80	24.84	27.13	29.68	32.42
	C_p	2.946	3.778	4.766	6.029	7.519
	T	35.37	38.62	42.41	46.77	51.37
	C_p	9.272	11.326	13.853	16.912	20.175
	T	55.69	60.31	65.69	67.93	74.47
	C_p	23.255	26.464	30.087	31.531	35.510
	T	81.12	88.26	95.95	99.69	105.55
	C_p	39.342	43.074	46.468	47.982	50.250
	T	113.22	121.26	129.48	137.71	146.21
	C_p	52.974	55.497	57.773	59.877	61.777
	T	152.68	161.27	170.28	179.68	189.01
	C_p	63.18	64.85	66.40	67.99	69.33
	T	198.36	208.09	217.17	226.11	234.95
	C_p	70.67	71.96	73.14	74.22	75.31
	T	243.59	252.95	262.52	271.95	281.27
	C_p	76.27	77.36	78.49	79.58	80.58
	T	290.46	294.47	298.72	302.09	307.85
	C_p	81.96	82.22	82.68	83.14	83.72
	T	317.34	326.78	336.11	345.33	
	C_p	84.85	86.02	87.36	88.12	
Na ethanoate	T	13.36	15.09	16.91	17.96	18.84
(Strelkov, 1955)	C_p	3.10	4.48	5.73	6.49	7.24
	T	19.60	20.65	20.83	20.83	21.03
	C_p	8.41	9.62	9.83	9.87	11.92
	T	21.04	21.13	21.22	21.27	21.28
	C_p	11.21	16.23	14.14	12.59	13.47
	T	21.30	21.31	21.63	21.69	21.76
	C_p	11.76	11.42	10.38	10.00	9.92

TABLE 9 (Continued); Na ethanoate (Continued)

T	22.06	22.42	23.07	26.80	31.33
C_p	9.96	10.17	10.42	13.64	17.32
T	34.87	37.76	40.38	42.75	43.03
C_p	20.00	22.93	24.10	25.77	24.43
T	44.51	47.20	48.95	50.83	52.81
C_p	26.86	28.74	30.50	31.92	33.60
T	60.96	63.97	66.85	69.80	73.06
C_p	38.24	39.66	40.04	42.38	42.84
T	75.88	77.46	78.61	81.10	83.83
C_p	45.14	45.81	46.61	46.78	46.94
T	93.30	95.28	101.26	103.28	108.50
C_p	52.17	52.13	54.27	54.60	57.11
T	112.07	121.27	127.84	132.54	138.88
C_p	58.66	60.84	60.50	62.68	63.93
T	140.87	150.96	154.57	165.54	189.38
C_p	64.10	65.61	67.99	68.53	70.37
T	191.21	193.53	206.46	220.43	237.12
C_p	70.25	72.38	72.05	72.84	76.94
T	249.27	269.95	274.72	276.03	284.18
C_p	77.70	76.32	76.69	77.91	80.42
T	291.18				
C_p	88.07				

Na butanedioate (Na succinate) (Strelkov, Ganenko & Alapina, 1955)					
T	69.77	74.46	78.75	83.86	86.60
C_p	60.96	67.32	72.17	76.90	79.96
T	89.50	93.65	97.04	97.77	99.07
C_p	82.55	87.57	90.79	91.63	93.60
T	100.04	100.43	103.97	108.79	127.05
C_p	92.59	93.72	96.27	102.68	115.02
T	136.10	139.51	143.88	151.56	155.43
C_p	122.01	125.65	126.15	132.72	135.77
T	160.40	161.31	164.74	169.06	173.56
C_p	137.32	138.87	139.16	143.39	144.01
T	178.33	193.60	207.58	212.03	213.80
C_p	147.32	153.18	156.61	156.86	158.62
T	229.10	239.50	248.75	263.07	268.08
C_p	161.96	166.36	167.90	168.24	170.21
T	300.43				
C_p	172.55				

Tetramethylammonium bromide (Chang & Westrum, 1962)					
T	4.99	5.61	6.63	7.64	8.52
C_p	0.159	0.276	0.473	0.841	1.280
T	9.45	10.51	11.67	12.92	14.28
C_p	1.845	2.628	3.648	4.904	6.468
T	15.86	17.66	19.61	21.66	23.85
C_p	8.494	10.899	13.627	16.606	19.790
T	26.27	28.96	32.01	35.41	39.11
C_p	23.271	26.978	30.999	35.158	39.208
T	43.17	47.63	49.88	55.49	61.62
C_p	43.145	47.041	48.83	52.84	56.82
T	68.24	75.40	78.21	84.56	91.67
C_p	60.58	64.31	65.86	69.41	72.97

TABLE   9 (Continued); Tetramethylammonium bromide (Continued)

	T	99.41	107.32	115.39	123.78	132.27
	C_P	76.65	80.63	84.64	88.87	92.89
	T	140.75	149.39	157.32	158.40	161.70
	C_P	96.82	100.83	104.43	105.10	106.44
	T	166.60	175.28	179.59	183.44	188.32
	C_P	108.45	111.92	113.76	115.19	117.19
	T	197.02	205.94	215.07	224.17	233.37
	C_P	120.58	124.10	127.65	131.21	135.19
	T	242.63	248.63	257.46	266.26	275.30
	C_P	139.12	141.75	145.73	148.03	151.88
	T	284.39	293.43	302.49	311.61	318.12
	C_P	155.85	159.79	163.72	167.61	170.37
	T	327.36	336.57	345.60		
	C_P	174.18	178.36	182.05		

Tetramethylammonium chloride (Chang & Westrum, 1962)	T	5.09	6.00	6.94	7.92	8.03
	C_P	0.088	0.172	0.272	0.427	0.460
	T	8.97	9.00	10.04	11.01	12.11
	C_P	0.686	0.695	1.029	1.397	1.925
	T	12.93	13.29	14.52	15.83	17.21
	C_P	2.339	2.515	3.314	4.289	5.477
	T	17.25	18.94	20.69	22.62	24.92
	C_P	5.515	7.150	9.042	11.318	14.196
	T	27.63	30.59	33.88	37.38	41.18
	C_P	17.711	21.656	25.966	30.326	34.736
	T	45.46	50.01	55.00	60.54	65.86
	C_P	39.288	43.72	48.03	52.59	56.27
	T	66.89	71.75	71.88	72.47	72.96
	C_P	56.94	60.12	60.33	60.38	60.92
	T	74.85	75.37	75.45	75.56	75.67
	C_P	62.22	64.31	135.73	144.98	315.26
	T	75.75	75.76	75.84	75.89	76.00
	C_P	327.23	348.03	270.24	245.22	163.80
	T	76.20	76.26	76.58	76.65	76.71
	C_P	111.46	94.27	71.46	70.63	*104.98*
	T	76.90	77.75	79.62	81.47	83.43
	C_P	65.69	64.06	*60.92*	65.81	67.03
	T	85.97	93.55	101.21	109.64	118.31
	C_P	68.58	72.76	76.99	81.55	86.15
	T	127.00	135.92	144.90	153.99	162.90
	C_P	90.71	94.81	99.96	104.43	108.83
	T	163.64	171.46	172.96	175.44	177.14
	C_P	109.12	113.55	113.93	115.02	115.69
	T	178.29	179.43	179.98	180.55	181.66
	C_P	116.69	117.44	118.24	118.49	119.66
	T	182.70	182.77	183.20	183.65	183.89
	C_P	120.75	120.88	121.88	122.72	124.18
	T	184.00	184.25	184.45	184.64	184.84
	C_P	124.39	126.27	129.08	131.34	137.70
	T	184.99	185.03	185.23	185.43	185.64
	C_P	*126.61*	129.03	117.44	115.94	115.48
	T	185.90	186.09	186.31	187.22	187.24
	C_P	115.23	*114.56*	115.19	115.27	*115.23*

TABLE 9 (Continued); Tetramethylammonium chloride (Continued)

T	189.01	191.35	193.64	198.47	200.15
C_p	115.65	116.27	117.03	118.87	119.20
T	209.45	218.72	228.08	237.19	246.30
C_p	122.72	126.23	129.24	133.18	136.73
T	255.35	264.32	273.13	281.78	290.36
C_p	140.33	143.80	146.86	150.62	153.64
T	298.94	302.14	311.87	321.43	330.84
C_p	157.32	158.41	162.26	166.02	169.79
T	339.21	346.70			
C_p	173.13	176.31			

The values in italic belong to measurement series covering the transition regions by large temperature steps

Tetramethylammonium iodide (Coulter, Pitzer & Latimer, 1940)	T	14.10	15.68	17.41	19.23	21.20
	C_p	7.171	9.146	11.535	14.129	17.393
	T	23.83	27.32	30.68	34.17	37.89
	C_p	21.489	26.514	30.832	35.510	40.489
	T	41.71	45.88	50.22	54.82	60.27
	C_p	44.39	48.03	51.59	54.85	58.66
	T	62.33	63.86	68.09	75.10	82.77
	C_p	60.54	61.17	62.97	66.78	71.17
	T	90.82	99.26	107.66	116.50	125.83
	C_p	75.48	79.91	84.18	88.53	93.01
	T	132.54	135.17	140.60	149.60	158.59
	C_p	96.36	97.28	99.54	103.34	107.70
	T	166.95	175.87	185.34	195.32	205.27
	C_p	111.29	115.06	118.87	123.01	126.78
	T	214.53	224.28	234.53	244.75	245.59
	C_p	130.16	133.89	137.19	143.01	141.50
	T	253.37	253.42	262.18	266.49	274.54
	C_p	143.93	144.22	148.70	150.71	162.09
	T	282.09	289.99	297.53		
	C_p	156.86	158.16	161.25		

ENTHALPIES OF FORMATION

Standard enthalpies of formation for salts with organic anion or cation have been so far evaluated only in a limited number of cases.

Some fifty data, partly drawn from the classical thermochemical work by Berthelot, may be found in the well known compilations on "Selected Values of Chemical Thermodynamic Properties" by Rossini et al. (1952) and by Wagman et al. (1968): such data are reported without further comment in the table hereafter under Ref.s 1 and 11, respectively.

Other sources supply some seventy more data, based either on calorimetric measurements of different kinds (Bousquet et al.; Kapustinskii et al.; Kirpichev & Rubtsov; Le Van & Perinet; Lobanov; etc.) or on emf determinations (Chen & Pan; Ostannii, Zharkova, Erofeev & Kazakova; etc.).

Also tabulated are the standard enthalpies of formation of some rare earth formates, calculated by Plyushchev et al. from the pertinent values of crystal lattice energy, whereas figures obtained through a variety of empirical correlations, such as those proposed e.g. by Zharkova & Nissenbaum, have been omitted.

TABLE 10. Free energies and enthalpies of formation at 298 K

Salt	$-\Delta G_f^{\circ}$ kJ mol^{-1}	$-\Delta H_f^{\circ}$ kJ mol^{-1}	Year	Ref.	Notes
Salts with organic anion					
$(CHO_2)_2Ba$		1366.1	1952	1	
$(CHO_2)_2Ca$		1353.5	1952	1	
$(CHO_2)_2Co$		877.2	1948	2	
		879.5±17.2	1969	3	
$(CHO_2)_2Cu$		751.4±16.3	1969	3	
		743.1	1970	4	*a*
		754.8	1970	4	*b*
$(CHO_2)_3Dy$		1875.3	1965	5	
$(CHO_2)_3Gd$		1881.5	1965	5	
$(CHO_2)_2Hg_2$	610.0	679.1	1971	6	
	591.1±1.2	750.9±1.6	1974	7	
$(CHO_2)_3Ho$		1872.3	1965	5	
$(CHO_2)K$		661.1	1952	1	
$(CHO_2)_3La$		2005.0	1965	5	
$(CHO_2)Li$		689.9	1947	8	*c*
		654.4	1947	8	*d*
$(CHO_2)_3Lu$		1823.0	1965	9	
$(CHO_2)_2Mn$		1013.4	1952	1	
$(CHO_2)Na$		648.6	1952	1	
$(CHO_2)_3Nd$		1892.0	1965	5	
$(CHO_2)NH_4$		575.3	1897	10	
		555.6	1952	1	
		567.5	1968	11	
$(CHO_2)_2Ni$		872.3	1948	2	

TABLE 10 (Continued); Salts with organic anion (Continued)

		875.7±18.4	1969	3	
$(CHO_2)_2Pb$		837.2	1952	1	
		878.6	1968	11	
$(CHO_2)_3Pr$		1919.2	1965	5	
$(CHO_2)_3Sm$		1884.5	1965	5	
$(CHO_2)_2Sr$		1361.9	1952	1	
$(CHO_2)_3Tb$		1872.8	1965	5	
$(CHO_2)_3Tm$		1844.7	1965	9	
$(CHO_2)_2UO_2$		1870.8±0.5	1975	12	
		1870.7±0.4	1976	13	
$(CHO_2)_3Yb$		1800.0	1965	9	
$(CHO_2)_2Zn$		941.4	1952	1	
		986.6	1968	11	
$(C_2H_3O_2)Ag$		390.8	1952	1	
$(C_2H_3O_2)_3Al$		1892.4	1968	11	
$(C_2H_3O_2)_2Ba$		1485.7	1952	1	
$(C_2H_3O_2)_2Be$		1292.0±3.1	1969	14	
$(C_2H_3O_2)_2Ca$		1485.3	1952	1	
$(C_2H_3O_2)_2Co$		965.5±2.8	1967	15	
		972.8±16.7	1969	3	
$(C_2H_3O_2)_2Cu$		896.6	1952	1	
		899.1±17.6	1969	3	
$(C_2H_3O_2)_2Hg$		834.3	1952	1	
$(C_2H_3O_2)_2Hg_2$		841.4	1952	1	
	646.4	846.8	1968	16	
$(C_2H_3O_2)K$		724.7	1952	1	
$(C_2H_3O_2)Li$		754.3	1947	8	*c*
		712.1	1947	8	*d*
$(C_2H_3O_2)_2Mn$		1142.2	1952	1	
$(C_2H_3O_2)Na$		710.4	1952	1	
$(C_2H_3O_2)_3Nd$		2196.1±1.8	1972	17	
$(C_2H_3O_2)NH_4$		628.6	1897	10	
		618.4	1952	1	
		616.1	1968	11	
$(C_2H_3O_2)_2Ni$		966.5±17.2	1969	3	
$(C_2H_3O_2)_2Pb$		964.4	1952	1	
		963.8	1968	11	
$(C_2H_3O_2)_2Sr$		1492.4	1952	1	
$(C_2H_3O_2)Tl$		527.6	1968	11	
$(C_2H_3O_2)_2Zn$		1079.9	1952	1	
		1078.6	1968	11	
$(C_3H_5O_2)_2Co$		1012.1±18.0	1969	3	
$(C_3H_5O_2)_2Cu$		937.6±18.4	1969	3	
$(C_3H_5O_2)_2Ni$		1004.2±17.6	1969	3	

TABLE 10 (Continued); Salts with organic anion (Continued)

$(C_4H_7O_2)_2Co$		1047.3	1969	18
$(C_5H_9O_2)_2Ni$		1091.6	1969	18
$(C_7H_5O_2)K$		574.5±41.8	1967	19
		617.1±0.8	1971	20
$(C_8H_{15}O_2)_2Cu$		1123.8	1969	18
$(C_2O_4)Ag_2$		665.7	1952	1
$(C_2O_4)Ba$		1370.7	1950	21
$(C_2O_4)Ca$		1393.7	1936	22
		1360.6	1950	21
$(C_2O_4)Cd$		909.6	1961	23
$(C_2O_4)Co$		869.4±18.0	1969	3
		868.6	1969	18
$(C_2O_4)Cu$		759.8±16.3	1969	3
		773.2	1969	18
$(C_2O_4)_3Ga_2$		2154.8	1952	1
$(C_2O_4)Hg$		677.0	1952	1
$(C_2O_4)K_2$		1342.2	1952	1
$(C_2O_4)Mg$		1265.7	1950	21
$(C_2O_4)Mn$		1080.3	1952	1
$(C_2O_4)Na_2$		1315.0	1952	1
$(C_2O_4)(NH_4)_2$		1124.3	1952	1
$(C_2O_4)Ni$		861.9	1956	24
		853.5±19.2	1969	3
$(C_2O_4)Pb$		858.1	1952	1
		850.2	1956	25
	756.9	862.7±2.1	1956	26
	756.9	862.9	1960	27
$(C_2O_4)Sr$		1371.1	1950	21

Salts with organic cation

$(CH_3NH_3)Br$		258.9	1976	27
$(CH_3NH_3)Cl$	146.8	285.8	1952	1
	159.0	297.9	1968	11
		298.3	1976	28
$(CH_3NH_3)I$		200.7	1976	28
$(CH_3NH_3)NO_3$		337.2	1951	29
		354.4	1968	11
$(CH_3NH_3)_2SO_4$		1155.2±0.8	1971	30
$(CH_3NH_3)_2SeO_4$		863.6±0.8	1971	30
$\{(CH_3)_2NH_2\}Br$		257.3	1976	28
$\{(CH_3)_2NH_2\}Cl$		291.1	1976	28
$\{(CH_3)_2NH_2\}I$		201.7	1976	28
$\{(CH_3)_2NH_2\}NO_3$		330.5	1951	29
		350.2	1968	11
$\{(CH_3)_3NH\}Br$		253.7	1976	28

e

TABLE 10 (Continued); Salts with organic cation (Continued)

$\{(CH_3)_3NH\}Cl$	281.5	1968	11	
	283.3	1976	28	
$\{(CH_3)_3NH\}I$	203.0	1976	28	
$\{(CH_3)_3NH\}NO_3$	305.9	1951	29	
	343.9	1968	11	
$\{(C_2H_5)_2NH_2\}Cl$	358.2	1968	11	
$\{(C_2H_5)_2NH_2\}NO_3$	413.0	1951	29	
	418.8	1968	11	
$\{(C_2H_5)_3NH\}Cl$	385.8	1968	11	
$\{(C_2H_5)_3NH\}NO_3$	407.1	1951	29	
	447.7	1968	11	
$(C_6H_5NH_3)NO_3$	177.8	1951	29	*f*
$(C_6H_5CH_2NH_3)NO_3$	239.3	1951	29	*g*

a: light blue anhydrous salt from the tetrahydrate
b: anhydrous intensely blue salt
c: from heats of solution
d: from reticular energies and heat of formation of the ions
e: $C_7H_5O_2$ = benzoate
f: C_6H_5 = phenyl
g: $C_6H_5CH_2$ = benzyl

1. Rossini, Wagman, Evans, Levine & Jaffe
2. Yatsimirskii (1948)
3. Le Van & Perinet (1969 a)
4. Ostannii, Zharkova & Erofeev (1970)
5. Plyushchev, Shklover, Shkol'nikova, Kuznetsova & Nadezhdina
6. Ostannii, Zharkova & Erofeev (1971)
7. Ostannii, Zharkova, Erofeev & Kazakova
8. Yatsimirskii (1947)
9. Shklover, Plyushchev, Kuznetsova & Trushina
10. Berthelot
11. Wagman, Evans, Parker, Halow, Bayley & Schumm
12. Bousquet, Bonnetot, Claudy, Mathurin & Turck (1975)
13. Bousquet, Bonnetot, Claudy, Mathurin & Turck (1976)
14. Kirpichev & Rubtsov
15. Nguen Zui Tkhin, Zharkova & Erofeev
16. Chen & Pan
17. Usubaliev & Batkibekova
18. Le Van & Perinet (1969 b)
19. Furuyama & Ebara
20. Lobanov
21. Kapustinskii & Samoilov
22. Bichowsky & Rossini
23. Kapustinskii & Samplavskaya
24. Kornienko & Petrenko
25. Scott, Good & Waddington
26. Kapustinskii, Selivanova & Stakhanova
27. Kapustinskii, Strelkov, Ganenko, Alapina, Stakhanova & Selivanova
28. Wilson
29. Cottrell & Gill
30. Selivanova & Zalogina

GLASS-FORMING ORGANIC SALTS

Among salts with organic anion or cation, the ability of some melts to supercool enough to produce glasses (i.e., non-crystalline solids the structure of which does not offer long range order) was put into evidence only in recent years, the tendency to glass formation being more frequent in multicomponent mixtures (see Chapter 2.1) than in single salts.

Acetate glasses have been the most widely investigated so far (Onodera, Suga & Seki, 1968; Duffy & Ingram, 1969; Bartholomew, 1970, 1972; Bartholomew & Lewek, 1970; Van Uitert, Bonner & Grodkiewicz, 1971), and it can be here stressed that one-component glasses were obtained, e.g., by Bartholomew & Lewek by quenching either the hydrated or the anhydrous melts of lithium, zinc and lead acetates.

Getting glassy acetate materials through ways different from supercooling of fused systems seems also to be possible. Thus Onodera, Suga & Seki by dehydration under vacuum of solid magnesium acetate tetrahydrate were able to obtain a sample of anhydrous $(C_2H_3O_2)_2Mg$ which: (i) was proved to be amorphous by X-ray diffraction patterns, and (ii) became crystalline when heated, after having undergone a "glass transition".

The latter phenomenon, which is peculiar to usual glasses, can be observed, e.g , by means of DTA or DSC analysis. At the glass transition temperature, T_g/K (which, however, is not to be taken as a thermodynamic quantity, inasmuch as it tends to increase with increasing heating rate), the recorded trace exhibits a characteristically shaped endothermic anomaly, soon followed by a sharp exothermic peak due to crystallization (see Fig. 14). As an example, for lithium acetate Bartholomew (1970) found by DTA a T_g value of 394 K when a heating rate of 10 K min^{-1} was employed.

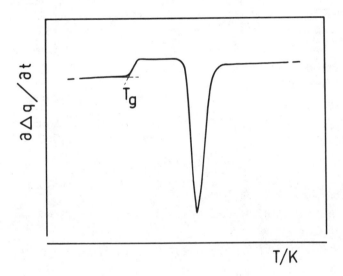

Fig. 14. Typical DSC trace for a glassy material, showing the glass transition temperature, T_g, followed by crystallization.

Concerning salts with organic cation, the case of the hydrochlorides of pyridine and substituted pyridines is worthy of mention. Although no tendency to produce glasses is exhibited by pyridinium chloride in the pure state, the introduction of a small sized substituent group into the pyridine ring may be sufficient to give rise to a remarkable glass-forming ability, as that occurring in α-picolinium chloride (Hodge, 1974).

A comprehensive survey of glass-forming molten salt systems was recently given by Angell & Tucker (1974).

REFERENCES

ADEOSUN, S.O., and SIME, S.J., *Thermochim. Acta*, 17, 351 (1976)

ALBERT, S., GUTOWSKY, H.S., and RIPMEESTER, J.A., *J. Chem. Phys.*, 56, 3672 (1972)

ANDREWS, J.T.S., and GORDON, J.E., *JCS Faraday Trans. I* , 69, 546 (1973)

ANGELL, C.A., and TUCKER, J.C., *"Physical Chemistry of Process Metallurgy - The Richardson Conference"*, Ed. J.H.E. Jeffes and R.J. Tait, Inst. Mining Metallurgy Publ., 1974, 207

ASTON, J.G., and ZIEMER, C.W., *J. Amer. Chem. Soc.*, 68, 1405 (1946)

BAKUNIN, M., and VITALE, E., *Gazz. Chim. Ital.*, 65, 593 (1935)

BALAREW, C., and STOILOVA, D., *J. Thermal Anal.*, 7, 561 (1975)

BARR, M.R., and DUNELL, B.A., *Can. J. Chem.*, 42, 1098 (1964)

BARSKII, Y.P., CHECHIK, E.Y., VENDEL'SHTEIN, E.G., ANDREEV, Y.K., and KON'KOV, K.D., *Inform. Soobshch. Gos. Nauch.- Issled. Proekt. Inst. Azotn. Prom. Prod. Org. Sin.*, 24, 36 (1969); taken from *CA* 75, 48188

BARTHOLOMEW, R.F., *J. Phys. Chem.*, 74, 2507 (1970)

BARTHOLOMEW, R.F., *U.S. Patent 3,649,551* (1972)

BARTHOLOMEW, R.F., and LEWEK, S.S., *J. Amer. Ceram. Soc.*, 53, 445 (1970)

BASKOV, A., *Zh. Russk. Fiz. Khim. Obshch.*, 47, 1533 (1915)

BAUM, E., DEMUS, D., and SACKMANN, H., *Wiss. Z. Univ. Halle XIX '70*, 37

BENTON, D.P., HOWE, P.G., FARNAND, R., and PUDDINGTON, I.E., *Can. J. Chem.*, 33, 1798 (1955)

BENTON, D.P., HOWE, P.G., and PUDDINGTON, I.E., *Can. J. Chem.*, 33, 1384 (1955)

BERCHIESI, G., and LAFFITTE, M., *J. Chim. Phys.*, 1971, 877

BERGMAN, A.G., and EVDOKIMOVA, K.A., *Izv. Sekt. Fiz.-Khim. Anal., Inst. Obshch. i Neorg. Khim. Akad. Nauk SSSR*, 27, 296 (1956)

BERNARD, M.A., and BUSNOT, F., *Bull. Soc. Chim.*, 12, 4649 (1967)

BERTHELOT, M.P.E., *"Thermochimie"*, Paris, 1897

BICHOWSKY, F.R., and ROSSINI, F.D., *"Thermochemistry of Chemical Substances"*, New York, 1936; taken from KAPUSTINSKII & SAMOILOV

BLOOM, H., and REINSBOROUGH, V.C., *Austr. J. Chem.*, 20, 2583 (1967)

BORNWATER, J.T., *Rec. Trav. Chim.*, 26, 413 (1907)

BOUAZIZ, R., and BASSET, J.Y., *Compt. Rend.*, 263, 581 (1966)

BOUSQUET, J., BONNETOT, B., CLAUDY, P., MATHURIN, D., and TURCK, G., *Conf. Int. Thermodyn. Chim. (C.R.), 4th*, 1, 169 (1975)

BOUSQUET, J., BONNETOT, B., CLAUDY, P., MATHURIN, D., and TURCK, G., *Thermochim. Acta*, 14, 357 (1976)

BRAGHETTI, M., and BERCHIESI, G., *Annual Meeting, Chimica Inorganica*, 1969, 101

BRAGHETTI, M., BERCHIESI, G., and FRANZOSINI, P., *Ric. Sci.*, 39, 576 (1969)

BRAGHETTI, M., LEONESI, D., and FRANZOSINI, P., *Ric. Sci.*, 38, 116 (1968)

BUERGER, M.J., SMITH, L.B., RYER, F.V., and SPIKE, J.E., Jr., *Proc. Nat. Acad. Sci., U.S.*, 31, 226 (1945)

BURNS, J.A., and VERRALL, R.E., *Thermochim. Acta*, 9, 277 (1974)

CHANG, S.S., and WESTRUM, E.F., Jr., *J. Chem. Phys.*, 36, 2420 (1962)

CHEN, L.S., and PAN, K., *J. Chin. Chem. Soc. (Taipei)*, 15, 106 (1968); taken from *CA*, 70, 51274

CHERBOV, S.I., and CHERNYAK, E.L., *Zh. Prikl. Khim.*, 10, 1220 (1937); taken from Gmelins Handbuch, Syst. N. 21, Na Erg (1967), 1443

CHESLEY, F.G., *J. Chem. Phys.*, 8, 643 (1940)

CHRISTIE, G.H., and MENZIES, R.C., *J. Chem. Soc.*, 127, 2369 (1925)

COKER, T.G., AMBROSE, J., and JANZ, G.J., *J. Amer. Chem. Soc.*, 92, 5293 (1970)

COKER, T.G., WUNDERLICH, B., and JANZ, G.J., *Trans. Faraday Soc.*, 65, 3361 (1969)

COTTRELL, T.L., and GILL, J.E., *J. Chem. Soc.*, 1951, 1798

COULTER, L.V., PITZER, K.S., and LATIMER, W.M., *J. Amer. Chem. Soc.*, 62, 2845 (1940)

COX, D.B., and McGLYNN, J.F., *Anal. Chem.*, 29, 960 (1957)

DAVIDSON, A.W., and McALLISTER, W.H., *J. Amer. Chem. Soc.*, 52, 507 (1930)

DAVIDSON, A.W., SISLER, H.H., and STOENNER, R., *J. Amer. Chem. Soc.*, 66, 779 (1944)

DIOGENOV, G.G., *Russ. J. Inorg. Chem.*, 1(4), 199 (1956); translated from *Zh. Neorg. Khim.*, 1,
        799 (1956)

DIOGENOV, G.G., BORZOVA, L.L., and SARAPULOVA, I.F., *Russ. J. Inorg. Chem.*, 10, 948 (1965);
        translated from *Zh. Neorg. Khim.*, 10, 1738 (1965)

DIOGENOV, G.G., BRUK, T.I., and NURMINSKII, N.N., *Russ. J. Inorg. Chem.*, 10, 814 (1965); trans-
        lated from *Zh. Neorg. Khim.*, 10, 1496 (1965)

DIOGENOV, G.G., and CHUMAKOVA, V.P., *Fiz.-Khim. Issled. Rasplavov Solei*, 1975, 7

DIOGENOV, G.G., and ERLYKOV, A.M., *Nauch. Doklady Vysshei Shkoly, Khim. i Khim. Tekhnol.*,
        1958, No. 3, 413

DIOGENOV, G.G., ERLYKOV, A.M., and GIMEL'SHTEIN, V.G., *Russ. J. Inorg. Chem.*, 19, 1069 (1974);
        translated from *Zh. Neorg. Khim.*, 19, 1955 (1974)

DIOGENOV, G.G., and GIMEL'SHTEIN, V.G., *Russ. J. Inorg. Chem.*, 10, 1395 (1965); translated
        from *Zh. Neorg. Khim.*, 10, 2567 (1965)

DIOGENOV, G.G., and MORGEN, L.T., *Nekotorye Vopr. Khimii Rasplavlen. Solei i Produktov Des-
        truktsii Sapropelitov*, 1974, 32

DIOGENOV, G.G., and MORGEN, L.T., *Fiz.-Khim. Issled. Rasplavov Solei*, 1975, 59

DIOGENOV, G.G., NURMINSKII, N.N., and GIMEL'SHTEIN, V.G., *Russ. J. Inorg. Chem.*, 2(7), 237
        (1957); translated from *Zh. Neorg. Khim.*, 2, 1596 (1957)

DIOGENOV, G.G., and SARAPULOVA, I.F., *Russ. J. Inorg. Chem.*, 9(2), 265 (1964); translated from
        Zh. Neorg. Khim., 9, 482 (1964)

DIOGENOV, G.G., and SARAPULOVA, I.F., *Russ. J. Inorg. Chem.*, 9(5), 704 (1964); translated from
        Zh. Neorg. Khim., 9, 1292 (1964)

DIOGENOV, G.G., and SARAPULOVA, I.F., *Russ. J. Inorg. Chem.*, 9(6), 814 (1964); translated from
        Zh. Neorg. Khim., 9, 1499 (1964)

DMITREVSKAYA, O.I., *J. Gen. Chem. USSR*, 28, 295 (1958); translated from *Zh. Obshch. Khim.*, 28,
        299 (1958)

DMITREVSKAYA, O.I., and SOKOLOV, N.M., *J. Gen. Chem. USSR*, 30, 19 (1960); translated from *Zh.
        Obshch. Khim.*, 30, 20 (1960)

DUFFY, J.A., and INGRAM, M.D., *J. Amer. Ceram. Soc.*, 52, 224 (1969)

DUNELL, B.A., and JANZEN, W.R., *Wiss. Z. Friedrich-Schiller Univ. Jena*, 14, 191 (1965); taken
        from CYR, T.J.R., JANZEN, W.R., and DUNELL, B.A., *"Ordered Fluids and Liquid Crystals"*,
        Adv. Chem. Ser., No. 63, Amer. Chem. Soc., Washington, 1967, 13

DURUZ, J.J., MICHELS, H.J., and UBBELOHDE, A.R., *Chem. & Ind.*, 1969, 1386

DURUZ, J.J., MICHELS, H.J., and UBBELOHDE, A.R., *Proc. R. Soc. London*, A322, 281 (1971)

ESCALES, R., and KOEPKE, H., *J. Prakt. Chem.*, 87, 258 (1913)

FERLONI, P., and FRANZOSINI, P., *Gazz. Chim. Ital.*, 105, 391 (1975)

FERLONI, P., SANESI, M., and FRANZOSINI, P., *Z. Naturforsch.*, 30a, 1447 (1975)

FERLONI, P., SANESI, M., and FRANZOSINI, P., *Z. Naturforsch.*, 31a, 679 (1976)

FERLONI, P., SANESI, M., TONELLI, P.L., and FRANZOSINI, P., *Z. Naturforsch.*, 33a, 240 (1978)

FERLONI, P., SPINOLO, G., ZANGEN, M., and FRANZOSINI, P., *Z. Naturforsch.*, 32a, 329 (1977)

FERLONI, P., ZANGEN, M., and FRANZOSINI, P., *Z. Naturforsch.*, 32a, 627 (1977)

FERLONI, P., ZANGEN, M., and FRANZOSINI, P., *Z. Naturforsch.*, 32a, 793 (1977)

FRANCHIMONT, N., *Ber.*, 12, 11 (1879)

FURUYAMA, S., and EBARA, N., *Scient. Coll. Gen. Educ. Univ. Tokyo*, 17, 81 (1967); taken from
        LOBANOV

GALLOT, B., and SKOULIOS, A., *Compt. Rend.*, 252, 142 (1961)

GALLOT, B., and SKOULIOS, A., *Acta Cryts.*, 15, 826 (1962)

GALLOT, B., and SKOULIOS, A., *Mol. Crystals*, 1, 263 (1966)

GALLOT, B., and SKOULIOS, A., *Kolloid-Z. Polym.*, 209, 164 (1966)

GALLOT, B., and SKOULIOS, A., *Kolloid-Z. Polym.*, 210, 143 (1966)

GALLOT, B., and SKOULIOS, A., *Kolloid-Z. Polym.*, 213, 143 (1966)

GIMEL'SHTEIN, V.G., and DIOGENOV, G.G., *Russ. J. Inorg. Chem.*, 3(7), 230 (1958); translated
        from *Zh. Neorg. Khim.*, 3, 1644 (1958)

GOBERT-RANCHOUX, E., and CHARBONNIER, F., *J. Thermal Anal.*, 12, 33 (1977)

GOC, R., PAJAK, Z., and SZAFRANSKA, B., *Magn. Res. Relat. Phenom., Proc. Congr. AMPERE, 18th*,
        2, 405 (1974)(Publ. 1975)

GODBOLE, N., and JOSHI, K., *Allgem. Oel u. Fett-Ztg.*, 27, 77 (1930); taken from VOLD, M.J.,
        MACOMBER & VOLD

GOLUBEVA, M.S., BERGMAN, A.G., and GRIGOR'EVA, E.A., *Uch. Zap. Rostovsk.-na-Donu Gos. Univ.*,
        41, 145 (1958)

GORDON, J.E., *J. Amer. Chem. Soc.*, 87, 4347 (1965)

GORDON, M., STENHAGEN, E., and VAND, V., *Acta Cryst.*, 6, 739 (1953)

GRANT, R.F., and DUNELL, B.A., *Can. J. Chem.*, 38, 1951 (1960)

GRANT, R.F., and DUNELL, B.A., *Can. J. Chem.*, 38, 2395 (1960)

GRANT, R.F., and DUNELL, B.A., *Can. J. Chem.*, 39, 359 (1961)

GRANT, R.F., HEDGECOCK, N., and DUNELL, B.A., *Can. J. Chem.*, 34, 1514 (1956)

GREEN, W.F., *J. Phys. Chem.*, 12, 655 (1908); taken from *C* 1909 I, 837

GROSCHUFF, E., *Ber.*, 36, 1783 (1903)

HALMOS, Z., MEISEL, T., SEYBOLD, K., and ERDEY, L., *Talanta*, 17, 1191 (1970)

HARRIES, C., and THIEME, C., *Ann.*, 343, 354 (1905)

HATIBARUA, J.M., and PARRY, G.S., *Acta Cryst.*, B28, 3099 (1972)

HATTIANGDI, G.S., VOLD, M.J., and VOLD, R.D., *Ind. Eng. Chem.*, 41, 2320 (1949)

HAWORTH, R.D., *J. Chem. Soc.*, 1929, 1456

HAZLEWOOD, F.J., RHODES, E., and UBBELOHDE, A.R., *Trans. Faraday Soc.*, 62, 3101 (1966)

HERRMANN, K., *Trans. Faraday Soc.*, 29, 972 (1933)

HODGE, I.M., *Ph.D. Thesis*, Purdue University, Lafayette, Indiana (1974)

HOLDE, D., and SELIM, M., *Ber.*, 58, 523 (1925)

HOLDE, D., and TAKEHARA, K., *Ber.*, 58, 1788 (1925)

IVANOFF, D., *Bull. Soc. Chim.*, 43, 441 (1928)

JACOBSON, C.A., and HOLMES, A., *J. Biol. Chem.*, 25, 29 (1916)

JANZEN, W.R., and DUNELL, B.A., *Trans. Faraday Soc.*, 59, 1260 (1963)

JUDD, M.D., PLUNKETT, B.A., and POPE, M.I., *J. Thermal Anal.*, 6, 555 (1974)

JURGA, S., DEPIREUX, J., and PAJAK, Z., *Magn. Res. Relat. Phenom., Proc. Congr. AMPERE, 18th*, 2, 403 (1974)(Publ. 1975)

KAPUSTINSKII, A.F., and SAMOILOV, O.Y., *Izv. Akad. Nauk SSSR, Otdel. Khim. Nauk*, 1950, 337

KAPUSTINSKII, A.F., and SAMPLAVSKAYA, K.K., *Zh. Neorg. Khim.*, 6, 2241 (1961)

KAPUSTINSKII, A.F., SELIVANOVA, N.M., and STAKHANOVA, M.S., *Trudy Moskov. Khim.-Tekhnol. Inst. im. D.I. Mendeleeva*, 1956, No. 22, 30; taken from *CA* 51, 16051

KAPUSTINSKII, A.F., STRELKOV, I.I., GANENKO, V.E., ALAPINA, A.V., STAKHANOVA, M.S., and SELIV-ANOVA, N.M., *Russ. J. Phys. Chem.*, 34, 517 (1960); translated from *Zh. Fiz. Khim.*, 34, 1088 (1960)

KENCH, J.E., and MALKIN, T., *J. Chem. Soc.*, 1939, 230

KENDALL, J., and ADLER, H., *J. Amer. Chem. Soc.*, 43, 1470 (1921)

KENNER, J., and MORTON, F., *Ber.*, 72, 452 (1939)

KIRPICHEV, E.P., and RUBTSOV, Y.I., *Russ. J. Phys. Chem.*, 43, 1137 (1969); translated from *Zh. Fiz. Khim.*, 43, 2029 (1969)

KORNIENKO, V.P., and PETRENKO, V., *Uch. Zap. Khar'kov Gos. Univ.*, 71, 77 (1956); taken from LE VAN & PERINET (1969 a)

KRAFFT, F., *Ber.*, 32, 1597 (1899)

LAWRENCE, A.S.C., *Trans. Faraday Soc.*, 34, 660 (1938)

LAWSON, K.D., and FLAUTT, T.J., *J. Phys. Chem.*, 69, 4256 (1965)

LEHRMAN, A., and LEIFER, E., *J. Amer. Chem. Soc.*, 60, 142 (1938)

LEHRMAN, A., and SKELL, P., *J. Amer. Chem. Soc.*, 61, 3340 (1939)

LEONESI, D., CINGOLANI, A., and BERCHIESI, G., *Z. Naturforsch.*, 31a, 1609 (1976)

LEONESI, D., PIANTONI, G., BERCHIESI, G., and FRANZOSINI, P., *Ric. Sci.*, 38, 702 (1968)

LE VAN, M., and PERINET, G., *Bull. Soc. Chim.*, 1969, 2681

LE VAN, M., and PERINET, G., *Bull. Soc. Chim.*, 1969, 3421

LEVKOV, J., KOHR, W., and MACKAY, R.A., *J. Phys. Chem.*, 75, 2066 (1971)

LIPPMAN, R., and RUDMAN, R., *J. Thermal Anal.*, 9, 229 (1976)

LOBANOV, G.A., *Izv. Vyssh. Uch. Zaved. Khim. Khim.-Tekhnol.*, 1971, 14(4), 638

LOMER, T.R., *Acta Cryst.*, 5, 11 (1952)

MADELMONT, C., and PERRON, R., *Bull. Soc. Chim.*, 1973, 3259

MARTIN, E.P., and PINK, R.C., *J. Chem. Soc.*, 1948, 1750

McADIE, H.G., *J. Inorg. Nucl. Chem.*, 28, 2801 (1966)

McBAIN, J.W., BROCK, G.C., VOLD, R.D., and VOLD, M.J., *J. Amer. Chem. Soc.*, 60, 1870 (1938)

McBAIN, J.W., and FIELD, M.C., *J. Phys. Chem.*, 30, 1545 (1926)

McBAIN, J.W., and FIELD, M.C., *J. Phys. Chem.*, 37, 675 (1933)

McBAIN, J.W., LAZARUS, L.H., and PITTER, A.V., *Z. physik. Chem.*, 147A, 87 (1930)

McBAIN, J.W., and STEWART, A., *J. Chem. Soc.*, 1933, 924

McBAIN, J.W., VOLD, R.D., and FRICK, M., *J. Phys. Chem.*, 44, 1013 (1940)

MEHLIS, T., *Ann.*, 185, 358 (1877)

MEIGEN, W., and NEUBERGER, A., *Chem. Umschau*, 29, 337 (1922); taken from HOLDE & SELIM

MEISEL, T., SEYBOLD, K., HALMOS, Z., ROTH, J., and MELYKUTI, C., *J. Thermal Anal.*, 10, 419 (1976)

MENZIES, R.C., and WILKINS, E.M., *J. Chem. Soc.*, 125, 1148 (1924)

MICHELS, H.J., and UBBELOHDE, A.R., *JCS Perkin II*, 1972, 1879

NADIROV, E.G., and BAKEEV, M.I., *Trudy Khim.-Metall. Inst. Akad. Nauk Kaz. SSR*, 25, 115 (1974)

NADIROV, E.G., and BAKEEV, M.I., *Trudy Khim.-Metall. Inst. Akad. Nauk Kaz. SSR*, 25, 129 (1974)

NAVES, Y.R., *Helv. Chim. Acta*, <u>32</u>, 2306 (1949)

NEAVE, G.B., *Analyst*, <u>37</u>, 399 (1912)

NESTEROVA, A.K., and BERGMAN, A.G., *J. Gen. Chem. USSR*, <u>30</u>, 339 (1960); translated from *Zh. Obshch. Khim.*, <u>30</u>, 317 (1960)

NGUEN ZUI TKHIN, ZHARKOVA, L.A., and EROFEEV, B.V., *Russ. J. Phys. Chem.*, <u>41</u>, 631 (1967); translated from *Zh. Fiz. Khim.*, <u>41</u>, 1187 (1967)

NOGTEVA, V.V., LUK'YANOVA, I.G., NAUMOV, V.N., and PAUKOV, I.E., *Russ. J. Phys. Chem.*, <u>48</u>, 1112 (1974); translated from *Zh. Fiz. Khim.*, <u>48</u>, 1873 (1974)

NURMINSKII, N.N., and DIOGENOV, G.G., *Russ. J. Inorg. Chem.*, <u>5</u>, 1011 (1960); translated from *Zh. Neorg. Khim.*, <u>5</u>, 2084 (1960)

ONODERA, N., SUGA, H., and SEKI, S., *Bull. Chem. Soc. Japan*, <u>41</u>, 2222 (1968)

OSTANNII, N.I., ZHARKOVA, L.A., and EROFEEV, B.V., *Russ. J. Phys. Chem.*, <u>44</u>, 798 (1970); translated from *Zh. Fiz. Khim.*, <u>44</u>, 1427 (1970)

OSTANNII, N.I., ZHARKOVA, L.A., and EROFEEV, B.V., *Uch. Zap. Mosk. Gos. Pedagog. Inst.*, No. 464, 114 (1971)

OSTANNII, N.I., ZHARKOVA, L.A., EROFEEV, B.V., and KAZAKOVA, G.D., *Russ. J. Phys. Chem.*, <u>48</u>, 209 (1974); translated from *Zh. Fiz. Khim.*, <u>48</u>, 358 (1974)

PACOR, P., and SPIER, H.L., *J. Amer. Oil Chem. Soc.*, <u>45</u>, 338 (1968)

PAVLOV, V.L., and GOLUBKOVA, V.V., *Vestn. Kiev. Politekhn. Inst. Ser. Khim. Mashinostr. Tekhnol.*, <u>1969</u>, No. 6, 76

PELZL, G., and SACKMANN, H., *Mol. Crystals and Liquid Crystals*, <u>15</u>, 75 (1971)

PENTHER, C.J., ABRAMS, S.T., and STROSS, F.H., *Anal. Chem.*, <u>23</u>, 1459 (1951)

PETERSEN, J., *Z. Elektrochem.*, <u>20</u>, 328 (1914)

PIANTONI, G., LEONESI, D., BRAGHETTI, M., and FRANZOSINI, P., *Ric. Sci.*, <u>38</u>, 127 (1968)

PIPER, J.D., FLEIGER, A.G., SMITH, C.C., and KERSTEIN, N.A., *Ind. Eng. Chem.*, <u>31</u>, 307 (1939)

PLYUSHCHEV, V.E., SHKLOVER, L.P., SHKOL'NIKOVA, L.M., KUZNETSOVA, G.P., and NADEZHDINA, G.V., *Dokl. Akad. Nauk SSSR*, <u>160</u>, 366 (1965); taken from *CA* <u>62</u>, 12727

POCHTAKOVA, E.I., *J. Gen. Chem. USSR*, <u>29</u>, 3149 (1959); translated from *Zh. Obshch. Khim.*, <u>29</u>, 3183 (1959)

POCHTAKOVA, E.I., *Zh. Obshch. Khim.*, <u>33</u>, 342 (1963)

POCHTAKOVA, E.I., *Zh. Obshch. Khim.*, <u>44</u>, 241 (1974)

POPPL, L., *Proc. Eur. Symp. Thermal Anal.*, *1st*, <u>1976</u>, 237

POTEMIN, S.S., TARASOV, A.A., and PANIN, O.B., *Vestn. Leningr. Univ. Fiz. Khim.*, <u>1973</u> (1), 86

RALSTON, A.W., and HOERR, C.W., *J. Amer. Chem. Soc.*, <u>64</u>, 772 (1942)

RALSTON, A.W., HOFFMAN, E.J., HOERR, C.W., and SELBY, W.M., *J. Amer. Chem. Soc.*, <u>63</u>, 1598 (1941)

RAVICH, G.B., and NECHITAYLO, N.A., *Dokl. Akad. Nauk SSSR*, <u>83</u>, 117 (1952)

REIK, R., *Monatsh.*, <u>23</u>, 1033 (1902)

RENARD, A., *Compt. Rend.*, <u>104</u>, 913 (1887)

RICE, M.J., Jr., and KRAUS, C.A., *Proc. Nat. Acad. Sci. U.S.*, <u>39</u>, 802 (1953)

RIES, A., *Z. Kryst.*, <u>55</u>, 454 (1915-1920)

RIPMEESTER, J.A., and DUNELL, B.A., *Can. J. Chem.*, <u>49</u>, 731 (1971)

RIPMEESTER, J.A., and DUNELL, B.A., *Can. J. Chem.*, <u>49</u>, 2906 (1971)

ROBINSON, G.M., *J. Chem. Soc.*, <u>125</u>, 226 (1924)

ROSSINI, F.D., WAGMAN, D.D., EVANS, W.H., LEVINE, S., and JAFFE, I., *"Selected Values of Chemical Thermodynamic Properties"*, Nat. Bur. Stand., Circ. 500, U.S. Dept. Commerce, Washington, D.C., 1952

ROTH, J., HALMOS, Z., and MEISEL, T., *Thermal Anal.*, Vol. 2, 343, Proc. 4th ICTA, Budapest, 1974

ROTH, J., MEISEL, T., SEYBOLD, K., and HALMOS, Z., *J. Thermal Anal.*, 10, 223 (1976)

SANESI, M., FERLONI, P., and FRANZOSINI, P., *Z. Naturforsch.*, 32a, 1173 (1977)

SANESI, M., FERLONI, P., SPINOLO, G., and TONELLI, P.L., *Z. Naturforsch.*, 33a, 386 (1978)

SANESI, M., FERLONI, P., ZANGEN, M., and FRANZOSINI, P., *Z. Naturforsch.*, 32a, 285 (1977)

SARAPULOVA, I.F., KASHCHEEV, G.N., and DIOGENOV, G.G., *Nekotorye Vopr. Khimii Rasplavlen. Solei i Produktov Destruktsii Sapropelitov*, 1974, 3

SCHAFFGOTSCH, F., *Pog.Ann.*, 102, 295 (1857); taken from Gmelins Handbuch, System N. 21, 821 (1928)

SCHWAB, G.M., PAPAROS, S., and TSIPOURIS, I., *Z. Naturforsch.*, 6a, 387 (1951)

SCOTT, D.W., GOOD, W.D., and WADDINGTON, G., *J. Phys. Chem.*, 60, 1090 (1956)

SELIVANOVA, N.M., and ZALOGINA, E.A., *Russ. J. Phys. Chem.*, 45, 904 (1971); translated from *Zh. Fiz. Khim.*, 45, 1592 (1971)

SHAW, D.J., and DUNELL, B.A., *Trans. Faraday Soc.*, 58, 132 (1962)

SHKLOVER, L.P., PLYUSHCHEV, V.E., KUZNETSOVA, G.P., and TRUSHINA, T.A., *Russ. J. Inorg. Chem.*, 10, 607 (1965); translated from *Zh. Neorg. Khim.*, 10, 1121 (1965)

SHMIDT, N.E., *Izv. Sekt. Fiz.-Khim. Anal., Inst. Obshch. i Neorg. Khim., Akad. Nauk SSSR*, 25, 381 (1954)

SIDGWICK, N.V., and GENTLE, J.A.H.R., *J. Chem. Soc.*, 121, 1837 (1922)

SIROTKIN, G.D., *Zh. Prikl. Khim.*, 26, 340 (1953); taken from Gmelins Handbuch, System N. 21, Na Erg (1967), 1401

SKODA, W., *Kolloid-Z. Polym.*, 234, 1128 (1969)

SKOULIOS, A., and LUZZATI, V., *Acta Cryst.*, 14, 278 (1961)

SOKOLOV, N.M., *Zh. Obshch. Khim.*, 24, 1150 (1954)

SOKOLOV, N.M., *Zh. Obshch. Khim.*, 24, 1581 (1954)

SOKOLOV, N.M., *Tezisy Dokl. X Nauch. Konf. S.M.I.* (1956)

SOKOLOV, N.M., and KHAITINA, M.V., *Zh. Obshch. Khim.*, 42, 2121 (1972)

SOKOLOV, N.M., and MINICH, M.A., *Russ. J. Inorg. Chem.*, 6, 1293 (1961); translated from *Zh. Neorg. Khim.*, 6, 2258 (1961)

SOKOLOV, N.M., and POCHTAKOVA, E.I., *J. Gen. Chem. USSR*, 28, 1449 (1958); translated from *Zh. Obshch. Khim.*, 28, 1391 (1958)

SOKOLOV, N.M., and POCHTAKOVA, E.I., *Zh. Obshch. Khim.*, 28, 1397 (1958)

SOKOLOV, N.M., and POCHTAKOVA, E.I., *J. Gen. Chem. USSR*, 28, 1741 (1958); translated from *Zh. Obshch. Khim.*, 28, 1693 (1958)

SOKOLOV, N.M., and POCHTAKOVA, E.I., *J. Gen. Chem. USSR*, 30, 1433 (1960); translated from *Zh. Obshch. Khim.*, 30, 1405 (1960)

SOKOLOV, N.M., TSINDRIK, N.M., and DMITREVSKAYA, O.I., *J. Gen. Chem. USSR*, 31, 971 (1961); translated from *Zh. Obshch. Khim.*, 31, 1051 (1961)

SOUTHARD, J.C., MILNER, R.T., and HENDRICKS, S.B., *J. Chem. Phys.*, 1, 95 (1933)

SPÄTH, E., *Monatsh.*, 33, 235 (1912)

SPEGT, P., and SKOULIOS, A., *Compt. Rend.*, 251, 2199 (1960)

SPEGT, P., and SKOULIOS, A., *Compt. Rend.*, 254, 4316 (1962)

STAMMLER, M., *J. Inorg. Nucl. Chem.*, 29, 2203 (1967)

STAMMLER, M., BRUENNER, R., SCHMIDT, W., and ORCUTT, D., *Adv. X-Ray Anal.*, 9, 170 (1966)

STORONKIN, A.V., VASIL'KOVA, I.V., and POTEMIN, S.S., *Vestn. Leningr. Univ. Fiz. Khim.*, 1974 (10), 84

STORONKIN, A.V., VASIL'KOVA, I.V., and POTEMIN, S.S., *Vestn. Leningr. Univ. Fiz. Khim.*, <u>1974</u> (16), 73

STORONKIN, A.V., VASIL'KOVA, I.V., and TARASOV, A.A., *Vestn. Leningr. Univ. Fiz. Khim.*, <u>1977</u> (4), 80

STRELKOV, I.I., *Ukr. Khim. Zh.*, <u>21</u>, 551 (1955)

STRELKOV, I.I., GANENKO, V.E., and ALAPINA, A.V., *Ukr. Khim. Zh.*, <u>21</u>, 291 (1955)

STROSS, F.H., and ABRAMS, S.T., *J. Amer. Chem. Soc.*, <u>72</u>, 3309 (1950)

STROSS, F.H., and ABRAMS, S.T., *J. Amer. Chem. Soc.*, <u>73</u>, 2825 (1951)

SUGDEN, S., *J. Chem. Soc.*, <u>1929</u>, 316

SYMONS, G.E., and BUSWELL, A.M., *Ind. Eng. Chem.*, <u>24</u>, 460 (1932)

TAKAGI, S., *J. Chem. Soc. Japan*, <u>60</u>, 625 (1939); taken from *CA* <u>36</u>, 6401

TAMAMUSHI, B., *Rheol. Acta*, <u>13</u>, 247 (1974); taken from *CA* <u>81</u>, 83127

TAMAMUSHI, B., and MATSUMOTO, M., *"Liquid Crystals and Ordered Fluids"*, Vol. 2, Ed. by Johnson, J.F., and Porter, R.S., Plenum Pub. Corp., New York (1974), 711

THIESSEN, P.A., and EHRLICH, E., *Z. physik. Chem.*, <u>B19</u>, 299 (1932)

THIESSEN, P.A., and EHRLICH, E., *Z. physik. Chem.*, <u>A165</u>, 453 (1933)

THIESSEN, P.A., and von KLENCK, J., *Z. physik. Chem.*, <u>A174</u>, 335 (1935)

THIESSEN, P.A., and STAUFF, J., *Z. physik. Chem.*, <u>A176</u>, 397 (1936)

TRZEBOWSKI, N., *Wiss. Z. d. Friedrich-Schiller-Universität Jena, Mat.-Naturwiss. Reihe*, <u>14</u>, 207 (1965)

TSAU, J., and GILSON, D.F.R., *J. Phys. Chem.*, <u>72</u>, 4082 (1968)

TSAU, J., and GILSON, D.F.R., *Can. J. Chem.*, <u>51</u>, 1990 (1973)

TSINDRIK, N.M., *Zh. Obshch. Khim.*, <u>28</u>, 830 (1958)

TSINDRIK, N.M., and SOKOLOV, N.M., *J. Gen. Chem. USSR*, <u>28</u>, 1462 (1958); translated from *Zh. Obshch. Khim.*, <u>28</u>, 1404 (1958)

TSINDRIK, N.M., and SOKOLOV, N.M., *J. Gen. Chem. USSR*, <u>28</u>, 1775 (1958); translated from *Zh. Obshch. Khim.*, <u>28</u>, 1728 (1958)

UBBELOHDE, A.R., *Chem. & Ind.*, <u>1968</u>, 313

UBBELOHDE, A.R., MICHELS, H.J., and DURUZ, J.J., *Nature*, <u>228</u>, 50 (1970)

USUBALIEV, D., and BATKIBEKOVA, M.V., *Fiz.-Khim. Issled. Redkozemel. Elem.*, <u>1972</u>, 148

VAN RENESSE, J.J., *Ann.*, <u>171</u>, 380 (1874)

VAN UITERT, L.G., BONNER, W.A., and GRODKIEWICZ, W.H., *Mat. Res. Bull.*, <u>6</u>, 513 (1971)

VOLD, M.J., *J. Amer. Chem. Soc.*, <u>63</u>, 160 (1941)

VOLD, M.J., *J. Amer. Chem. Soc.*, <u>65</u>, 465 (1943)

VOLD, M.J., FUNAKOSHI, H., and VOLD, R.D., *J. Phys. Chem.*, <u>80</u>, 1753 (1976)

VOLD, M.J., MACOMBER, M., and VOLD, R.D., *J. Amer. Chem. Soc.*, <u>63</u>, 168 (1941)

VOLD, M.J., and VOLD, R.D., *J. Amer. Oil Chem. Soc.*, <u>26</u>, 520 (1949)

VOLD, M.J., and VOLD, R.D., *J. Coll. Sci.*, <u>5</u>, 1 (1950)

VOLD, R.D., *J. Phys. Chem.*, <u>43</u>, 1213 (1939)

VOLD, R.D., *J. Amer. Chem. Soc.*, <u>63</u>, 2915 (1941)

VOLD, R.D., *J. Phys. Chem.*, <u>49</u>, 315 (1945)

VOLD, R.D., GRANDINE, J.D., Jr., and SCHOTT, H., *J. Phys. Chem.*, <u>56</u>, 128 (1952)

VOLD, R.D., GRANDINE, J.D., Jr., and VOLD, M.J., *J. Coll. Sci.*, <u>3</u>, 339 (1948)

VOLD, R.D., REIVERE, R., and McBAIN, J.W., *J. Amer. Chem. Soc.*, <u>63</u>, 1293 (1941)

VOLD, R.D., ROSEVEAR, F.B., and FERGUSON, R.H., *Oil & Soap*, <u>16</u>, 48 (1939)

VOLD, R.D., and VOLD, M.J., *J. Phys. Chem.*, <u>49</u>, 32 (1945)

VORLÄNDER, D., *Ber.*, <u>43</u>, 3120 (1910)

WAGMAN, D.D., EVANS, W.H., PARKER, W.B., HALOW, I., BAYLEY, S.M., and SCHUMM, H., *"Selected Values of Chemical Thermodynamic Properties"*, Nat. Bur. Stand., Tech. Note 270-3, U.S. Dept. Commerce, Washington, D.C., 1968

WAGNER, H., MUESMANN, J., and LAMPART, J.B., *Z. f. Unters. Nahrgs.-u.-Genussmittel*, <u>28</u>, 244 (1914); taken from *C* <u>1914</u>(II), 1458

WAGNER, L., *Z. Kryst.*, <u>43</u>, 148 (1907)

WALTER, R., *Ber.*, <u>59</u>, 962 (1926)

WASIK, S.P., *J. Chromatogr. Sci.*, <u>14</u>, 516 (1976)

WESTRUM, E.F., Jr., CHANG, S.S., and LEVITIN, N.E., *J. Phys. Chem.*, <u>64</u>, 1553 (1960)

WHITMORE, W.F., and LAURO, M., *Ind. Eng. Chem.*, <u>22</u>, 646 (1930)

WILSON, J.W., *JCS Dalton Trans.*, <u>1976</u>, 890

WIRTH, H.E., and WELLMAN, W.W., *J. Phys. Chem.*, <u>60</u>, 921 (1956)

YATSIMIRSKII, K.B., *Izv. Akad. Nauk SSSR, Otdel. Khim. Nauk*, <u>1947</u>, 453; taken from Gmelins Handbuch, System N. 20, Li Erg (1960) 505, 507

YATSIMIRSKII, K.B., *Izv. Akad. Nauk SSSR, Otdel. Khim. Nauk*, <u>1948</u>, 590; taken from LE VAN & PERINET (1969 a)

ZHARKOVA, L.A., and NISSENBAUM, V.D., *Uch. Zap. Mosk. Gos. Pedagog. Inst.*, <u>464</u>, 137 (1971)

ZINCKE, T., *Ann.*, <u>152</u>, 1 (1869)

ZUFFANTI, S., *J. Amer. Chem. Soc.*, <u>63</u>, 3123 (1941)

ADDENDA

Addendum *to*: TABLE 1. Temperatures and enthalpies of phase transitions of alkali n.alkanoates

Cation	Anion	Transition	T/K	$\dfrac{\Delta H}{kJ\ mol^{-1}}$	Method	Year	Ref.	Notes
Li	$C_2H_3O_2$	fusion nstr	565 405		DTA	1971	1	
Na	$C_2H_3O_2$	fusion nstr	601 543		DTA	1971	1	
K	$C_2H_3O_2$	fusion nstr	585 428		DTA	1971	1	
Rb	$C_2H_3O_2$	fusion nstr	508 479		DTA	1971	1	
Cs	$C_2H_3O_2$	fusion nstr	458 308		DTA	1971	1	

1. GIMEL'SHTEIN, V.G., *Tr. Irkutsk. Politekh. Inst.*, <u>66</u>, 80 (1971)

Addendum *to*: TABLE 6. Temperatures and enthalpies of phase transitions of Zn, Cd, Hg and Pb n.alkanoates

Cation	Anion	Transition	T/K	$\dfrac{\Delta H}{kJ\ mol^{-1}}$	Method	Year	Ref.	Notes
Hg(II)	$C_8H_{15}O_2$	solid → liquid	387.2±0.1	61.5±2 %	DTA	1978	1	
Hg(II)	$C_{10}H_{19}O_2$	phase I → liquid solid → phase I	389.3±0.1 380.8±0.1	70.2±2 % 5.3±2 %	DTA	1978	1	a
Hg(II)	$C_{12}H_{23}O_2$	solid → liquid	394.2±0.1	94.8±2 %	DTA	1978	1	
Hg(II)	$C_{14}H_{27}O_2$	phase I → liquid solid → phase I	387.0±0.1 382.4±0.1	40.0±2 % 57.9±2 %	DTA	1978	1	b
Hg(II)	$C_{16}H_{31}O_2$	phase I → liquid solid → phase I	390.3±0.1 383.4±0.1	59.5±2 % 49.5±2 %	DTA	1978	1	b
Hg(II)	$C_{18}H_{35}O_2$	phase I → liquid solid → phase I	393.2±0.1 355.2±0.1	116.5±2 % 4.4±2 %	DTA	1978	1	a

a: phase I is probably a solid phase
b: phase I is probably a G (smectic) phase

1. ADEOSUN, S.O., *J. Thermal Anal.*, <u>14</u>, 235 (1978)

Addendum *to*: TABLE 8. Temperatures and enthalpies of phase transitions of alkyl-substituted ammonium salts

Cation	Anion	Transition [a]	T/K	$\dfrac{\Delta H}{\text{kJ mol}^{-1}}$	Method	Year	Ref.	Notes
$(CH_3)_2(C_2H_5)(C_{16}H_{33})N$	ClO_4	I → liquid	429	19.2±10 %	DSC	1978	1	
		II → I	359	18.4±10 %				
		III → II	352.7	6.4±10 %				
$(CH_3)(C_3H_7)_2(C_{13}H_{27})N$	I	F	345	37.7±10 %	DSC	1978	1	
$(C_2H_5)_3(C_{14}H_{29})N$	ClO_4	I → liquid	425	50.2±10 %	DSC	1978	1	*b*
$(C_2H_5)_2(C_8H_{17})_2N$	ClO_4	I → liquid	333	50.2±10 %	DSC	1978	1	*c*
$(C_2H_5)(C_6H_{13})_3N$	ClO_4	F	320	18.8±10 %	DSC	1978	1	
$(C_3H_7)_3(C_{11}H_{23})N$	ClO_4	F	335	25.1±10 %	DSC	1978	1	
$(C_4H_9)_3(C_8H_{17})N$	ClO_4	F	338	28.9±10 %	DSC	1978	1	
$(C_4H_9)_2(C_6H_{13})_2N$	ClO_4	F	355	34.3±10 %	DSC	1978	1	
$(C_4H_9)(C_5H_{11})_2(C_6H_{13})N$	ClO_4	F	355.7	26.4±10 %	DSC	1978	1	
$(C_5H_{11})_4N$	ClO_4	I → liquid	391	18.0±10 %	DSC	1978	1	
		II → I	364	36.8±10 %				

a: F, fusion; II → I, solid-solid transition; I → liquid, fusion of modification I

b: modification I crystallizes from solution; the cooled melt crystallizes as modification III

c: modification I grows from solution, II from the melt held at 318 K, and III from the melt below 317 K; no solid-solid transitions could be observed

1. GORDON, J.E., and SUBBA RAO, G.N., *J. Amer. Chem. Soc.*, <u>100</u>, 7445 (1978)

Chapter 1.3

MELTING MECHANISM OF PURE SALTS

A. R. Ubbelohde

INTRODUCTION

Interionic forces are long range, which entrains distinctive properties for crystalline or=
ganic salts and even more so for organic ion melts. It is useful to discuss these with refer=
ence to the enthalpy ΔH_F and the entropy ΔS_F of fusion.

Quite generally, mechanisms of chemical decomposition of organic ions can be greatly promoted
or facilitated by the positional disordering on melting, because of the increase in free volume and because of the long range electrostatic forces which become partly uncompensated
through the disorder. Unless the ratio of the two quantities which determines the thermodynamic melting point, i.e.,

$$T_F = \Delta H_F / \Delta S_F$$

is sufficiently low, intrinsic chemical instability becomes so pronounced in the melt at T_F
that some melts can hardly be studied, so that many organic salts are only of very limited
interest in the liquid state. Probably for this reason, the variety of pure salts with organic cation or anion whose properties have been investigated in any detail in the molten
state has hitherto been restricted to very few types. This is a pity, in view of the versatil-
ity of organic synthesis that is feasible, and in view of potentially valuable properties of
organic ionic melts for many applications. To develop those possibilities more systematically,
it is quite generally desirable to achieve salts whose melting points are low, by contriving
low enthalpies and high entropies of fusion, through appropriate selection of the mechanisms
of melting that operate with any particular type of salt. These may be discussed in terms of
molecular and crystal structure, with the remark that determinations of crystal structures of
organic salts have unfortunately hitherto been extremely scanty.

Obviously, the intrinsic chemical stability of any organic ion will be determined by the temperature for thermal cracking as well as for other bond rearrangements. These matters of physical organic chemistry will not be discussed here. However, it is essential to point out that
chemical instability of organic ion melts can often arise from causes that are extrinsic. For
example, the group of alkali carboxylates discussed below are rapidly attacked in the molten
state by molecular oxygen, or by ionic impurities such as OH^- or ions of transitional metals.
But when these extrinsic sources of chemical instability are suppressed, as is often readily
done (Duruz, Michels & Ubbelohde, 1971), the organic ionic melts have quite a long temperature
range over which they remain quite stable, so that they can be studied with precision and used
in a number of applications.

ENTHALPIES IN RELATION TO MECHANISM OF MELTING

Familiar contributions to the overall enthalpy of fusion of inorganic crystals can be conveniently discussed (Ubbelohde, 1978, pages 126,130) in terms of *positional defects* and of *orientational defects* such as may already be incipient in the crystal lattice on approaching T_F.
These defects introduce terms

$$\Delta H_{positional} \quad \text{and} \quad \Delta H_{orientational}$$

in the enthalpy, which can be approximately treated as additive. A jump in these enthalpy
terms is found, corresponding with the jump in structure on passing from the crystal lattice
to the melt. Additional contributions to the melt enthalpy of fusion may arise from various
anticrystalline changes on rearranging the molecules, so that the overall increase can be regarded at least conceptually as resulting from a sequence of operations on the lattice structure, i.e.,

crystal $\rightarrow$ quasi-crystalline melt $\rightarrow$ actual melt after rearrangements

$$\Delta H_F = \Delta H \text{ positional defects} + \Delta H \text{ orientational defects} +$$

$$\left\{ \begin{array}{l} \Delta H \text{ configurational} \\ \text{molecular changes} \end{array} \right\} + \Delta H \text{ association of ions} + \ldots \quad (1)$$

each of which may be conveniently discussed in turn for the melting of organic salts.

POSITIONAL DEFECTS

Positional defects of the simplest types involve the creation of lattice vacancies (Schottky defects) or interstitial insertions (Frenkel defects). Both types are known e.g. from studies with salts of the inert gas type (Ubbelohde, 1978, pages 126, 130). In salts both require quite high energies of creation per defect, so that if these are the principal contributors to the overall mechanism of melting, they inevitably tend to lead to high melting points. However two other kinds of positional defect seem likely to be at least equally significant for certain classes of salts.

(i) Sub-lattice positional disorder

Whenever anion and cation have sufficiently dissimilar radii of their repulsion envelopes, the sub-lattice with the smaller of the two ions may show positional disordering akin to melting at a transition temperature T_c considerably below the final and overall melting temperature T_F. The concept of repulsion radius for organic ions which are polyatomic needs special consideration, as is further discussed below, but to some extent it is comparable with inorganic ions. For these, when

$$r_+/r_- \gg 1 \quad \text{or conversely} \quad r_-/r_+ \gg 1$$

a lambda transition is found at T_c. Well known examples include silver salts such as AgI and KAg_4I_5 (Ubbelohde, 1978, pages 88, 218), in which the cations have a much smaller repulsion envelope than the anion. Above the lambda peak at T_c the small cation migrates easily within the rigid skeleton lattice maintained by the large anions, which does not itself "melt" until T_F. Clear-cut analogous examples of sub-lattice melting in organic ionic crystals do not appear to have been described as yet. Subject to the reservation (see below) about the concept of ionic size in organic salts, the numerical ratio r_+/r_- may well favour sub-lattice melting in them. Since the mobility of the smaller ion in the lattice above T_c can be even greater than above T_F in the melt, more intensive search may *inter alia* reveal useful novel solid electrolytes amongst organic salts, with possible applications.

(ii) Cooperative positional disorder

The simplest kinds of positional defects are at lattice sites which can be treated as independent of one another. However, *cooperative* positional defects can be just as important for melting in some crystal lattices. Elementary kinds of cooperative positional defects are types of crystal dislocations (Ubbelohde, 1978, page 296) as often discussed for inorganic compounds because of their role in the mechanical strength of the solids, and in crystal growth from vapour or solution. For organic salts, cooperative positional disordering as a mechanism of melting could follow a greater diversity of possibilities, because the crystal structures are more diverse. Unfortunately, as stated, the number of crystal structures that are well-determined is as yet very scanty amongst organic salts.

One class that has been studied fairly extensively comprises the alkane carboxylates of mono-valent cations (Duruz, Michels & Ubbelohde, 1971; Ubbelohde, 1976). In these, when the alkane is fairly short (C_nH_{2n+1} with $n \leq 7$) the organic ions in the crystal are anchored into rafts with a structure illustrated diagrammatically in Fig. 1. The alkane chains are held parallel by nearest neighbour Van der Waals repulsion forces, together with strong attachment of the carboxyl group at one end to the ionic rafts which also contain cations. Whilst electrostatic forces are strongest in these rafts, electrostriction throughout the crystal remains high since the polymethylene chains themselves heve only low polarisability and low dielectric constant (Ubbelohde, 1938; Ubbelohde & Woodward, 1952).

As the temperature of the crystal rises, thermal vibrations of all kinds induce a kind of "Dervish dance" in the alkane chains which remain anchored at one end to the ionic raft. Usually this would generate a large thermal expansion, but owing to the strong electrostatic forces pulling the layers of paired ions together in the direction normal to them, the principal thermal effect of the enthalpy increase is that the lattice expands in directions parallel to the layers because of the increase of vibrational energy. Some positional defects may also be generated at the sites of alkali cations, and it is not known whether the chains actually undergo rotational disordering of their carboxyl anions whilst remaining anchored in the ionic rafts. At a first melting point T_F, the main disordering mechanism appears to be the break up of the ion layers into smaller uncorrelated rafts. X-ray data (Duruz & Ubbelohde, 1972) show that above the first melting point the molten salt contains highly anisotropic "striped domains" (Fig. 2) with good parallelism between the layers in any one domain. In fact, a kind of cooperative positional disordering occurs. Positional dislocation melting would appear to account for the remarkably low heat of fusion ($\Delta H_F = \Delta S_F\, T_F$).

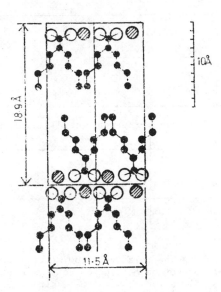

Fig. 1. Alternative configurations of n.alkane chains between "electrostatic sandwiches" for solid potassium caproate (ions and atoms not to scale)
(Michels & Ubbelohde, 1972; by kind permission of *The Chemical Society*)

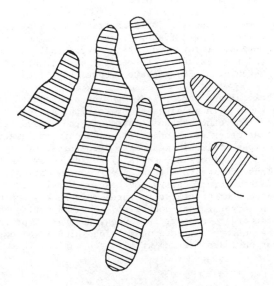

Fig. 2. "Striped domains": possible texture of stacked electrostatic sandwiches in the ionic mesophases
(Duruz & Ubbelohde, 1972 ; by kind permission of *The Royal Society*)

TABLE 1. Comparative thermal parameters (T/K; ΔS/J K^{-1}mol^{-1}) for melting mechanisms of salts

	------Melting------				
	T	ΔS			
Sodium chloride	1074	28.0			
Sodium nitrate	591	25.5			

	---Solid → Solid---		------Melting------		
	T	ΔS	T	ΔS	
(C$_4$H$_9$)$_4$N iodide	665	71.5	692	23.0	
(C$_5$H$_{11}$)$_4$N iodide	628	42.7	685	95.8	

	----Solid → Mesophase Melt----		--Mesophase → Isotropic Melt--	
	T	ΔS	T	ΔS
Sodium butyrate	525	11.7	597	2.5
Sodium *iso*valerate	461	9.2	553	1.3
p-azoxyanisole	391	75.3	409	1.4

The low values of ΔH_F are one helpful factor in achieving the relatively low melting point of this class of organic salts. However, the structural changes on melting do not involve very large entropy increases so that the ratio $\Delta H_F/\Delta S_F$ is not as low as might be desired. Increasing positional disordering in the rafts themselves probably accompanies further rises of temperature since the specific heat capacity of the anisotropic melts remains high (Michels & Ubbelohde, unpublished observations). At a second melting point, T_{Cl}, the molten salt passes into an isotropic melt, with only a small further jump in entropy (see Table 1). These remarkable anisotropic organic ionic melts also show a very high viscosity (about 3 P) as may be expected from their structure; above the second melting point, when the melt is isotropic, the viscosity drops to values normal for ionic melts (about 0.04 P).

It is customary to use the all-embracing term "mesophase" for anisotropic melts of the well-known liquid crystals. However, ionic melt mesophases differ in important particulars from molecular "liquid crystals". For these, the entropy of formation of the crystals from the melts is much higher (see examples in Table 1) and the viscosity of even the anisotropic molten phase though anomalous remains relatively low (McLaughlin, Shakespeare & Ubbelohde, 1964).

The sequence of melting mechanisms of alkali alkane carboxylates has been discussed in some detail since the structural features are known with considerable precision. As yet, corresponding structural information is not available for other organic salts, in the same detail. In this context, there can be little doubt that other positional melting mechanisms await discovery, and would point to other unusual types of liquid mesophase properties. By way of example, it may be stressed that in another class of organic salts discussed below the thermodynamic parameters are quite different. Thus, for salts with alkane ammonium cations, low melting points are achieved because the positional defect energy increase on passing from crystal to melt is quite low; this might be expected from crystal structures more nearly analogous to those of the alkali halides, but with much larger (quasi-spherical) ions. As yet, no statistical thermodynamic theory has been put forward to support this view, however. Experimentally, for this group of organic salts, it is the entropy increase on melting which is exceptionally large, through the operation of configurational disordering within each cation. This favours low values of $\Delta H_F/\Delta S_F$.

SIZES OF IONS IN ORGANIC SALTS

In relation to diverse mechanisms of positional melting of ionic crystals, the relative sizes

as well as shapes of anion and cation play an important role, which is as yet by no means fully investigated. Unlike what applies with many inorganic ions, even those that are polyatomic,
the charge density distribution and thus electrical properties over the entire repulsion envelope of many organic ions is probably far from uniform. These ions cannot be simply regarded as
quasi-spherical, or even quasi-elliptical. The two principal types of organic ions as yet investigated at all fully with regard to melting mechanisms illustrate the differences which may
arise. Substituted ammonium salts with cation $(R_1, R_2, R_3, R_4)N^+$ may achieve approximately
spherical charge density distribution at the ion surfaces in the crystals, if the number of
carbon atoms in the alkyl groups is not too dissimilar. If, however, the organic substituents
R_1, R_2, R_3, R_4 have configurational flexibility, the lattice forces may push the substituents
into the most nearly spherical envelope in the crystals, but on melting the greater freedom in
the melt would certainly permit many more stretched out configurations of these organisations
to appear and make the ionic envelope of the cations less even. This factor could no doubt be
controlled by special choice of the substituents R_1, R_2, R_3, R_4 so as to enhance or to minimize irregularities within the repulsion envelope. By contrast, n.alkane carboxylates with the
organic anion $C_nH_{2n+1}COO^-$ (which normally assumes the most stretched out form in the crystals
because of electrostrictive forces) must in all states have maximum charge density at the carboxyl end. It is not known as a refinement whether there is resonance between the two C-O bonds
of the carboxyl group; if they behave differently a further mechanism of local disordering by
bond switching is potentially feasible in the crystal lattice. However this may be, the broad
general difference of geometry of the organic ions in these two classes of organic salts must
result in striking differences of electrostrictive fields operating in them. These fields are
likely to show fairly high symmetry in the substituted ammonium salts, but only layered peri-
odicity in the alkane salts. Electrostriction (which pulls the ions together) has consequences
analogous to applying external pressure on the crystals. For example, external pressure can
influence the feasibility of orientational randomisation of molecular axes at transformation
temperatures T_c below the first melting point T_F. With simpler inorganic salts, external pressure may merely push T_c above T_F (see following sections). For many types of organic ions their
shape and charge distribution must however lead to electrostriction that is markedly non-uni-
form. This is what may lead to melting by stages, involving various kinds of cooperative disorder, and passing through mesophase "liquids" with novel properties.

Deliberate synthesis of organic salts in which the molecular skeletons of the ions are even
less uniform than the contrasting examples cited, but which are still capable of melting to
organic liquids which are intrinsically stable, has not yet received the attention warranted.
As one example, particularly interesting local electrical peculiarities could be associated
with organic ions when their molecular skeletons have conjugated bonds. Such skeletons could
exhibit a kind of internal "microcharge transport" within each ion, with unusual consequences
for "macrotransport" in ionic melts.

ORIENTATIONAL DISORDERING IN IONIC CRYSTALS

Preceding sections mainly deal with aspects of positional disordering in a salt lattice, when
this contains organic ions. Orientational randomisation of molecular axes of polyatomic ions
in a crystal lattice at a transformation temperature (lambda peak) T_c well below the melting
point T_F is well known for a number of ammonium salts (Ubbelohde, 1978, page 95). As the cation
is made progressively larger in substituted ammonium salts, the increasing radius ratio r_+/r_-
must influence the potential energy minima for different orientations of the cation in the
crystal. This could complicate the melting mechanisms whenever the increase of r_+/r_- lowers
the temperature of positional melting T_{pos} well below that of any orientational lambda peak
T_{or}. The simplest thing that can then happen is that at the positional melting point the
expansion in free volume and overall loosening up lowers barriers to orientational randomisation simultaneously, permitting random orientation so that the melt is effectively quasi-
crystalline and disordered in all respects. Such simultaneous disordering is known, e.g., for
some inorganic crystals (Ubbelohde, 1978, page 130).

Room to rotate in the melt

The simple combination at T_F of randomisation with respect to orientation as well as positional
correlation is in fact fairly common. For salts with organic ions some orientational correlations may however persist well above the first melting point, whenever positional melting is
not accompanied by a sufficient fractional increase in molecular volume $\Delta V_F/V_s$ to allow com-

plete uncorrelation of molecular axes of every organic ion to occur in the melt, with respect to its neighbouring ions. This can usually be diagnosed from calculations on the molecular volume required for "free" rotation. Some long range though imperfect correlation of molecular axes may well persist in a variety of organic ionic melts, in order to pack the ions in these liquids economically.

Spread of melting mechanisms over a temperature range

One of the features of melting mechanisms that is receiving increasing attention is the role of precursor effects on either side of the discontinuous jump in entropy which characterizes identifiable melting. Premelting effects in the crystals may well be prominent in certain organic salts, but no clear reports appear to be available. Prefreezing effects in the melts seem likely to be quite general, and are more immediately important. Organic ions are seldom like simple spherical molecules, and various kinds of packing into anticrystalline clusters or domains must be common in the melt above T_F.

In one way, this can be regarded as a spread of the overall entropy change for configurational disordering, ΔS_{config}, over a range of temperatures, instead of being practically complete at a definable melting point as is usual, e.g., for positional randomisation.

Spread of configurational entropy can have marked consequential influences on first derivatives such as the specific heat capacity, the thermal expansion and the compressibility of these ionic melts. Although data are still far too scanty, studies of these quantities form an important part of studies of mechanisms of melting generally, and will assist in the elucidation of organic ionic melts as a novel class of liquids (Ubbelohde, 1976).

Another way of regarding anticrystalline features in the structure and topology of a melt is that when these are prominent they contribute to the overall enthalpy jump and even more to the overall entropy jump on fusion, usually in such a way that the first melting point is lower than in the absence of such association into anticrystalline clusters. This lowering effect could be particularly helpful in order to obtain organic ionic melts that are still intrinsically stable. Thermodynamic reasons have been discussed to account for the low melting points of certain inorganic salts with polyatomic ions on this basis; similar arguments may provide a useful guide to obtaining melts of organic salts at temperatures where these are still intrinsically stable.

REFERENCES

DURUZ, J.J., MICHELS, H.J., and UBBELOHDE, A.R., *Proc. Roy. Soc. London*, A322, 281 (1971)

DURUZ, J.J., and UBBELOHDE, A.R., *Proc. Roy. Soc. London*, A330, 1 (1972)

McLAUGHLIN, E., SHAKESPEARE, M.A., and UBBELOHDE, A.R., *Trans. Faraday Soc.*, 60, 25, 1884 (1964)

MICHELS, H.J., and UBBELOHDE, A.R., *J.C.S. Perkin II*, 1972, 1879

UBBELOHDE, A.R., *Trans. Faraday Soc.*, 34, 282 (1938)

UBBELOHDE, A.R., *Rev. int. Htes Temp. et Refract.*, 13, 5 (1976)

UBBELOHDE, A.R., *"The Molten State of Matter"*, John Wiley & Sons, 1978

UBBELOHDE, A.R., and WOODWARD, I., *Trans. Faraday Soc.*, 48, 113 (1952)

Chapter 1.4

TRANSPORT PROPERTIES OF PURE MOLTEN SALTS WITH ORGANIC IONS

A. Kisza

INTRODUCTION

Molten salts with organic ions form a group of ionic liquids of special interest. Contrary to inorganic salts whose physicochemical properties are far better understood, salts with organic anion or cation have low melting points and offer a possibility of tayloring their size and shape by a proper choice of the organic ion. Applications of fused salts in organic chemistry were carefully reviewed by Gordon (1969). The study of the transport properties may yield valuable information both on structure and mechanism of motion. Most data on such properties have been collected for molten inorganic salts (Sundheim, 1964; Klemm, 1966; Janz et al., 1968). An important feature of the melts of salts with organic anion or cation is their frequent structural complexity. Beside the translational degrees of freedom, which are mostly pronounced in molten inorganic salts, the internal ones – vibrational and rotational – become also important in many fused salts with organic ions . The number of possible modes of momentum transfer is thus increased in the latter liquids: as a consequence, the simple relationships usually expressing the temperature dependence of transport properties only seldom hold. Melts formed from organic salt hydrates have recently received considerable attention. In view of the low melting points and the (comparatively) simple structure of these melts, these form a useful intermediate between organic ionic melts, and more dilute solutions of organic salts in water.

In the following, attention is restricted to viscosity and electrical conductivity, whereas other transport properties such as diffusion and migration (transference numbers) will not be considered due to the complete lack of data for this kind of liquids.

VISCOSITY OF PURE MOLTEN SALTS WITH ORGANIC IONS

DEFINITIONS

Viscosity, which is a measure of the frictional resistance a fluid in motion offers to an applied shearing force, is defined by the phenomenological equation (De Groot & Mazur, 1962)

$$\Pi^s = -2\eta \, (\overset{o}{\mathrm{grad}} \, v)^s \tag{1}$$

in which Π^s is the symmetric part of the viscous pressure tensor, Π, and η is the ordinary or shear viscosity. For a symmetric tensor with zero trace three independent components are possible:

$$\overset{o}{\Pi}{}^s_{xy} = -\eta \left(\frac{\partial v_y}{\partial x} + \frac{\partial v_x}{\partial y} \right) \tag{2}$$

$$\overset{o}{\Pi}{}^s_{xz} = -\eta \left(\frac{\partial v_z}{\partial x} + \frac{\partial v_x}{\partial z} \right) \tag{3}$$

$$\overset{o}{\Pi}{}^s_{yz} = -\eta \left(\frac{\partial v_z}{\partial y} + \frac{\partial v_y}{\partial z} \right) \tag{4}$$

Here v_x, v_y and v_z are the cartesian components of the local mass center velocity, v. Eq.s (1-4) represent the Newton's relations (Baranowski, 1974) and a liquid obeying them is called newtonian. When $v_x = 0$, Eq. (2) takes the familiar form

$$F_{xy} = \eta \, A \left(\frac{\partial v_y}{\partial x} \right) \tag{5}$$

where F_{xy} is the force in the y direction exerted by a fluid in motion on a surface A normal to the x axis.

The most striking feature of the viscosity of liquids is its temperature dependence. The Arrhenius type equation

$$\eta = \eta_\infty \exp(E_\eta / RT) \tag{6}$$

first adopted by Guzman (1913) has been frequently used for this purpose. However, even for simple liquids (Harrap & Heyman, 1951) the plot of $\ln \eta$ *vs* $1/T$ is non-linear.

Several models have been proposed to account for this temperature dependence. That developed by Frenkel (1946), known as the "hole" model of viscous flow, emphasizes the similarity between liquid and solid rather than liquid and gas. In addition to the lattice vacancies in the crystal, the melting process increases the free volume within the liquid, which has been assumed by McLeod as being equal to the difference between the total volume and the volume of the spheres of influence of all the molecular species. The equation derived by Frenkel is similar to the Arrhenius type Eq. (6), but the pre-exponential factor is about 1,000 times larger than that determined experimentally and strongly temperature dependent.

The justification of Eq. (6) was also given by Eyring (1936) in his rate theory, according to which

$$\eta = \frac{hN}{V} \exp\{-\frac{\Delta S^{\bullet}}{R}\} \exp\{\frac{\Delta H^{\bullet}}{RT}\} \tag{7}$$

V, $\Delta S^{\bullet}$ and $\Delta H^{\bullet}$ being the molar volume, and the activation entropy and enthalpy, respectively.

In the model developed by Batchinski (1913) the specific volume, v, of the liquid has been taken as the proper independent variable for the viscosity:

$$\eta = \frac{B}{v-b} \tag{8}$$

According to McLeod (1923), for molecular liquids the viscosity should be related to the free space within the liquid. His result was essentially the same as that by Batchinski (for a=1):

$$\eta = \frac{B}{x^a} = \frac{B}{(v-b')^a} \tag{9}$$

Here x is the free space per unit volume and a, B, b and b' are constants. Constant b in Eq. (8) may bear some analogy to covolume in Van der Waals' equation. Although both Batchinski and McLeod attempted to relate the constant B to the molecular weight, McLeod suggested rather its proportionality to the "size of the unit flow".

The rate theory of liquid viscosity has difficulties in accounting for the non-Arrhenius behaviour of liquids.

Macedo & Litovitz (1965) reconsidered the relative roles of free volume and activation energy. The hybrid equation

$$\eta = A_{\circ} \exp\{\frac{\gamma V_{\circ}}{V_f} + \frac{E_v^{\bullet}}{RT}\} \tag{10}$$

combines both the Eyring rate theory and the Cohen & Turnbull free volume theory. The constant $A_{\circ}$ varies with the temperature, but far less than the exponential term; γ is a numerical factor ($0.5 < \gamma < 1$) needed to correct for the overlap of free volume; $V_{\circ}$ is the close packed volume; V_f is the molar free volume, and $E_v^{\bullet}$ is the activation energy measured at constant volume. Eq. (10) has been obtained assuming that two events must occur simultaneously before a molecule can undergo a diffusive jump, i.e., (i) the molecule must attain a sufficient energy to break away from neighbours, and (ii) it must have at disposal an empty site large enough to jump into. In the light of this theory the Arrhenius behaviour of viscosity can occur if either of the following facts is true: (i) the liquid has a zero or a very small thermal expansion coefficient, i.e., V_f is temperature independent; (ii) V_f is directly proportional to temperature.

In those liquids where $E_v^{\bullet}/RT \ll V_{\circ}/V_f$ Eq. (10) is reduced to the Williams, Landel & Ferry expression:

$$\eta = A_{\circ} \exp\{\gamma \frac{V_{\circ}}{V_f}\} \tag{11}$$

If one assumes that

$$V_f = \alpha V (T - T_{\circ}) \tag{12}$$

the Cohen-Turnbull equation is obtained:

$$\eta = A \exp\{\frac{1}{\alpha'(T - T_{\circ})}\} \tag{13}$$

Here $\alpha' = \alpha V/\gamma V_{\circ}$ and $T_{\circ}$ is the apparent temperature where the free volume becomes zero.

If the term $E_v^{\bullet}/RT$ is not negligible the above condition leads to Dienes' (1953) equation:

$$\eta = A \exp\{\frac{E}{RT} + \frac{U}{R(T - T_o)}\} \tag{14}$$

where $U/R(T - T_o)$ represents the free volume contribution to the exponent.

The Eyring significant structure theory of liquids (1969) leads for the viscosity to the expression

$$\eta = (\pi m k T)^{\frac{1}{2}} N \frac{\ell_f}{2(v - v_s)K} \exp\{\frac{a'v_s Z \phi(a)}{(v - v_s) 2kT}\} + \frac{v - v_s}{v} \frac{2}{3d^2} \left(\frac{mkT}{\pi^3}\right)^{\frac{1}{2}} \tag{15}$$

where m : mass of the molecule;
        v : specific volume of the liquid;
        v_s: specific volume of the solid-like structure
        d : diameter of the molecule
        ℓ_f: free length
        K : transmission coefficient
        a': constant
        Z : number of nearest molecules
        $\phi(a)$: intermolecular potential
and k, T, N have their usual meaning.

Using the concept of vanishing mobility, Angell (1967) derived the equation

$$\eta = A_\eta T^{\frac{1}{2}} \exp\{\frac{k_{o\eta}}{T - T_{o\eta}}\} \tag{16}$$

where A_η, $k_{o\eta}$ and $T_{o\eta}$ are constants. The temperature $T_{o\eta}$ reflects the cohesion forces in the liquid and has been interpreted as the theoretical glass temperature at which the configurational entropy of the liquid vanishes (temperature of zero mobility).

EXPERIMENTAL RESULTS

The viscosities measured for pure molten salts with organic ions range from a few Pa s 10^{-3} to several tens Pa s 10^{-3}. Typical plots of viscosity vs temperature for different salts families are presented in Fig. 1.

Among molten alkylammonium chlorides Gatner & Kisza (1969) found non-linear Arrhenius behaviour for $C_2H_5NH_3^+Cl^-$ and $i.C_3H_7NH_3^+Cl^-$. This may be seen as a curvature in the plots of $\log \eta$ vs $1/T$ presented in Fig. 2.

Among molten alkylammonium fluoborates non-linear Arrhenius behaviour was found by Gatner (1978) in the case of $C_2H_5NH_3^+BF_4^-$, $i.C_3H_7NH_3^+BF_4^-$, $i.C_4H_9NH_3^+BF_4^-$ and $(C_2H_5)_3NH^+BF_4^-$.

Similar plots were also established by Lind, Abdel-Rehim & Rudich (1966) for the following quaternary ammonium salts: $(C_6H_{13})_4N^+BF_4^-$, $(C_4H_9)_4N^+BF_4^-$ and $(C_3H_7)_4N^+BF_4^-$.

Anyway, by least-squares fitting of the experimental data Eq. (6) allows to obtain an average "activation energy of viscous flow". For a given salts family, the latter quantity changes with the number of $-CH_2-$ groups in the alkyl chain (Fig. 3) or with the radius of the alkali cation (Fig. 4).

The dependence of E_η on the number of $-CH_2-$ groups or the alkali cation radius does not lead to any structural conclusion unless a detailed model of mass transport is known. It is obvious that the increase of E_η in pure molten alkylammonium picrates in comparison to fused alkylammonium chlorides is not only due to the increase in size of the picrate in comparison with the chloride or tetrafluoborate ions. Molten alkylammonium picrates (except quaternary) are much more strongly associated by hydrogen bonds, e.g., of the type $R_3N^+-H...O^-$(picrate ion) than molten alkylammonium chlorides or fluoborates in which hydrogen bonds are much weaker. The abnormally high "activation energy of viscous flow" in molten $CH_3COO^-Li^+$ as compared with larger cations in alkali metal acetates must be due to very strong polarization of the acetate ion by the small Li^+ cation.

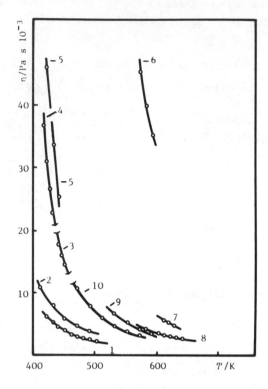

Fig. 1. Viscosity *vs* temperature for different salts families.
1: $C_2H_5NH_3^+BF_4^-$; 2: $C_2H_5NH_3^+Cl^-$; 3: $C_2H_5NH_3^+Pic^-$; 4: $C_3H_7NH_3^+Pic^-$; 5: $C_{16}H_{33}NH_3^+Pic^-$;
6: $CH_3COO^-Li^+$; 7: $CH_3COO^-Na^+$; 8: $CH_3COO^-K^+$; 9: $CH_3COO^-Rb^+$; 10: $CH_3COO^-Cs^+$

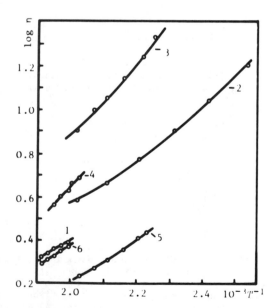

Fig. 2. Plots of $\log \eta$ *vs* $1/T$ for molten alkylammonium chlorides.
1: $CH_3NH_3^+Cl^-$; 2: $C_2H_5NH_3^+Cl^-$; 3: $i.C_3H_7NH_3^+Cl^-$; 4: $C_4H_9NH_3^+Cl^-$; 5: $(CH_3)_2NH_2^+Cl^-$;
6: $(C_2H_5)_2NH_2^+Cl^-$ (η in Cp, Gatner & Kisza, 1969)

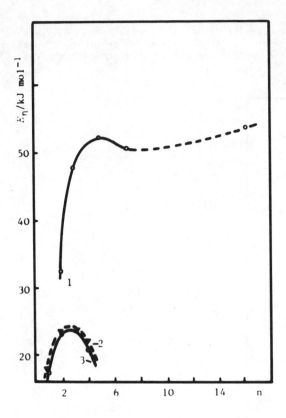

Fig. 3. "Activation energy of
viscous flow" vs number of $-CH_2-$
groups in the alkyl chain.
1: $RNH_3^+Pic^-$
2: $RNH_3^+BF_4^-$
3: $RNH_3^+Cl^-$

Fig. 4. "Activation energy of
viscous flow" as a function of
ionic radius for molten alkali
acetates.

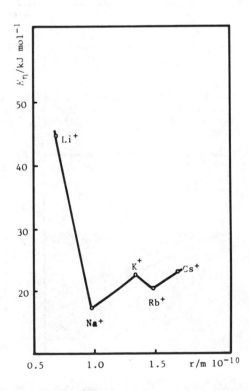

The Batchinski equation (Eq. 8) was seldom used for quantitative description of viscous flow. It is possible to correlate the parameter B of this equation - which should be proportional to the size of the unit flow - with the average "activation energy of viscous flow" (Fig. 5).

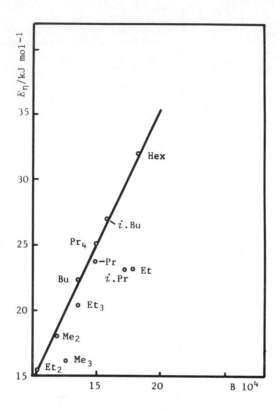

Fig. 5. Correlation between "activation energy of viscous flow" and para= meter B of the Batchinski equation for molten alkylammonium fluoborates.

Contrary to the behaviour of the activation energy, the plot of covolume, b, vs cationic radius for molten alkali acetates (Fig. 6) is not sensitive to the polarization interactions which are assumed to exist in molten $CH_3COO^-Li^+$: a smooth decay of the covolume is observed when passing from Li^+ to Cs^+.

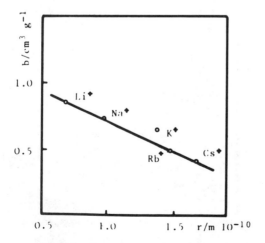

Fig. 6. Trend of covolume, b, as a function of the cationic radius in molten alkali acetates.

The three-parameter Angell equation (Eq. 16) describes the experimental temperature dependence of the viscosity very well, provided that $T_{0\eta}$ is also determined experimentally. In most cases the experimental plots of viscosity vs temperature in molten salts with organic ions are not obtained in conditions of sufficient accuracy and over a temperature range large enough to allow the parameter $T_{0\eta}$ to be calculated by the best-fit method. Lampreia & Barreira (1976) observed a regular decrease of the best-fit $T_{0\eta}$ values in the series of molten quaternary alkylammonium picrates: $Pr_4N^+Pic^-$ (244 K), $Bu_4N^+Pic^-$ (221 K) and $Am_4N^+Pic^-$ (194 K). This is, however, not the case with the acidic alkylammonium cations which form hydrogen bonds with the anion. In such cases no regularity is observed.

CONDUCTANCE OF PURE MOLTEN SALTS WITH ORGANIC IONS

DEFINITIONS

Electrical conductivity is a measure of ability for transporting charge by electrolytic conductors, e.g., molten salts. The conductivity, κ, is obtained from the measured resistance, R, and the cell constant, k,

$$\kappa = k/R \tag{17}$$

The molar conductivity is obtained from the conductivity and the molar volume, V,

$$\Lambda = \kappa V \tag{18}$$

Klemm (1964), Sundheim (1964) and Janz et al. (1968) gave extensive tables of conductance data in pure molten inorganic salts and their solutions.

The conductivity measurements provide a method of distinguishing between molecular liquids and molten salts. The theory of electrical conductivity in molten salts, however, is not well developed. Several authors treated the conductivity of pure salts as a rate process; the equation

$$\Lambda = \Lambda_\infty \exp\{-E_\Lambda/RT\} \tag{19}$$

where Λ_∞ is a constant for a given salt, and E_Λ is the energy of activation of the conductance process, is frequently used. It was pointed out by Sundheim that the rate process approach assumes tacitly that the temperature dependence is measured at constant volume, whereas most experimental measurements are performed at constant pressure.

By applying the free volume model to electrical conductance in molten salts, Angell (1964, 1967) derived for the temperature dependence of the molar conductance the expression

$$\Lambda = A_\Lambda T^{-\frac{1}{2}} \exp\{-\frac{k_{0\Lambda}}{T - T_{0\Lambda}}\} \tag{20}$$

in which A_Λ, $k_{0\Lambda}$ and $T_{0\Lambda}$ are constants. The constant $T_{0\Lambda}$ represents the zero free volume temperature in the liquid.

The conductance of molten salts can be put in a comprehensive and rigorous framework with the aid of the thermodynamics of irreversible processes. The application of the latter to transport processes in molten salts was presented by Sundheim and by Klemm (1964). Following the notations by Laity (1959), the phenomenological equations relating generalized forces to fluxes for a molten salt with N ionic species are given as

$$K_i = -(\nabla\mu_i \pm z_i F\phi) = \sum_{k=1}^{N} r_{ik} x_k (v_i - v_k) \tag{21}$$

where K_i: generalized force;
μ_i: chemical potential;

v_i, v_k: average velocity of the i-th and k-th species;
x_k : mole fraction of the k-th species;
r_{ik} : friction coefficient to be determined from the experiment;
ϕ : electric potential.

The friction coefficients obey the Onsager relation

$$r_{ik} = r_{ki} \tag{22}$$

and the relation

$$\sum_{i=1}^{N} x_i K_i = 0 \tag{23}$$

allows to reduce the number of equations from N (as given in 21) to N-1. The solution for a pure molten salt (two ionic species: cation (+) and anion (-)) yield the molar conductance

$$\Lambda = F^2 \frac{z_+ + z_-}{r_{+-}} \tag{24}$$

EXPERIMENTAL RESULTS

The conductivities of pure molten salts with organic ions range from 0.001 S cm^{-1} to approximately 1 S cm^{-1}, or from 1 S cm^2 mol^{-1} to $\approx$30 S cm^2 mol^{-1} in terms of molar conductivity. Typical molar conductivity vs temperature plots are presented in Fig. 7.

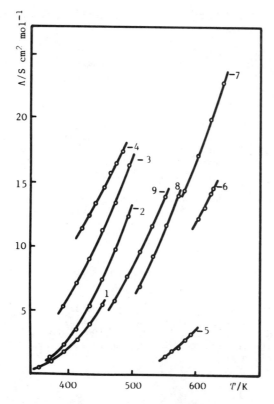

Fig. 7. Molar conductivity vs temperature for different molten salts families.
1: $(C_2H_5)_2NH_2^+Pic^-$; 2: $(C_4H_9)_4N^+Pic^-$; 3: $C_2H_5NH_3^+Cl^-$; 4: $C_2H_5NH_3^+BF_4^-$; 5: $CH_3COO^-Li^+$; 6: $CH_3COO^-Na^+$; 7: $CH_3COO^-K^+$; 8: $CH_3COO^-Rb^+$; 9: $CH_3COO^-Cs^+$.

Non-linear Arrhenius plots have been found for molten $(C_3H_7)_4N^+BPh_4^-$, $(C_6H_{13})_4N^+BF_4^-$, and $(C_4H_9)_4N^+BF_4^-$ by Lind, Abdel-Rehim & Rudich (1966); for molten $C_2H_5NH_3^+BF_4^-$, $i.C_3H_7NH_3^+BF_4^-$, $i.C_4H_9NH_3^+BF_4^-$ and $(C_2H_5)_3NH^+BF_4^-$ by Gatner (1978), and for molten $C_2H_5NH_3^+Cl^-$ by Kisza & Hawranek (1968). Some results are shown in Fig. 8.

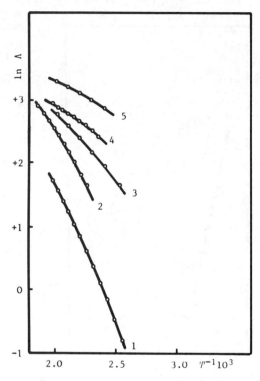

Fig. 8. Plots of $\ln \Lambda$ vs $1/T$ for molten alkylammonium salts.
1: $(C_6H_{13})_4N^+BF_4^-$
2: $(C_4H_9)_4N^+BF_4^-$
3: $C_2H_5NH_3^+Cl^-$
4: $C_2H_5NH_3^+BF_4^-$
5: $(C_2H_5)_3NH^+BF_4^-$

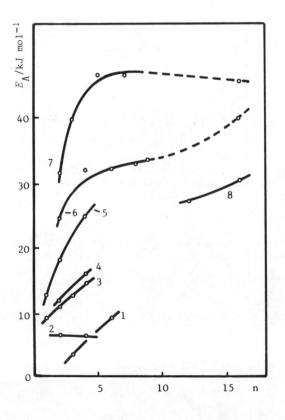

Fig. 9. "Activation energy of the conductance process" vs number of carbon atoms in the alkyl chain of the alkylammonium cation.
1: $R_3NH^+BF_4^-$
2: $R_2NH_2^+BF_4^-$
3: $RNH_3^+BF_4^-$
4: $R_2NH_2^+Cl^-$
5: $RNH_3^+Cl^-$
6: $R_2NH_2^+Pic^-$
7: $RNH_3^+Pic^-$
8: $R_4N^+Pic^-$

The least-squares fit of the experimental molar conductivity to the Arrhenius type Eq. (19) yields an average "activation energy of the conductance process". The latter quantity depends, for a given salts family, on the number of $-CH_2-$ groups in the alkyl chain (Fig. 9) or on the radius of the alkali metal cation (Fig. 10).

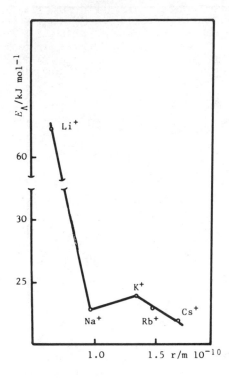

Fig. 10. "Activation energy of the conductance process" *vs* ionic radius of cation in molten alkali acetates.

According to Fig.s 9, 10 the average "activation energy of the conductance process" is apparently a function of the molten salt molecular structure, being a measure of the potential barrier that must be overcome by the ions to move from one position to another.

THE WALDEN PRODUCT AND THE STRUCTURE OF MOLTEN SALTS WITH ORGANIC IONS

Another approach to the transport properties of molten salts relates conductivity to viscosity. From the earliest measurements of the electrical conductivity and the viscosity of quaternary alkylammonium picrates, Walden (1931) proposed that

$$\Lambda\eta = \text{constant} \tag{25}$$

Eq. (25) is known as the "Walden rule", and is obeyed by molten quaternary alkylammonium picrates and also by molten alkali nitrates and chlorides (Lind, 1973) at their melting point. Frenkel (1946) proposed the relation

$$\Lambda^m\eta = \text{constant} \tag{26}$$

where m is to be determined from the temperature dependence of both viscosity and molar conductivity of a given pure molten salt.

Lind, Abdel-Rehim & Rudich (1966) computed the values of the Walden product according to two theoretical models, i.e., the hard spheres model and Rice's (1961, 1962) acoustical modification of Kirkwood's theory. Both of them yield essentially the same expression for the Walden product, although the latter theory better accounts for the actual viscosity of the melt. It was found that the expression for the Walden product is insensitive to the energy terms and depends primarily upon the structure of the melt. These relations were fully discussed by Lind (1973).

The experimental results presented in Tables A, B and C essentially confirm these conclusions. In the light of the rate theory, the constancy of the Walden product is only possible if the activation energy of the conductance process is the same as that of viscous flow, and this in turn is possible only in those molten salts in which the same molecular species are responsible for both charge transport and momentum transfer. Fully dissociated tetraalkylammonium salts with Arrhenius behaviour are good examples although even here the activation energy of viscous flow is higher than that of the conductance process.

A considerable difference between the Arrhenius average activation energies of the viscous flow and of the conductance process is observed in acidic alkylammonium salts. As a result of acidic dissociation of the alkylammonium cation (Kisza, 1968)

$$(CH_3)_2NH_2^+ \; \rightleftharpoons \; (CH_3)_2NH + H^+ \tag{27}$$

the conductance process is governed by other ionic species than those taking part in the momentum transfer.

It was believed that for most molten salts the Walden product should either be constant or decrease with increasing temperature. It can be seen, however, in Fig. 11 that this is not the case. Molten lithium acetate and tributylammonium picrate are examples of an opposite behaviour.

In molten rubidium and cesium acetates a minimum in the plot of Walden product vs temperature is observed. The latter case seems to represent the most general type of dependence of the Walden product upon temperature, while both decrease and increase of the Walden product with increasing temperature are only particular cases of the most general one.

If the ion-like interactions are responsible for the decrease of the Walden product with the

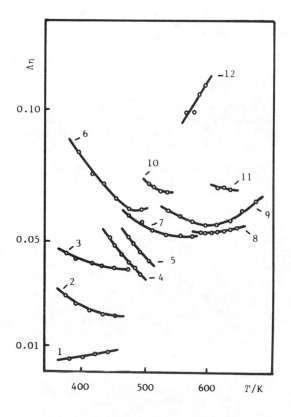

Fig. 11. Dependence of the Walden product on temperature for molten salts with organic ions.
1: $(C_4H_9)_3 NH^+Pic^-$
2: $(C_4H_9)_2 NH_2^+Pic^-$
3: $(C_4H_9)_4 N^+Pic^-$
4: $C_3H_7NH_3^+BF_4^-$
5: $CH_3NH_3^+BF_4^-$
6: $C_2H_5NH_3^+Cl^-$
7: $CH_3COO^-Cs^+$
8: $CH_3COO^-K^+$
9: $CH_3COO^-Rb^+$
10: $CH_3NH_3^+Cl^-$
11: $CH_3COO^-Na^+$
12: $CH_3COO^-Li^+$

increase of the temperature (Lind et al., 1966), then the polarization interactions should be responsible for the increase. This very interesting problem, however, needs more theoretical consideration.

TABLES

In the following tables the data available on the transport properties of pure molten salts with organic ions are presented.

The experimental values of density, ρ, conductivity, κ, and viscosity, η, at several temperatures have been taken from the references given at the bottom of each table. If in the original paper other than SI units were used, the experimental values have been converted.

For numerical purposes the temperature dependence of density is presented in the linear form

$$\rho = a + b\,T \tag{28}$$

whereas quadratic equations have been used for conductivity and viscosity:

$$\kappa = a_1 + b_1 T + c_1 T^2 \tag{29}$$

$$\eta = a_2 + b_2 T + c_2 T^2 \tag{30}$$

The coefficients of Eq.s (28-30) have been evaluated by least-squares analysis of the experimental data.

In a limited number of cases (when justified by the experimental data in the original papers) linear instead of quadratic equations have been used for the temperature dependence of conductivity (e.g., Tables B1, B2, B3).

The validity of these equations does not extend beyond the temperature ranges given in the tables. In most cases the linear equation (28) represents the experimental density fairly well over the whole temperature range of its applicability. With Eq.s (29) and (30) this is only true if the temperature range is rather narrow. The latter representations are somewhat poor for molten salts with non-Arrhenius behaviour of either conductance or viscosity. The correlation coefficient, r^2, is given for the Arrhenius type equations, for which it represents a good criterion of applicability (in the case of ideal fit, $r^2 = 1$).

Tables Ai (i = 1, 2, ..., 26) present transport properties of molten salts with organic cation, Tables Bi (i = 1, 2, ..., 8) contain transport properties for molten salts with organic anion, and Tables Ci (i = 1, 2, ..., 27) present the same properties for molten salts with both organic anion and cation.

For comparison and review, the more compact Tables A, B and C have been prepared, which summarize the relevant transport properties in correspondence with a "characteristic temperature" about 5 % higher than the melting point, T_F/K.

TABLE A. Transport properties of molten salts with organic cation at the characteristic temperature, T_{ch}

Salt	T_{ch} K	Λ S cm² mol⁻¹	η 10³ Pa s	$\Lambda\eta$	E_Λ J mol⁻¹	E_η J mol⁻¹	B 10⁴	b cm³ g⁻¹
$CH_3NH_3^+Cl^-$	523	33.52	2.08	0.0698	12598	17287	2.61	0.865
$C_2H_5NH_3^+Cl^-$	401	5.98	13.96	0.0835	18213	23042	-0.81	1.034
$i.C_3H_7NH_3^+Cl^-$	447	3.41	19.88	0.0678	28129	36254	2.97	1.051
$C_4H_9NH_3^+Cl^-$	508	8.66	4.01	0.0347	21327	25654	1.34	1.066
$(CH_3)_2NH_2^+Cl^-$	463	31.13	2.27	0.0705	11801	18816	1.03	0.998
$(C_2H_5)_2NH_2^+Cl^-$	523	12.59	1.94	0.0243	15832	21277	1.13	1.057
$CH_3NH_3^+BF_4^-$	493	16.20	3.05	0.0465	9191	18045	3.00	0.727
$C_2H_5NH_3^+BF_4^-$	443	13.42	4.59	0.0615	11195	23267	1.79	0.757
$C_3H_7NH_3^+BF_4^-$	459	12.11	3.89	0.0471	12516	23872	1.49	0.825
$i.C_3H_7NH_3^+BF_4^-$	413	10.21	7.74	0.0790	11511	23196	1.72	0.826
$C_4H_9NH_3^+BF_4^-$	493	13.38	2.73	0.0364	14553	22476	1.35	0.883
$i.C_4H_9NH_3^+BF_4^-$	443	9.31	5.71	0.0531	17047	26945	1.57	0.871
$(CH_3)_2NH_2^+BF_4^-$	413	19.21	3.78	0.0727	6413	18169	1.19	0.763
$(C_2H_5)_2NH_2^+BF_4^-$	463	22.81	1.72	0.0391	6111	15439	1.02	0.863
$(CH_3)_3NH^+BF_4^-$	513	26.27	1.69	0.0445	3456	16082	1.26	0.827
$(C_2H_5)_3NH^+BF_4^-$	413	17.35	4.50	0.0781	9016	20420	1.36	0.903
$(C_3H_7)_4N^+BF_4^-$	548	33.76	1.70	0.0472	20947	25237	1.50	1.029
$(C_4H_9)_4N^+BF_4^-$	453	6.57	6.90	0.0453	28638	33700		
$(C_6H_{13})_4N^+BF_4^-$	383	0.401	86.9	0.0348	38030	41903	6.99	1.089
$C_6H_{11}NH_3^+BF_4^-$	473	12.29	7.21	0.0885	19500	32056	1.82	0.859
$(C_4H_9)_4N^+Br^-$	408	0.407			32349			
$(C_4H_9)_4N^+I^-$	438	2.77	1.69	0.0443	38444	44096		
$(i.C_5H_{11})_3NH^+I^-$	393	0.308	33.8	0.0104	40234	44263	5.69	0.876
$(C_4H_9)_4N^+ClO_4^-$	493	9.52	4.76	0.0454	30222	36401	1.89	1.027
$(C_3H_7)_4N^+PF_6^-$	533	24.74	2.56	0.0633	25514	27032	6.49	0.907
$(C_4H_9)_4N^+PF_6^-$	533	17.16	2.73	0.0469	25101	29954	1.23	0.971

TABLE A1. Methylammonium chloride, $CH_3NH_3^+Cl^-$

T_F = 498 K   Ref. 1

ρ = 1.3131 − 5.70 10⁻⁴T Ref. 1

κ = − 2.6110 + 9.486 10⁻³T − 6.75 10⁻⁶T^2 Ref. 1

η = 0.0362 − 1.145 10⁻⁴T + 9.43 10⁻⁸T^2 Ref. 2

T	ρ	κ	Λ	η	$\Lambda\eta$	κ: Ref. 1
K	g cm^{-3}	S cm^{-1}	S cm^2 mol^{-1}	Pa s		η: Ref. 2
503	1.0264	0.4526	29.78	2.438 10^{-3}	0.0726	
513	1.0207	0.4789	31.68	2.259 10^{-3}	0.0715	
523	1.0150	0.5038	33.52	2.085 10^{-3}	0.0698	
533	1.0093	0.5274	35.28	1.961 10^{-3}	0.0691	

$$\Lambda = 606.5 \exp\left\{-\frac{12598}{R\,T}\right\} \qquad\qquad r^2 = 0.9993 \qquad \text{Note (a)}$$

$$\eta = 0.039\ 10^{-3}\ \exp\left\{\frac{17287}{R\,T}\right\} \qquad\qquad r^2 = 0.9978 \qquad \text{Note (a)}$$

$$\eta = \frac{2.991\ 10^{-3}}{v - 0.8501} \qquad\qquad\qquad\qquad\qquad\qquad \text{Note (a)}$$

Ref. 1: Kisza & Hawranek (1968)

Ref. 2: Gatner & Kisza (1969)   } data in the table given according to Ref.s 1, 2

Note (a): equations obtained by least-squares analysis of the data given in Ref.s 1, 2

TABLE A2. Ethylammonium chloride, $C_2H_5NH_3^+Cl^-$

$T_F = 382$ K   Ref. 1

$\rho = 1.1758 - 4.343\ 10^{-4}T$ Ref. 1

$\kappa = 0.02571 - 8.320\ 10^{-4}T + 2.370\ 10^{-6}T^2$ Ref. 1

$\eta = 0.30755 - 1.2415\ 10^{-3}T + 1.2699\ 10^{-6}T^2$ Ref. 2

T	ρ	κ	Λ	η	$\Lambda\eta$	κ: Ref. 1
K	g cm^{-3}	S cm^{-1}	S cm^2 mol^{-1}	Pa s		η: Ref. 2
393	1.0051	0.0648	5.256	16.04 10^{-3}	0.0843	
413	0.9965	0.0864	7.067	10.85 10^{-3}	0.0766	
433	0.9878	0.1098	9.067	7.988 10^{-3}	0.0724	
453	0.9791	0.1352	11.26	5.940 10^{-3}	0.0668	
473	0.9704	0.1625	13.65	4.613 10^{-3}	0.0629	
493	0.9617	0.1916	16.25	3.884 10^{-3}	0.0631	

$$\Lambda = 1367.8 \exp\left\{-\frac{18113}{R\,T}\right\} \qquad\qquad r^2 = 0.9989 \qquad \text{Note (a)}$$

$$\eta = 1.3442\ 10^{-3}\ \exp\left\{\frac{23042}{R\,T}\right\} \qquad\qquad r^2 = 0.9967 \qquad \text{Note (a)}$$

$$\eta = \frac{-\ 8.197\ 10^{-5}}{v - 1.0342} \qquad\qquad\qquad\qquad\qquad\qquad \text{Note (a)}$$

Ref. 1: Kisza & Hawranek (1968)

Ref. 2: Gatner & Kisza (1969)   } data in the table given according to Ref.s 1, 2

Note (a): equations obtained by least-squares analysis of the data given in Ref.s 1, 2

TABLE A3. Isopropylammonium chloride, $i.C_3H_7NH_3^+Cl^-$

T_F = 426 K	Ref. 1
$\rho = 1.1048 - 3.742 \ 10^{-4}T$	Ref. 1
$\kappa = 0.29412 - 1.7431 \ 10^{-3}T + 2.595 \ 10^{-6}T^2$	Ref. 1
$\eta = 1.1613 - 4.657 \ 10^{-3}T + 4.707 \ 10^{-6}T^2$	Ref. 2

T	ρ	κ	Λ	η	$\Lambda\eta$	
K	g cm^{-3}	S cm^{-1}	S cm^2 mol^{-1}	Pa s		κ: Ref. 1 η: Ref. 2
433	0.9427	0.0259	2.62			
443	0.9390	0.0312	3.17	21.64 10^{-3}	0.0685	
453	0.9353	0.0370	3.78	17.26 10^{-3}	0.0652	
463	0.9317	0.0434	4.44	13.78 10^{-3}	0.0611	
473	0.9280	0.0502	5.17	11.25 10^{-3}	0.0581	
483	0.9234	0.0576	5.95	9.69 10^{-3}	0.0576	
493	0.9206	0.0655	6.79			

$\Lambda = 6564.8 \ \exp\{-\dfrac{28129}{R\ T}\}$ $r^2 = 0.9993$ Note (a)

$\eta = 1.1371 \ 10^{-6} \ \exp\{\dfrac{36254}{R\ T}\}$ $r^2 = 0.9970$ Note (a)

$\eta = \dfrac{2.978 \ 10^{-4}}{v - 1.0514}$ Note (a)

Ref. 1: Kisza & Hawranek (1968)

Ref. 2: Gatner & Kisza (1969)   } data in the table given according to Ref.s 1, 2

Note (a): equations obtained by least-squares analysis of the data given in Ref.s 1, 2

TABLE A4. Butylammonium chloride, $C_4H_9NH_3^+Cl^-$

T_F = 489 K	Ref. 1
$\rho = 1.052 - 2.8 \ 10^{-4}T$	Ref. 1
$\kappa = -9.535 + 3.770 \ 10^{-2}T - 3.7 \ 10^{-5}T^2$	Ref. 1
$\eta = 0.20771 - 7.5803 \ 10^{-4}T + 7.029 \ 10^{-7}T^2$	Ref. 2

T	ρ	κ	Λ	η	$\Lambda\eta$	
K	g cm^{-3}	S cm^{-1}	S cm^2 mol^{-1}	Pa s		ρ: Ref. 1 κ: Ref. 1 η: Ref. 2
493	0.914	0.0619	7.42	4.82 10^{-3}	0.0358	
498	0.913	0.0672	8.06	4.52 10^{-3}	0.0364	
503	0.911	0.0704	8.45	4.24 10^{-3}	0.0358	
508	0.910	0.0720	8.66	4.01 10^{-3}	0.0347	

$\Lambda = 1367.2 \ \exp\{-\dfrac{21322}{R\ T}\}$ $r^2 = 0.9427$ Note (a)

$\eta = 9.2127 \ 10^{-6} \ \exp\{\dfrac{25654}{R\ T}\}$ $r^2 = 0.9993$ Note (a)

$$\eta = \frac{1.336 \ 10^{-4}}{v - 1.0661}$$ Note (a)

Ref. 1: Kisza & Hawranek (1968)

Ref. 2: Gatner & Kisza (1969)   } data in the table given according to Ref.s 1, 2

Note (a): equations obtained by least-squares analysis of the data given in Ref.s 1, 2

TABLE A5. Dimethylammonium chloride, $(CH_3)_2NH_2^+Cl^-$

$T_F = 440$ K   Ref. 1
$\rho = 1.1712 - 4.62 \ 10^{-4}T$ Ref. 1
$\kappa = -1.4215 + 5.414 \ 10^{-3}T - 3.357 \ 10^{-6}T^2$ Ref. 1
$\eta = 0.059754 - 2.229 \ 10^{-4}T + 2.133 \ 10^{-7}T^2$ Ref. 2

T	ρ	κ	Λ	η	$\Lambda\eta$	
K	g cm^{-3}	S cm^{-1}	S cm^2 mol^{-1}	Pa s		κ: Ref. 1 η: Ref. 2
443	0.9667	0.3183	26.85			
453	0.9621	0.3422	29.00	2.543 10^{-3}	0.0737	
463	0.9574	0.3655	31.13	2.266 10^{-3}	0.0705	
473	0.9528	0.3883	33.23	2.028 10^{-3}	0.0673	
483	0.9482	0.4105	35.30	1.860 10^{-3}	0.0656	
493	0.9436	0.4321	37.35	1.692 10^{-3}	0.0631	
503	0.9390	0.4523	39.37			

$\Lambda = 665.0 \ \exp\{-\frac{11801}{R \ T}\}$ $r^2 = 0.9991$ Note (a)

$\eta = 1.7114 \ 10^{-5} \ \exp\{\frac{18816}{R \ T}\}$ $r^2 = 0.9985$ Note (a)

$\eta = \frac{1.0371 \ 10^{-4}}{v - 0.9985}$ Note (a)

Ref. 1: Kisza & Hawranek (1968)

Ref. 2: Gatner & Kisza (1969)   } data in the table given according to Ref.s 1, 2

Note (a): equations obtained by least-squares analysis of the data given in Ref.s 1, 2

TABLE A6. Diethylammonium chloride, $(C_2H_5)_2NH_2^+Cl^-$

$T_F = 497$ K   Ref. 1
$\rho = 1.1195 - 4.28 \ 10^{-4}T$ Ref. 1
$\kappa = -2.3359 + 8.88 \ 10^{-3}T - 8.07 \ 10^{-6}T^2$ Ref. 1
$\eta = 0.07445 - 2.6117 \ 10^{-4}T + 2.3428 \ 10^{-7}T^2$ Ref. 2

$\dfrac{T}{K}$	$\dfrac{\rho}{g\ cm^{-3}}$	$\dfrac{\kappa}{S\ cm^{-1}}$	$\dfrac{\Lambda}{S\ cm^2\ mol^{-1}}$	$\dfrac{\eta}{Pa\ s}$	$\Lambda\eta$	κ: Ref. 1 η: Ref. 2
498	0.906	0.0868	10.49			
503	0.904	0.0909	11.01	$2.360\ 10^{-3}$	0.0259	
508	0.902	0.0944	11.46	$2.227\ 10^{-3}$	0.0255	
513	0.899	0.0977	11.90	$2.124\ 10^{-3}$	0.0252	
518	0.897	0.1005	12.28	$2.035\ 10^{-3}$	0.0249	
523	0.895	0.1029	12.59	$1.936\ 10^{-3}$	0.0243	

$\Lambda = 483.9\ \exp\{-\dfrac{15832}{R\,T}\}$ $\qquad\qquad\qquad\qquad r^2 = 0.9911 \qquad$ Note (a)

$\eta = 1.45117\ 10^{-5}\ \exp\{\dfrac{21277}{R\,T}\}$ $\qquad\qquad\qquad r^2 = 0.9980 \qquad$ Note (a)

$\eta = \dfrac{1.1299\ 10^{-4}}{v - 1.0578}$

Ref. 1: Kisza & Hawranek (1968)

Ref. 2: Gatner & Kisza (1969) $\qquad\qquad\qquad$ } data in the table given according to Ref.s 1, 2

Note (a): equations obtained by least squares analysis of the data given in Ref.s 1, 2

TABLE A7. Methylammonium fluoborate, $CH_3NH_3^+BF_4^-$

$T_F = 467.6$ K $\qquad\qquad\qquad\qquad\qquad\qquad\qquad\qquad\qquad$ Ref. 1

$\rho = 1.7115 - 7.4809\ 10^{-4}T$ $\qquad\qquad\qquad\qquad\qquad$ Note (a)

$\kappa = -1.2417 + 5.108\ 10^{-3}T - 4.50\ 10^{-6}T^2$ $\qquad\quad$ Note (a)

$\eta = 0.09440 - 3.412\ 10^{-3}T + 3.156\ 10^{-7}T^2$ $\qquad\quad$ Note (a)

$\dfrac{T}{K}$	$\dfrac{\rho}{g\ cm^{-3}}$	$\dfrac{\kappa}{S\ cm^{-1}}$	$\dfrac{\Lambda}{S\ cm^2\ mol^{-1}}$	$\dfrac{\eta}{Pa\ s}$	$\Lambda\eta$
468	1.3616				
473	1.3569	0.1676	14.68	$3.846\ 10^{-3}$	0.0523
478	1.3528			$3.563\ 10^{-3}$	
483	1.3519	0.1755	15.43	$3.394\ 10^{-3}$	0.0493
488	1.3461			$3.200\ 10^{-3}$	
493	1.3421	0.1829	16.20	$3.048\ 10^{-3}$	0.0465
498	1.3396			$2.873\ 10^{-3}$	
503		0.1890	16.86	$2.733\ 10^{-3}$	0.0438
508				$2.598\ 10^{-3}$	

$\Lambda = 152.07\ \exp\{-\dfrac{9191}{RT}\}$ $\qquad\qquad\qquad\qquad r^2 = 0.9988 \qquad$ Note (a)

$\eta = 3.4609\ 10^{-5}\ \exp\{\dfrac{18045}{R\,T}\}$ $\qquad\qquad\qquad r^2 = 0.987 \qquad$ Note (a)

$\eta = \dfrac{3.0029\ 10^{-4}}{v - 0.7275}$ $\qquad\qquad\qquad\qquad\qquad\qquad\qquad$ Note (a)

Ref. 1: Gatner (1978) $\qquad\qquad\qquad$ data in the table given according to Ref. 1

Note (a): equations obtained by least-squares analysis of the data given in Ref. 1

TABLE A8. Ethylammonium fluoborate, $C_2H_5NH_3^+BF_4^-$

T_F= 422.9			Ref. 1
$\rho = 1.5837 - 7.575 \ 10^{-4}T$		$r^2 = 0.988$	Note (a)
$\kappa = -0.8668 + 3.580 \ 10^{-3}T - 3.021 \ 10^{-6}T^2$			Note (a)
$\eta = 0.13625 - 5.2384 \ 10^{-4}T + 5.1193 \ 10^{-7}T^2$			Note (a)

$\dfrac{T}{K}$	$\dfrac{\rho}{g \ cm^{-3}}$	$\dfrac{\kappa}{S \ cm^{-1}}$	$\dfrac{\Lambda}{S \ cm^2 \ mol^{-1}}$	$\dfrac{\eta}{Pa \ s}$	$\Lambda\eta$
423	1.2632	0.1074	11.30	6.383 10^{-3}	0.0721
433		0.1169	12.37	5.311 10^{-3}	0.0656
443	1.2491	0.1261	13.42	4.589 10^{-3}	0.0615
453		0.1351	14.48	3.953 10^{-3}	0.0572
463	1.2316	0.1438	15.52	3.455 10^{-3}	0.0536
473		0.1505	16.33	3.078 10^{-3}	0.0502
483	1.2180	0.1583	17.27	2.745 10^{-3}	0.0474
493		0.1635	17.95	2.455 10^{-3}	0.0440
503	1.2030	0.1699	18.77	2.197 10^{-3}	0.0412

$\Lambda = 278.52 \exp\{-\dfrac{11195}{R \ T}\}$	$r^2 = 0.992$	Note (a)
$\eta = 8.341 \ 10^{-6} \exp\{\dfrac{23267}{R \ T}\}$	$r^2 = 0.998$	Note (a)
$\eta = \dfrac{1.799 \ 10^{-4}}{v - 0.7571}$		Note (a)

Ref. 1: Gatner (1978)   data in the table given according to Ref. 1

Note (a): equations obtained by least-squares analysis of the data given in Ref. 1

TABLE A9. Propylammonium fluoborate, $C_3H_7NH_3^+BF_4^-$

T_F= 437 K			Ref. 1
$\rho = 1.4982 - 7.454 \ 10^{-4}T$		$r^2 = 0.997$	Note (a)
$\kappa = -0.2901 + 1.047 \ 10^{-3}T - 4.482 \ 10^{-7}T^2$			Note (a)
$\eta = 0.12622 - 4.786 \ 10^{-4}T + 4.619 \ 10^{-7}T^2$			Note (a)

$\dfrac{T}{K}$	$\dfrac{\rho}{g \ cm^{-3}}$	$\dfrac{\kappa}{S \ cm^{-1}}$	$\dfrac{\Lambda}{S \ cm^2 \ mol^{-1}}$	$\dfrac{\eta}{Pa \ s}$	$\Lambda\eta$
443	1.1683	0.08574	10.78	4.859 10^{-3}	0.0524
453	1.1602	0.09227	11.68	4.164 10^{-3}	0.0486
463	1.1522	0.09841	12.55	3.619 10^{-3}	0.0454
473	1.1464	0.1053	13.50	3.188 10^{-3}	0.0430
483	1.1390	0.1106	14.27	2.836 10^{-3}	0.0404
493	1.1300	0.1173	15.25	2.504 10^{-3}	0.0381

$$\Lambda = 323.37 \, \exp\{-\frac{12516}{R\,T}\}$$ $r^2 = 0.999$ Note (a)

$$\eta = 7.3958 \; 10^{-6} \, \exp\{\frac{23872}{R\,T}\}$$ $r^2 = 0.999$ Note (a)

$$\eta = \frac{1.494 \; 10^{-4}}{v - 0.8256}$$ Note (a)

Ref. 1: Gatner (1978)   data in the table given according to Ref. 1

Note (a): equations obtained by least-squares analysis of the data given in Ref. 1

TABLE A10. *Iso*propylammonium fluoborate, $i.C_3H_7NH_3^+BF_4^-$

$T_F = 391.3$ K   Ref. 1

$\rho = 1.4408 - 6.3985 \; 10^{-4}T$ $r^2 = 0.999$ Note (a)

$\kappa = -0.7804 + 3.350 \; 10^{-3}T - 3.061 \; 10^{-6}T^2$ Note (a)

$\eta = 0.21240 - 8.544 \; 10^{-4}T + 8.704 \; 10^{-7}T^2$ Note (a)

T	ρ	κ	Λ	η	$\Lambda\eta$
K	g cm^{-3}	S cm^{-1}	S cm^2 mol^{-1}	Pa s	
393	1.1895	0.06328	7.82	11.24 10^{-3}	0.0878
413	1.1773	0.08185	10.21	7.738 10^{-3}	0.0790
433	1.1624	0.09737	12.31	5.413 10^{-3}	0.0666
453	1.1513	0.1088	13.88	4.261 10^{-3}	0.0591
473	1.1383	0.1190	15.36	3.159 10^{-3}	0.0485
493	1.1256	0.1281	16.72	2.571 10^{-3}	0.0429

$$\Lambda = 287.0 \, \exp\{-\frac{11511}{R\,T}\}$$ $r^2 = 0.969$ Note (a)

$$\eta = 8.811 \; 10^{-6} \, \exp\{\frac{23196}{R\,T}\}$$ $r^2 = 0.994$ Note (a)

$$\eta = \frac{1.717 \; 10^{-4}}{v - 0.8262}$$ Note (a)

Ref. 1: Gatner (1978)   data in the table given according to Ref. 1

Note (a): equations obtained by least-squares analysis of the data given in Ref. 1

TABLE A11. Butylammonium fluoborate, $C_4H_9NH_3^+BF_4^-$

$T_F = 468$ K   Ref. 1

$\rho = 1.4095 - 6.84 \; 10^{-4}T$ $r^2 = 0.998$ Note (a)

$\kappa = -0.3075 + 1.038 \; 10^{-3}T - 4.75 \; 10^{-7}T^2$ Note (a)

$\eta = 0.10740 - 3.912 \; 10^{-4}T + 3.628 \; 10^{-7}T^2$

T	ρ	κ	Λ	η	$\Lambda\eta$
K	g cm^{-3}	S cm^{-1}	S cm^2 mol^{-1}	Pa s	
473	1.0864	0.07747	11.48	3.545 10^{-3}	0.0406

483	1.0790	0.08314	12.40	$3.086 \ 10^{-3}$	0.0382
493	1.0722	0.08915	13.38	$2.726 \ 10^{-3}$	0.0364
503	1.0652	0.09463	14.30	$2.438 \ 10^{-3}$	0.0348
513	1.0591			$2.197 \ 10^{-3}$	

$$\Lambda = 464.68 \ \exp\{-\frac{14553}{R\,T}\} \qquad\qquad r^2 = 0.999 \qquad \text{Note (a)}$$

$$\eta = 1.1288 \ 10^{-5} \ \exp\{\frac{22476}{R\,T}\} \qquad\qquad r^2 = 0.9777 \qquad \text{Note (a)}$$

$$\eta = \frac{1.356 \ 10^{-4}}{v - 0.8831} \qquad\qquad\qquad \text{Note (a)}$$

Ref. 1: Gatner (1978)   data in the table given according to Ref. 1

Note (a): equations obtained by least-squares analysis of the data given in Ref. 1

TABLE A12. Isobutylammonium fluoborate, $i.C_4H_9NH_3^+BF_4^-$

$$T_F = 420.6 \ K \qquad\qquad\qquad\qquad\qquad\qquad\qquad \text{Ref. 1}$$
$$\rho = 1.4166 - 6.9 \ 10^{-4}T \qquad\qquad r^2 = 0.995 \qquad \text{Note (a)}$$
$$\kappa = -0.3984 + 1.377 \ 10^{-3}T - 7.536 \ 10^{-7}T^2 \qquad\qquad \text{Note (a)}$$
$$\eta = 0.19563 - 7.591 \ 10^{-4}T + 7.459 \ 10^{-7}T^2 \qquad\qquad \text{Note (a)}$$

T	ρ	κ	Λ	η	$\Lambda\eta$
K	g cm^{-3}	S cm^{-1}	S cm^2 mol^{-1}	Pa s	
433	1.1190	0.05689	8.18	$6.853 \ 10^{-3}$	0.0560
443		0.06427	9.31	$5.709 \ 10^{-3}$	0.0531
453	1.1025	0.07107	10.38	$4.768 \ 10^{-3}$	0.0494
463		0.07788	11.43	$4.089 \ 10^{-3}$	0.0467
473	1.0902	0.08495	12.54	$3.543 \ 10^{-3}$	0.0444
483		0.09124	13.55	$3.047 \ 10^{-3}$	0.0412
493	1.0771	0.09776	14.61	$2.658 \ 10^{-3}$	0.0388

$$\Lambda = 947.72 \ \exp\{-\frac{17047}{R\,T}\} \qquad\qquad r^2 = 0.996 \qquad \text{Note (a)}$$

$$\eta = 3.8353 \ 10^{-6} \ \exp\{\frac{26945}{R\,T}\} \qquad\qquad r^2 = 0.972 \qquad \text{Note (a)}$$

$$\eta = \frac{1.578 \ 10^{-4}}{v - 0.8719} \qquad\qquad\qquad \text{Note (a)}$$

Ref. 1: Gatner (1978)   data in the table given according to Ref. 1

Note (a): equations obtained by least-squares analysis of the data given in Ref. 1

TABLE A13. Dimethylammonium fluoborate, $(CH_3)_2NH_2^+BF_4^-$

$$T_F = 360 \ K \qquad\qquad\qquad\qquad\qquad\qquad\qquad \text{Ref. 1}$$
$$\rho = 1.5570 - 7.346 \ 10^{-4}T \qquad\qquad r^2 = 0.999 \qquad \text{Note (a)}$$
$$\kappa = -0.3738 + 1.866 \ 10^{-3}T - 1.267 \ 10^{-6}T^2 \qquad\qquad \text{Note (a)}$$

$$\eta = 0.09512 - 3.834 \ 10^{-4}T + 3.935 \ 10^{-7}T^2 \qquad \text{Note (a)}$$

$\dfrac{T}{K}$	$\dfrac{\rho}{g \ cm^{-3}}$	$\dfrac{\kappa}{S \ cm^{-1}}$	$\dfrac{\Lambda}{S \ cm^2 \ mol^{-1}}$	$\dfrac{\eta}{Pa \ s}$	$\Lambda\eta$
373	1.2828			$7.009 \ 10^{-3}$	
393	1.2692			$5.031 \ 10^{-3}$	
413	1.2531	0.1811	19.21	$3.787 \ 10^{-3}$	0.0727
433	1.2389	0.1955	20.97	$2.885 \ 10^{-3}$	0.0604
453	1.2238	0.2115	22.93	$2.336 \ 10^{-3}$	0.0535
473	1.2094	0.2260	24.83	$1.905 \ 10^{-3}$	0.0473
493	1.1953	0.2376	26.42	$1.617 \ 10^{-3}$	0.0427

$$\Lambda = 126.11 \ \exp\{-\frac{6413}{RT}\} \qquad\qquad r^2 = 0.995 \qquad \text{Note (a)}$$

$$\eta = 1.8746 \ 10^{-5} \ \exp\{\frac{18169}{R \ T}\} \qquad\qquad r^2 = 0.994 \qquad \text{Note (a)}$$

$$\eta = \frac{1.191 \ 10^{-4}}{v - 0.7636} \qquad\qquad \text{Note (a)}$$

Ref. 1: Gatner (1978)   data in the table given according to Ref. 1

Note (a): equations obtained by least-squares analysis of the data given in Ref. 1

TABLE A14. Diethylammonium fluoborate, $(C_2H_5)_2NH_2^+BF_4^-$

$$T_F = 444.6 \ K \qquad\qquad\qquad\qquad\qquad\qquad\qquad\qquad \text{Ref. 1}$$

$$\rho = 1.3719 - 6.23 \ 10^{-4}T \qquad\qquad r^2 = 0.998 \qquad \text{Note (a)}$$

$$\kappa = -0.1232 + 7.648 \ 10^{-4}T - 3.571 \ 10^{-7}T^2 \qquad\qquad \text{Note (a)}$$

$$\eta = 0.02829 - 9.973 \ 10^{-5}T + 9.142 \ 10^{-8}T^2 \qquad\qquad \text{Note (a)}$$

$\dfrac{T}{K}$	$\dfrac{\rho}{g \ cm^{-3}}$	$\dfrac{\kappa}{S \ cm^{-1}}$	$\dfrac{\Lambda}{S \ cm^2 \ mol^{-1}}$	$\dfrac{\eta}{Pa \ s}$	$\Lambda\eta$
453	1.0896	0.1502	22.17	$1.875 \ 10^{-3}$	0.0415
463	1.0839	0.1536	22.81	$1.718 \ 10^{-3}$	0.0391
473	1.0768	0.1589	23.75	$1.582 \ 10^{-3}$	0.0375
483	1.0710	0.1631	24.51	$1.442 \ 10^{-3}$	0.0353
493	1.0649	0.1668	25.21	$1.351 \ 10^{-3}$	0.0340

$$\Lambda = 112.06 \ \exp\{-\frac{6111}{RT}\} \qquad\qquad r^2 = 0.9969 \qquad \text{Note (a)}$$

$$\eta = 3.1134 \ 10^{-5} \ \exp\{\frac{15439}{R \ T}\} \qquad\qquad r^2 = 0.998 \qquad \text{Note (a)}$$

$$\eta = \frac{1.020 \ 10^{-4}}{v - 0.8634} \qquad\qquad \text{Note (a)}$$

Ref. 1: Gatner (1978)   data in the table given according to Ref. 1

Note (a): equations obtained by least-squares analysis of the data given in Ref. 1

TABLE A15. Trimethylammonium fluoborate, $(CH_3)_3NH^+BF_4^-$

$$T_F = 489 \text{ K}$$ Ref. 1
$$\rho = 1.4411 - 6.46 \ 10^{-4}T \qquad\qquad r^2 = 0.995$$ Note (a)
$$\kappa = -2.1620 + 9.091 \ 10^{-3}T - 8.75 \ 10^{-6}T^2$$ Note (a)
$$\eta = 0.05951 - 2.140 \ 10^{-4}T + 1.975 \ 10^{-7}T^2$$ Note (a)

T	ρ	κ	Λ	η	$\Lambda\eta$
K	g cm^{-3}	S cm^{-1}	S cm^2 mol^{-1}	Pa s	
493	1.1225	0.1930	25.26	1.998 10^{-3}	0.0504
503	1.1161	0.1971	25.95	1.819 10^{-3}	0.0472
513	1.1106	0.1986	26.27	1.694 10^{-3}	0.0445
523	1.1028	0.1992	26.54	1.594 10^{-3}	0.0423

$$\Lambda = 58.947 \ \exp\{-\frac{3456}{RT}\} \qquad\qquad r^2 = 0.950$$ Note (a)
$$\eta = 3.924 \ 10^{-5} \ \exp\{\frac{16082}{R \ T}\} \qquad r^2 = 0.993$$ Note (a)
$$\eta = \frac{1.263 \ 10^{-4}}{v - 0.8271}$$ Note (a)

Ref. 1: Gatner (1978)   data in the table given according to Ref. 1
Note (a): equations obtained by least-squares analysis of the data given in Ref. 1

TABLE A16. Triethylammonium fluoborate, $(C_2H_5)_3NH^+BF_4^-$

$$T_F = 389 \text{ K}$$ Ref. 1
$$\rho = 1.3216 - 6.07 \ 10^{-4}T \qquad\qquad r^2 = 0.999$$ Note (a)
$$\kappa = -0.6471 + 2.841 \ 10^{-3}T - 2.507 \ 10^{-6}T^2$$ Note (a)
$$\eta = 0.10530 - 4.164 \ 10^{-4}T + 4.189 \ 10^{-7}T^2$$ Note (a)

T	ρ	κ	Λ	η	$\Lambda\eta$
K	g cm^{-3}	S cm^{-1}	S cm^2 mol^{-1}	Pa s	
393	1.0835			6.741 10^{-3}	
413	1.0708	0.09828	17.35	4.504 10^{-3}	0.0781
433	1.0587	0.1132	20.20	3.644 10^{-3}	0.0736
453	1.0462	0.1251	22.59	2.614 10^{-3}	0.0590
473	1.0345	0.1354	24.73	2.156 10^{-3}	0.0533
493	1.0228	0.1441	26.62	1.744 10^{-3}	0.0464

$$\Lambda = 243.77 \ \exp\{-\frac{9016}{RT}\} \qquad\qquad r^2 = 0.991$$ Note (a)
$$\eta = 1.1913 \ 10^{-5} \ \exp\{\frac{20420}{R \ T}\} \qquad r^2 = 0.989$$ Note (a)
$$\eta = \frac{1.364 \ 10^{-4}}{v - 0.9035}$$ Note (a)

Ref. 1: Gatner (1978)   data in the table given according to Ref. 1
Note (a): equations obtained by least-squares analysis of the data given in Ref. 1

TABLE A17. Tetrapropylammonium fluoborate, $(C_3H_7)_4N^+BF_4^-$

T_F= 521 K   Ref. 1

$\rho = 1.3116 - 7.637\ 10^{-4}T$ Note (a)

$\rho = 1.2397 - 6.286\ 10^{-4}T$ Note (b)

$\kappa = 0.2753 - 1.459\ 10^{-3}T + 2.114\ 10^{-6}T^2$ Note (c)

$\eta = 0.05599 - 1.812\ 10^{-4}T + 1.50\ 10^{-7}T^2$ Note (c)

T	ρ	κ	Λ	η	$\Lambda\eta$
K	g cm^{-3}	S cm^{-1}	S cm^2 mol^{-1}	Pa s	
523	0.9121	0.09049	27.10	2.218 10^{-3}	0.0601
528	0.9083	0.09432	28.36	2.099 10^{-3}	0.0595
533	0.9044	0.09821	29.66	1.989 10^{-3}	0.0589
538	0.9014	0.1023	30.99	1.886 10^{-3}	0.0584
543	0.8982	0.1064	32.36	1.791 10^{-3}	0.0579
548	0.8951	0.1106	33.76	1.702 10^{-3}	0.0574

$\Lambda = 3345\ \exp\{-\dfrac{20947}{R\ T}\}$ Note (d)

$\eta = 6.687\ 10^{-6}\ \exp\{\dfrac{25237}{R\ T}\}$ Note (e)

$\eta = \dfrac{1.5062\ 10^{-4}}{v - 1.0291}$ Note (c)

Ref. 1: Lind, Abdel-Rehim & Rudich (1966)

Note (a): equation given in Ref. 1, valid in the temperature range 525-533 K; standard deviation: 0.0001

Note (b): equation given in Ref. 1, valid in the temperature range 533-548 K; standard deviation: 0.0001

Note (c): equations obtained by least-squares analysis of the data given in Ref. 1

Note (d): equation obtained using $\ln \Lambda = 8.1154 - 2519.4/T$, as given in Ref. 1; standard deviation: 0.001

Note (e): equation obtained using $\ln \eta 100 = -5.0112 + 3038.41/T$ (η in cP), as given in Ref. 1; standard deviation: 0.0007

TABLE A18. Tetrabutylammonium fluoborate, $(C_4H_9)_4N^+BF_4^-$

T_F= 435 K   Ref. 1

$\rho = 1.1906 - 5.812\ 10^{-4}T$ Note (a)

$\kappa = 0.37497 - 1.8511\ 10^{-3}T + 2.3488\ 10^{-6}T^2$ Note (b)

$\eta = 0.2519 - 9.564\ 10^{-4}T + 9.177\ 10^{-7}T^2$ Note (b)

T	ρ	κ	Λ	η	$\Lambda\eta$
K	g cm^{-3}	S cm^{-1}	S cm^2 mol^{-1}	Pa s	
443	0.9330	0.01553	5.483	8.429 10^{-3}	0.0462

453	0.9272	0.01851	6.572	$6.897\ 10^{-3}$	0.0453
463	0.9214	0.02175	7.771	$5.730\ 10^{-3}$	0.0445
473	0.9156	0.02525	9.079	$4.823\ 10^{-3}$	0.0438
483	0.9098	0.02896	10.48	$4.110\ 10^{-3}$	0.0431
493	0.9040	0.03289	11.98	$3.541\ 10^{-3}$	0.0424
503	0.8982	0.03704	13.58	$3.082\ 10^{-3}$	0.0419
513	0.8924	0.04431	16.35	$2.707\ 10^{-3}$	0.0443

$\Lambda = 13123.2\ \exp\{-\dfrac{28638}{R\,T}\}$ $\qquad\qquad\qquad r^2 = 0.9982$ $\qquad$ Note (b)

$\eta = 9.26\ 10^{-7}\ \exp\{\dfrac{33700}{R\,T}\}$ $\qquad\qquad\qquad r^2 = 0.9950$ $\qquad$ Note (b)

$\eta = \dfrac{1.8279\ 10^{-4}}{v\ -\ 1.0522}$ $\qquad\qquad\qquad\qquad\qquad\qquad$ Note (b)

Ref. 1: Lind, Abdel-Rehim & Rudich (1966)

Note (a): equation given in Ref. 1, valid in the temperature range 436-539 K; standard deviation: 0.0003

Note (b): equations obtained by least-squares analysis of the data given in Ref. 1:

$\ln \Lambda = 3.8298 + 1811.0/T - 1.2199\ 10^6/T^2$; standard deviation: 0.001

$\ln \eta 100 = -0.59671 - 1661/T + 1.27194\ 10^6/T^2$; standard deviation: 0.002

TABLE A19. Tetrahexylammonium fluoborate, $(C_6H_{13})_4N^+BF_4^-$

$T_F = 364$ K $\qquad\qquad\qquad\qquad\qquad\qquad\qquad\qquad$ Ref. 1

$\rho = 1.1296 - 5.772\ 10^{-4}T$ $\qquad\qquad\qquad\qquad\qquad$ Note (a)

$\kappa = 0.083639 - 4.4926\ 10^{-4}T + 6.0813\ 10^{-7}T^2$ $\qquad$ Note (b)

$\eta = 2.8551 - 0.01230T + 1.326\ 10^{-5}T^2$ $\qquad\qquad$ Note (c)

T	ρ	κ	Λ	η	$\Lambda\eta$
K	g cm^{-3}	S cm^{-1}	S cm^2 mol^{-1}	Pa s	
373	0.9142	$6.930\ 10^{-4}$	0.3317	0.1197	0.0397
393	0.9027	$9.829\ 10^{-4}$	0.4765	0.05412	0.0258
413	0.8911	$1.801\ 10^{-3}$	0.8843	0.02772	0.0245
433	0.8796	$3.126\ 10^{-3}$	1.555	0.01570	0.0244
453	0.8680	$4.937\ 10^{-3}$	2.489	0.009668	0.0241
473	0.8565	$7.214\ 10^{-3}$	3.686	0.006370	0.0235
493	0.8450	$9.937\ 10^{-3}$	5.147	0.004443	0.0229

$\Lambda = 62879\ \exp\{-\dfrac{38030}{R\,T}\}$ $\qquad\qquad\qquad r^2 = 0.970$ $\qquad$ Note (d)

$\eta = 1.482\ 10^{-7}\ \exp\{\dfrac{41903}{R\,T}\}$ $\qquad\qquad\qquad r^2 = 0.996$ $\qquad$ Note (d)

$\eta = \dfrac{6.9936\ 10^{-4}}{v\ -\ 1.0891}$ $\qquad\qquad\qquad\qquad\qquad\qquad$ Note (d)

Ref. 1: Lind, Abdel-Rehim & Rudich (1966)

Note (a): equation given in Ref. 1, valid in the temperature range 375-491 K; standard deviation: 0.0003

Note (b): equation obtained by least-squares analysis using $\ln \Lambda = 3.3488 + 2042.6/T - 1.4287$
$10^6/T^2$ (standard deviation: 0.002), and densities as given in Ref. 1

Note (c): equation obtained by least-squares analysis using $\ln \eta 100 = -0.6263 - 1987.1/T +$
$1.49496\ 10^6/T^2$ (standard deviation: 0.004), as given in Ref. 1

Note (d): equations obtained by least-squares analysis of the data given in Ref. 1.
Data in the table given according to Ref. 1

TABLE A20. Cyclohexylammonium fluoborate, $C_6H_{11}NH_3^+BF_4^-$

$T_F = 452$ K		Ref. 1
$\rho = 1.4355 - 6.44\ 10^{-4}T$	$r^2 = 0.995$	Note (a)
$\kappa = -0.10219 + 1.296\ 10^{-4}T + 4.0\ 10^{-7}T^2$		Note (a)
$\eta = 0.22699 - 8.1468\ 10^{-4}T + 7.4\ 10^{-7}T^2$		Note (a)

T	ρ	κ	Λ	η	$\Lambda\eta$
K	g cm^{-3}	S cm^{-1}	S cm^2 mol^{-1}	Pa s	
473	1.1312	0.04859	12.29	7.209 10^{-3}	0.0885
483	1.1244	0.05375	13.68	6.139 10^{-3}	0.0839
493	1.1173	0.05889	15.08	5.211 10^{-3}	0.0785
503	1.1121	0.06421	16.52	4.437 10^{-3}	0.0732

$\Lambda = 1752.2\ \exp\{-\dfrac{19500}{R\ T}\}$	$r^2 = 0.999$	Note (a)
$\eta = 2.0892\ 10^{-6}\ \exp\{\dfrac{32056}{R\ T}\}$		Note (a)
$\eta = \dfrac{1.8215\ 10^{-4}}{v - 0.8592}$		Note (a)

Ref. 1: Gatner (1978)   data in the table given according to Ref. 1

Note (a): equations obtained by least-squares analysis of the data given in Ref. 1

TABLE A21. Tetrabutylammonium bromide, $(C_4H_9)_4N^+Br^-$

$T_F = 392.6$ K	Ref. 1
$\rho = 1.287 - 7.039\ 10^{-4}T$	Note (a)
$\kappa = 0.0011834 - 2.5828\ 10^{-5}T + 6.381\ 10^{-8}T^2$	Note (b)

T	ρ	κ	Λ	η	$\Lambda\eta$
K	g cm^{-3}	S cm^{-1}	S cm^2 mol^{-1}	Pa s	
393	1.0103	8.881 10^{-4}	0.2834		
398	1.0067	1.002 10^{-3}	0.3209		
403	1.0032	1.127 10^{-3}	0.3623		
408	0.9997	1.264 10^{-3}	0.4078		

$$\Lambda = 5653.3 \exp\{-\frac{32349}{R\,T}\}$$ Note (c)

Ref. 1: Lind, Abdel-Rehim & Rudich (1966)   data in the table given according to Ref. 1

Note (a): equation given in Ref. 1, valid in the temperature range 392–408 K; standard deviation 0.0002

Note (b): equation obtained by least-squares analysis of the data given in Ref. 1

Note (c): equation obtained using ln Λ = 8.64 – 3891/T, as given in Ref. 1; standard deviation: 0.07

TABLE A22. Tetrabutylammonium iodide, $(C_4H_9)_4N^+I^-$

T_F= 419 K   Ref. 1

ρ = 1.4460 – 8.388 $10^{-4}T$ Note (a)

κ = 0.0085927 – 1.6638 $10^{-4}T$ + 3.768 $10^{-7}T^2$ Note (b)

η = 27 10^{-3} Pa s at 422 K

T	ρ	κ	Λ	η	$\Lambda\eta$
K	g cm^{-3}	S cm^{-1}	S cm^2 mol^{-1}	Pa s	
418				30.30 10^{-3}	
423	1.0911	5.635 10^{-3}	1.907	26.08 10^{-3}	0.0497
428	1.0869	6.347 10^{-3}	2.167		
433	1.0827	7.196 10^{-3}	2.455		
438	1.0785	8.097 10^{-3}	2.773		

$$\Lambda = 1.0619\ 10^5 \exp\{-\frac{38444}{R\,T}\}$$ Note (c)

$$\eta = 9.35\ 10^{-8} \exp\{\frac{44096}{R\,T}\}$$ Note (d)

Ref. 1: Lind, Abdel-Rehim & Rudich (1966)   data in the table according to Ref. 1, except η

Note (a): equation given in Ref. 1, valid in the temperature range 420–435 K; standard deviation: 0.0003

Note (b): equation obtained by least-squares analysis of the data given in Ref. 1

Note (c): equation obtained using ln Λ = 11.573 – 4623.8/T, as given in Ref. 1; standard deviation: 0.005

Note (d): see Walden & Birr (1932); equation obtained from two temperature values only

TABLE A23. Tri*iso*amylammonium iodide, $(i.C_5H_{11})_3NH^+I^-$

T_F= 373 K   Ref. 1

ρ = 1.4027 – 7.24 $10^{-4}T$ r^2 = 0.9998   Note (a)

κ = 0.043156 – 2.4533 $10^{-4}T$ + 3.511 $10^{-7}T^2$ Note (a)

η = 4.17813 – 0.019715T + 2.3339 $10^{-5}T^2$ Note (a)

T	ρ	κ	Λ	η	$\Lambda\eta$
K	g cm^{-3}	S cm^{-1}	S cm^2 mol^{-1}	Pa s	
373	1.1328	4.914 10^{-4}	0.1541	0.07246	0.0112
383	1.1254	7.013 10^{-4}	0.2214	0.04893	0.0108
393	1.1182	9.699 10^{-4}	0.3082	0.03381	0.0104
403	1.1107	1.301 10^{-3}	0.4162	0.02432	0.0101
413	1.1036	1.721 10^{-3}	0.5541	0.01760	0.00975
423	1.0967	2.203 10^{-3}	0.7137	0.01353	0.00966

$\Lambda = 67606.2 \exp\{- \dfrac{40234}{R\,T}\}$ $\qquad\qquad\qquad r^2 = 0.9993$ $\qquad$ Note (a)

$\eta = 4.5\ 10^{-8} \exp\{\dfrac{44263}{R\,T}\}$ $\qquad\qquad\qquad r^2 = 0.9991$ $\qquad$ Note (a)

$\eta = \dfrac{5.6967\ 10^{-4}}{v - 0.8760}$ $\qquad\qquad\qquad\qquad\qquad\qquad$ Note (a)

Ref. 1: Walden & Birr (1932)   data in the table given according to Ref. 1

Note (a): equations obtained by least-squares analysis of the data given in Ref. 1

TABLE A24. Tetrabutylammonium perchlorate, $(C_4H_9)_4N^+ClO_4^-$

$T_F = 482$ K $\qquad\qquad\qquad\qquad\qquad\qquad\qquad\qquad\qquad$ Ref. 1

$\rho = 1.2277 - 5.9\ 10^{-4}T$ $\qquad\qquad\qquad r^2 = 0.9999$ $\qquad$ Note (a)

$\kappa = - 1.14903 + 4.4542\ 10^{-3}T + 4.2\ 10^{-6}T^2$ $\qquad\qquad$ Note (a)

$\eta = 1.1141 - 4.449\ 10^{-3}T + 4.46\ 10^{-6}T^2$ $\qquad\qquad$ Note (a)

T	ρ	κ	Λ	η	$\Lambda\eta$
K	g cm^{-3}	S cm^{-1}	S cm^2 mol^{-1}	Pa s	
483	0.9428	0.02253	8.17	5.725 10^{-3}	0.0468
488	0.9398	0.02441	8.88	5.133 10^{-3}	0.0456
493	0.9369	0.02608	9.52	4.764 10^{-3}	0.0454

$\Lambda = 15189.2 \exp\{- \dfrac{30222}{R\,T}\}$ $\qquad\qquad\qquad r^2 = 0.9977$ $\qquad$ Note (a)

$\eta = 6.588\ 10^{-7} \exp\{\dfrac{36401}{R\,T}\}$ $\qquad\qquad\qquad r^2 = 0.9895$ $\qquad$ Note (a)

$\eta = \dfrac{1.8979\ 10^{-4}}{v - 1.0273}$ $\qquad\qquad\qquad\qquad\qquad\qquad$ Note (a)

Ref. 1: Walden & Birr (1932)   data in the table given according to Ref. 1

Note (a): equations obtained by least-squares analysis of the data given in Ref. 1

TABLE A25. Tetrapropylammonium fluorophosphate, $(C_3H_7)_4N^+PF_6^-$

$T_F = 510$ K $\qquad\qquad\qquad\qquad\qquad\qquad\qquad\qquad\qquad$ Ref. 1

$\rho = 1.2433 - 3.224\ 10^{-4}T$ $\qquad\qquad\qquad\qquad\qquad\qquad$ Note (a)

$$\kappa = 0.47927 - 2.4082 \ 10^{-3}T + 3.1493 \ 10^{-6}T^2 \qquad \text{Note (b)}$$
$$\eta = 0.08254 - 2.703 \ 10^{-4}T + 2.2571 \ 10^{-7}T^2 \qquad \text{Note (b)}$$

T	ρ	κ	Λ	η	$\Lambda\eta$
K	g cm^{-3}	S cm^{-1}	S cm^2 mol^{-1}	Pa s	
513	1.0779	0.07264	19.77	3.247 10^{-3}	0.0642
523	1.0746	0.08119	22.16	2.876 10^{-3}	0.0637
533	1.0714	0.09036	24.74	2.560 10^{-3}	0.0633
543	1.0682	0.1002	27.50	2.288 10^{-3}	0.0629
553	1.0650	0.1106	30.46	2.053 10^{-3}	0.0625

$$\Lambda = 7821.2 \ \exp\{-\frac{25514}{R\,T}\} \qquad \text{Note (c)}$$

$$\eta = 5.7515 \ 10^{-6} \ \exp\{\frac{27032}{R\,T}\} \qquad \text{Note (d)}$$

$$\eta = \frac{6.4947 \ 10^{-5}}{v - 0.9078} \qquad \text{Note (d)}$$

Ref. 1: Lind, Abdel-Rehim & Rudich (1966)   data in the table given according to Ref. 1

Note (a): equation given in Ref. 1, valid in the temperature range 513-545 K; standard devi-
        ation : 0.0001

Note (b): equations obtained by least-squares analysis from the data given in Ref. 1

Note (c): equation obtained using $\ln \Lambda = 8.9646 - 3068.9/T$ (standard deviation: 0.0005), as
        given in Ref. 1

Note (d): equations obtained using $\ln \eta \ 100 = -5.1583 + 3251.4/T$ (standard deviation: 0.001),
        as given in Ref. 1

TABLE A26. Tetrabutylammonium fluorophosphate, $(C_4H_9)_4N^+PF_6^-$

$$T_F = 510 \text{ K} \qquad \text{Ref. 1}$$
$$\rho = 1.3252 - 6.557 \ 10^{-4}T \qquad \text{Note (a)}$$
$$\kappa = 0.17896 - 9.3948 \ 10^{-4}T + 1.285 \ 10^{-6}T^2 \qquad \text{Note (b)}$$
$$\eta = 0.095175 - 3.1202 \ 10^{-4}T + 2.6 \ 10^{-7}T^2 \qquad \text{Note (b)}$$

T	ρ	κ	Λ	η	$\Lambda\eta$
K	g cm^{-3}	S cm^{-1}	S cm^2 mol^{-1}	Pa s	
523	0.9822	0.03910	15.42	3.107 10^{-3}	0.0479
533	0.9756	0.04328	17.19	2.730 10^{-3}	0.0469
543	0.9691	0.04771	19.07	2.411 10^{-3}	0.0460
553	0.9625	0.05240	21.09	2.138 10^{-3}	0.0451

$$\Lambda = 4950.2 \ \exp\{-\frac{25101}{R\,T}\} \qquad \text{Note (c)}$$

$$\eta = 3.1726 \ 10^{-6} \ \exp\{\frac{29954}{R\,T}\} \qquad \text{Note (d)}$$

$$\eta = \frac{1.4383 \ 10^{-4}}{v - 0.9719} \qquad \text{Note (b)}$$

Ref. 1: Lind, Abdel-Rehim & Rudich (1966)   data in the table given according to Ref. 1

Note (a): equation given in Ref. 1, valid in the temperature range 529-548 K; standard deviation: 0.0003

Note (b): equations obtained by least-squares analysis of the data given in Ref. 1

Note (c): equation obtained using $\ln \Lambda = 8.5072 - 3019.2/T$ (standard deviation: 0.003), as given in Ref. 1

Note (d): equation obtained from $\ln \eta\ 100 = -5.7532 + 3602.95/T$ (standard deviation: 0.002), as given in Ref. 1

TABLE B. Transport properties of molten salts with organic anion at the characteristic temperature, T_{ch}

Salt	T_{ch} K	Λ S cm^2 mol^{-1}	$\eta\ 10^3$ Pa s	$\Lambda\eta$	E_{Λ} J mol^{-1}	E_{η} J mol^{-1}	$B\ 10^4$	b cm^3 g^{-1}
$CH_3COO^-Li^+$	583	2.689	40.0	0.107	62282	44925	13.0	0.853
$CH_3COO^-Na^+$	628	14.41	4.89	0.0705	22858	17360	2.57	0.742
$CH_3COO^-K^+$	603	16.95	3.50	0.0593	23889	22706	1.62	0.683
$CH_3COO^-Rb^+$	533	9.21	6.80	0.0626	22964	20451	1.16	0.492
$CH_3COO^-Cs^+$	493	7.53	5.80	0.0552	21714	23148	0.80	0.406
$C_6H_5SO_3^-K^+$	673	9.29	7.40	0.0697	37921	105855	1.61	0.679
$C_6H_5SO_3^-Rb^+$	633	8.85	5.30	0.0469	24617	26013	1.19	0.586
$C_6H_5SO_3^-Cs^+$	533	4.15	13.0	0.0540	26038	29189	1.17	0.502

TABLE B1. Lithium ethanoate, $CH_3COO^-Li^+$

$T_F = 553$ K   Ref. 1

$\rho = 1.5581 - 7.414\ 10^{-4}T$ $r^2 = 0.964$ Note (a)

$\kappa = -0.4204 + 8\ 10^{-4}T$ Note (a)

$\eta = 6.8129 - 0.02269T + 1.9\ 10^{-5}T^2$ Note (a)

T K	ρ g cm^{-3}	κ S cm^{-1}	Λ S cm^2 mol^{-1}	η Pa s	$\Lambda\eta$
553	1.1460	0.022	1.266		
563	1.1405	0.030	1.735	0.0578	0.100
573	1.1345	0.038	2.209	0.0454	0.100
583	1.1285	0.046	2.689	0.0400	0.107
593	1.1210	0.054	3.178	0.0352	0.111

$\Lambda = 1010887\ \exp\{-\dfrac{62282}{R\ T}\}$ $r^2 = 0.988$ Note (a)

$$\eta = 3.813 \ 10^{-6} \ \exp\{\frac{44915}{R\ T}\} \qquad\qquad r^2 = 0.975 \qquad\qquad \text{Note (a)}$$

$$\eta = \frac{1.3087 \ 10^{-3}}{v - 0.8537} \qquad\qquad\qquad\qquad\qquad\qquad \text{Note (a)}$$

Ref. 1: Meisel, Halmos & Pungor (1973) } data in the table given according to Ref.s 1, 2
Ref. 2: Meisel (private communication)

Note (a): equations obtained by least-squares analysis of the data given in Ref. 1

TABLE B2. Sodium ethanoate, $CH_3COO^-Na^+$

T_F= 602 K   Ref. 1
$\rho = 1.6529 - 6.2971 \ 10^{-4}T$ $\qquad\qquad\qquad r^2 = 0.999 \qquad$ Note (a)
$\kappa = -0.6449 + 1.3788 \ 10^{-3}T$ $\qquad\qquad\qquad\qquad\qquad$ Note (b)
$\eta = 0.24579 - 7.3165 \ 10^{-4}T + 5.5419 \ 10^{-7}T^2$ $\qquad\qquad$ Note (c)

T	ρ	κ	Λ	η	$\Lambda\eta$	ρ: Ref. 1
K	g cm^{-3}	S cm^{-1}	S cm^2 mol^{-1}	Pa s		κ: Ref. 2 η: Ref. 3
603	1.2729	0.1865	12.02			
608	1.2698	0.1934	12.49			
613	1.2666	0.2003	12.97	5.52 10^{-3}	0.0716	
618	1.2634	0.2071	13.45	5.31 10^{-3}	0.0714	
623	1.2603	0.2140	13.93	5.10 10^{-3}	0.0710	
628	1.2572	0.2209	14.41	4.89 10^{-3}	0.0705	

$$\Lambda = 1149.34 \ \exp\{-\frac{22858}{R\ T}\} \qquad\qquad r^2 = 0.9998 \qquad \text{Note (b)}$$

$$\eta = 1.7843 \ 10^{-4} \ \exp\{\frac{17360}{R\ T}\} \qquad\qquad r^2 = 0.932 \qquad \text{Note (c)}$$

$$\eta = \frac{2.5792 \ 10^{-4}}{v - 0.7428} \qquad\qquad\qquad\qquad\qquad \text{Note (c)}$$

Ref. 1: Hazlewood, Rhodes & Ubbelohde (1966)

Ref. 2: Leonesi, Cingolani & Berchiesi (1973)

Ref. 3: Meisel, Halmos & Pungor (1973); the conductivity values reported by these authors, extending up to 663 K, are in good agreement with those of Ref. 2

Note (a): equation obtained by least-squares analysis of the data given in Ref. 1

Note (b): equations obtained by least-squares analysis of the data given in Ref.s 1, 2

Note (c): equations obtained by least-squares analysis of the data given in Ref. 3

TABLE B3. Potassium ethanoate, $CH_3COO^-K^+$

T_F= 577 K   Ref. 1
$\rho = 1.7694 - 6.618 \ 10^{-4}T$ $\qquad\qquad\qquad r^2 = 0.9999 \qquad$ Note (a)
$\kappa = -0.8506 + 1.8032 \ 10^{-3}T$ $\qquad\qquad\qquad\qquad\qquad$ Note (b)
$\eta = 0.10411 - 3.0466 \ 10^{-4}T + 2.286 \ 10^{-7}T^2$ $\qquad\qquad$ Note (c)

T	ρ	κ	Λ	η	$\Lambda\eta$	
K	g cm^{-3}	S cm^{-1}	S cm^2 mol^{-1}	Pa s		ρ: Ref. 1
						κ: Ref. 2
						η: Ref. 3
583	1.3837	0.2006	14.23	4.20 10^{-3}	0.0598	
593	1.3770	0.2186	15.58	3.80 10^{-3}	0.0592	
603	1.3702	0.2367	16.95	3.50 10^{-3}	0.0593	
613	1.3636	0.2547	18.33	3.24 10^{-3}	0.0594	
623	1.3570	0.2727	19.72	3.04 10^{-3}	0.0599	
633	1.3505	0.2908	21.13	2.86 10^{-3}	0.0604	
643	1.3440	0.3088	22.55	2.70 10^{-3}	0.0609	

$\Lambda = 1979.93 \exp\{-\dfrac{23889}{R\,T}\}$ $r^2 = 0.9991$ Note (b)

$\eta = 3.8141\ 10^{-5} \exp\{\dfrac{22706}{R\,T}\}$ $r^2 = 0.9948$ Note (c)

$\eta = \dfrac{1.6130\ 10^{-4}}{v - 0.6839}$ Note (c)

Ref. 1: Hazlewood, Rhodes & Ubbelohde (1966)

Ref. 2: Leonesi, Cingolani & Berchiesi (1973)

Ref. 3: Meisel, Halmos & Pungor (1973); the conductivity values reported by these authors,
     extending up to 693 K, are in good agreement with those of Ref. 2

Note (a): equation obtained by least-squares analysis of the data given in Ref. 1

Note (b): equations obtained by least-squares analysis of the data given in Ref.s 1, 2

Note (c): equations obtained by least-squares analysis of the data given in Ref. 3

TABLE B4. Rubidium ethanoate, CH$_3$COO$^-$Rb$^+$

T_F= 514 K   Ref. 1

$\rho = 2.3851 - 7.893\ 10^{-4}T$ $r^2 = 0.9963$ Note (a)

$\kappa = -0.87258 + 2.2161\ 10^{-3}T - 6.466\ 10^{-7}T^2$ Note (a)

$\eta = 0.12491 - 3.758\ 10^{-4}T + 2.887\ 10^{-7}T^2$ Note (a)

T	ρ	κ	Λ	η	$\Lambda\eta$
K	g cm^{-3}	S cm^{-1}	S cm^2 mol^{-1}	Pa s	
513	1.979	0.094	6.87		
533	1.961	0.125	9.21	0.0068	0.0626
553	1.943	0.156	11.60	0.0052	0.0603
573	1.925	0.184	13.81	0.0042	0.0580
593	1.908	0.214	16.21	0.0035	0.0567
613	1.984	0.242	18.46	0.0031	0.0572
633	1.879	0.272	20.92	0.0028	0.0586
653	1.864	0.300	23.25	0.0027	0.0628
673	1.852	0.325	25.36	0.0026	0.0659

$\Lambda = 1633.4 \exp\{-\dfrac{22964}{R\,T}\}$ $r^2 = 0.987$ Note (a)

$$\eta = 6.0581\ 10^{-5}\ \exp\{\frac{20451}{R\,T}\} \qquad\qquad r^2 = 0.951 \qquad \text{Note (a)}$$

$$\eta = \frac{1.1558\ 10^{-4}}{v - 0.4924} \qquad\qquad \text{Note (a)}$$

Ref. 1: Meisel, Halmos & Pungor (1973)   data in the table given according to Ref. 1

Note (a): equations obtained by least-squares analysis of the data given in Ref. 1

TABLE B5. Cesium ethanoate, $CH_3COO^-Cs^+$

$T_F = 470$ K   Ref. 1

$\rho = 2.8221 - 8.6428\ 10^{-4}T$ $r^2 = 0.9977$ Note (a)

$\kappa = -0.32202 + 5.2226\ 10^{-4}T + 6.548\ 10^{-7}T^2$ Note (a)

$\eta = 0.18427 - 6.1215\ 10^{-4}T + 5.164\ 10^{-7}T^2$ Note (a)

T	ρ	κ	Λ	η	$\Lambda\eta$
K	g cm^{-3}	S cm^{-1}	S cm^2 mol^{-1}	Pa s	
473	2.416	0.072	5.72	0.0106	0.0606
493	2.396	0.094	7.53	0.0077	0.0580
513	2.378	0.118	9.52	0.0058	0.0552
533	2.360	0.142	11.54	0.0046	0.0531
553	2.341	0.168	13.77	0.0039	0.0537
573	2.326	0.192	15.84	0.0033	0.0523
593	2.310	0.218	18.11	0.0030	0.0543
613	2.295	0.244	20.40	0.0028	0.0571

$$\Lambda = 1501.6\{\exp - \frac{21714}{R\,T}\} \qquad\qquad r^2 = 0.9949 \qquad \text{Note (a)}$$

$$\eta = 2.6828\ 10^{-5}\ \exp\{\frac{23148}{R\,T}\} \qquad\qquad r^2 = 0.978 \qquad \text{Note (a)}$$

$$\eta = \frac{8.009\ 10^{-5}}{v - 0.4065} \qquad\qquad \text{Note (a)}$$

Ref. 1: Meisel, Halmos & Pungor (1973)   data in the table given according to Ref. 1

Note (a): equations obtained by least-squares analysis of the data given in Ref. 1

TABLE B6. Potassium benzenesulphonate, $C_6H_5SO_3^-K^+$

$T_F = 648$ K   Ref. 1

$\rho = 2.1575 - 1.0893\ 10^{-3}T$ $r^2 = 0.9959$ Note (a)

$\kappa = -0.15281 + 6.6556\ 10^{-6}T + 4.7619\ 10^{-7}T^2$ Note (a)

$\eta = 5.0367 - 1.4876\ 10^{-2}T + 1.1\ 10^{-5}T^2$ Note (a)

T	ρ	κ	Λ	η	$\Lambda\eta$
K	g cm^{-3}	S cm^{-1}	S cm^2 mol^{-1}	Pa s	
653	1.445	0.0545	7.40	0.0132	0.0977
663	1.435	0.0610	8.34	0.0092	0.0767
673	1.425	0.0675	9.29	0.0074	0.0697
683	1.415	0.0740	10.26		

693	1.405	0.0800	11.17
703	1.390	0.0875	12.35
713	1.380	0.0940	13.36

$\Lambda = 8087.0 \exp\{-\dfrac{37921}{R\,T}\}$ $r^2 = 0.9985$ Note (a)

$\eta = 4.3997 \ 10^{-11} \exp\{\dfrac{105855}{R\,T}\}$ $r^2 = 0.982$ Note (a)

$\eta = \dfrac{1.6104 \ 10^{-4}}{v - 0.6797}$ Note (a)

Ref. 1: Meisel, Halmos & Pungor (1973)   data in the table given according to Ref. 1
Note (a): equations obtained by least-squares analysis of the data given in Ref. 1

TABLE B7. Rubidium benzenesulphonate, $C_6H_5SO_3^-Rb^+$

$T_F = 598$ K   Ref. 1
$\rho = 2.0476 - 6.3828 \ 10^{-4}T$ $r^2 = 0.9828$ Note (a)
$\kappa = -0.018705 + 2.8711 \ 10^{-5}T + 1.6369 \ 10^{-8}T^2$ Note (a)
$\eta = 0.23359 - 6.5778 \ 10^{-4}T + 4.6981 \ 10^{-7}T^2$ Note (a)

T	ρ	κ	Λ	η	$\Lambda\eta$
K	g cm^{-3}	S cm^{-1}	S cm^2 mol^{-1}	Pa s	
613	1.655	0.0515	7.55	0.0070	0.0529
633	1.645	0.0600	8.85	0.0053	0.0469
653	1.630	0.0695	10.34	0.0044	0.0455
673	1.620	0.0790	11.83	0.0038	0.0450
693	1.605	0.0900	13.60	0.0034	0.0462
713	1.585	0.100	15.30		
733	1.575	0.110	16.94		
753	1.573	0.118	18.20		

$\Lambda = 957.64 \exp\{-\dfrac{24617}{R\,T}\}$ $r^2 = 0.9973$ Note (a)

$\eta = 3.8998 \ 10^{-5} \exp\{\dfrac{26013}{R\,T}\}$ $r^2 = 0.9267$ Note (a)

$\eta = \dfrac{1.1900 \ 10^{-4}}{v - 0.5866}$ Note (a)

Ref. 1: Meisel, Halmos & Pungor (1973)   data in the table given according to Ref. 1
Note (a): equations obtained by least-squares analysis of the data given in Ref. 1

TABLE B8. Cesium benzenesulphonate, $C_6H_5SO_3^-Cs^+$

$T_F = 505$ K   Ref. 1
$\rho = 2.3740 - 7.8363 \ 10^{-4}T$ $r^2 = 0.9993$ Note (a)
$\kappa = -0.09136 + 1.9252 \ 10^{-5}T - 3.8352 \ 10^{-7}T^2$ Note (a)
$\eta = 0.20513 - 5.9279 \ 10^{-4}T + 4.337 \ 10^{-7}T^2$ Note (a)

T / K	ρ / g cm^{-3}	κ / S cm^{-1}	Λ / S cm^2 mol^{-1}	η / Pa s	$\Lambda\eta$
533	1.955	0.0280	4.15	0.0130	0.0540
553	1.940	0.0365	5.45	0.0096	0.0523
573	1.925	0.0455	6.86	0.0075	0.0515
593	1.910	0.0545	8.27	0.0057	0.0471
613	1.895	0.0645	9.87	0.0046	0.0454
633	1.880	0.0755	11.64	0.0039	0.0454
653	1.863	0.0845	13.15	0.0034	0.0447
673	1.845	0.0950	14.93	0.0030	0.0448
693	1.830	0.1060	16.80	0.0027	0.0454
713	1.815	0.1175	18.77	0.0025	0.0464

$$\Lambda = 1578.1 \, \exp\left\{-\frac{26038}{R\,T}\right\} \qquad\qquad r^2 = 0.9952 \qquad \text{Note (a)}$$

$$\eta = 1.6345 \; 10^{-5} \, \exp\left\{\frac{29189}{R\,T}\right\} \qquad\qquad r^2 = 0.985 \qquad \text{Note (a)}$$

$$\eta = \frac{1.1741 \; 10^{-4}}{v - 0.5028} \qquad\qquad\qquad\qquad\qquad\qquad\quad \text{Note (a)}$$

Ref. 1: Meisel, Halmos & Pungor (1973)   data in the table given according to Ref. 1

Note (a): equations obtained by least-squares analysis of the data given in Ref. 1

TABLE C. Transport properties of molten salts with organic cation and anion at the characteristic temperature, T_{ch}

Salt	T_{ch} / K	Λ / S cm^2 mol^{-1}	$\eta \; 10^3$ / Pa s	$\Lambda\eta$	E_Λ / J mol^{-1}	E_η / J mol^{-1}	$B \; 10^4$	b / cm^3 g^{-1}
$C_2H_5NH_3^+Pic^-$	453	3.012	14.4	0.0435	31288	32616	3.05	0.724
$C_3H_7NH_3^+Pic^-$	433	1.690	22.7	0.0384	39449	48130	3.30	0.734
$C_4H_9NH_3^+Pic^-$	423	1.167	28.3	0.0331				
$i.C_4H_9NH_3^+Pic^-$	423	1.111	37.1	0.0412				
$C_5H_{11}NH_3^+Pic^-$	418	0.915	35.1	0.0321	46107	52684		
$i.C_5H_{11}NH_3^+Pic^-$	423	0.889	37.6	0.0335	47817	56173	6.65	0.772
$C_7H_{15}NH_3^+Pic^-$	418	0.742	35.5	0.0264	46058	50962	5.91	0.814
$C_{16}H_{33}NH_3^+Pic^-$	403	0.225	101.0	0.0227	44975	53995	1.03	0.929
$(CH_3)_2NH_2^+Pic^-$	443	5.138	10.2	0.0526	24367	35206	1.32	0.719
$(CH_3)(C_2H_5)NH_2^+Pic^-$	393	1.672	33.5	0.0560	32445	38854	3.60	0.722
$(C_2H_5)_2NH_2^+Pic^-$	363	0.676	45.2	0.0306	31806	31259	3.13	0.743
$(C_3H_7)_2NH_2^+Pic^-$	393	1.209	25.5	0.0308	31969	35239	3.87	0.794
$(C_4H_9)_2NH_2^+Pic^-$	393	1.900	29.9	0.0269	32731	37731	4.76	0.852
$(i.C_5H_{11})_2NH_2^+Pic^-$	383	0.443	48.0	0.0213	36652	39624	5.55	0.863
$(C_{16}H_{33})_2NH_2^+Pic^-$	343	0.0166	111.0	0.0018	39792	36729	8.00	1.041

$(C_2H_5)_3NH^+Pic^-$	468	4.457	5.13	0.0229	32700	21633	1.94	0.797
$(C_3H_7)_3NH^+Pic^-$	408	0.904	10.6	0.0096	36591	32660	2.40	0.846
$(C_4H_9)_3NH^+Pic^-$	393	0.359	15.0	0.0054	40357	31655	2.49	0.880
$(i.C_5H_{11})_3NH^+Pic^-$	418	0.548	10.6	0.0058	42417	36755	2.53	0.934
$(C_3H_7)_4N^+Pic^-$	413	4.219	10.9	0.0462	27150	29730	1.97	0.870
$(C_4H_9)_4N^+Pic^-$	383	1.583	29.1	0.0460	30048	31814	2.65	0.905
$(i.C_5H_{11})_4N^+Pic^-$	373	0.654	81.2	0.0531	35802	38747	4.95	0.951
$(CH_3)_2(C_3H_7)_2N^+Pic^-$	383	2.169	25.6	0.0543	28374	31869	2.52	0.818
$(C_3H_7)(C_2H_5)_3N^+Pic^-$	433	6.331	9.57	0.0501	24572	27302	1.92	0.830
$(CH_3)(C_3H_7)_3N^+Pic^-$	373	1.475	3.41	0.0503	31041	33092	2.66	0.840
$(C_2H_5)_2(C_3H_7)_2N^+Pic^-$	373	1.681	3.07	0.0516	30815	32681	2.98	0.818
$(C_2H_5)(C_3H_7)_3N^+Pic^-$	403	3.43	13.8	0.0476	28791	31507	2.21	0.853

TABLE C1. Ethylammonium picrate, $C_2H_5NH_3^+ C_6H_2(NO_2)_3O^-$

$T_F = 443$ K		Ref. 1
$\rho = 1.6450 - 6.7 \ 10^{-4}T$	$r^2 = 0.9993$	Note (a)
$\kappa = 0.14411 - 8.292 \ 10^{-4}T + 1.2 \ 10^{-6}T^2$		Note (b)
$\eta = 0.63742 - 2.4624 \ 10^{-3}T + 2.4 \ 10^{-6}T^2$		Note (c)

T	ρ	κ	Λ	η	$\Lambda\eta$	ρ: Ref. 1
K	g cm^{-3}	S cm^{-1}	S cm^2 mol^{-1}	Pa s		κ: Ref. 2 η: Ref. 3
443	1.3483	0.01228	2.497	0.01757	0.0439	
448	1.3448	o.01348	2.748	0.01595	0.0438	
453	1.3416	0.01474	3.012	0.01445	0.0435	

$\Lambda = 12211 \ \exp\{-\dfrac{31288}{R\,T}\}$	$r^2 = 0.9999$	Note (b)
$\eta = 2.5075 \ 10^{-6} \ \exp\{\dfrac{32616}{R\,T}\}$	$r^2 = 0.9998$	Note (c)
$\eta = \dfrac{3.0462 \ 10^{-4}}{v - 0.7243}$		Note (c)

Ref. 1: Walden, Ulich & Birr (1927a)

Ref. 2: Walden, Ulich & Birr (1927b)

Ref. 3: Walden, Ulich & Birr (1927c)

Note (a): equation calculated from the data given in Ref. 1

Note (b): equations obtained by least-squares analysis of the data given in Ref. 2

Note (c): equations obtained by least-squares analysis of the data given in Ref. 3

TABLE C2. Propylammonium picrate, $C_3H_7NH_3^+ C_6H_2(NO_2)_3O^-$

$T_F = 417$ K		Ref. 1
$\rho = 1.6216 - 6.6 \ 10^{-4}T$	$r^2 = 0.9998$	Note (a)
$\kappa = 0.11636 - 6.8364 \ 10^{-4}T + 1 \ 10^{-6}T^2$		Note (b)
$\eta = 4.0273 - 1.7865 \ 10^{-2}T + 1.99 \ 10^{-5}T^2$		Note (c)

T	ρ	κ	Λ	η	$\Lambda\eta$	
$\overline{}$						ρ: Ref. 1
						κ: Ref. 2
K	g cm^{-3}	S cm^{-1}	S cm^2 mol^{-1}	Pa s		η: Ref. 3
418	1.3457	0.005326	1.141	0.03668	0.0419	
423	1.3425	0.006106	1.311	0.0310	0.0406	
428	1.3392	0.006953	1.496	0.0264	0.0395	
433	1.3358	0.007833	1.690	0.0227	0.0384	

$\Lambda = 97306.4 \, \exp\{-\dfrac{39449}{R\,T}\}$ $\qquad\qquad$ $r^2 = 0.9996$ $\qquad$ Note (b)

$\eta = 3.54 \; 10^{-8} \, \exp\{\dfrac{48130}{R\,T}\}$ $\qquad\qquad$ $r^2 = 0.9997$ $\qquad$ Note (c)

$\eta = \dfrac{3.3053 \; 10^{-4}}{v - 0.7341}$ $\qquad\qquad\qquad\qquad\qquad\qquad\qquad$ Note (c)

Ref. 1: Walden, Ulich & Birr (1927a)

Ref. 2: Walden, Ulich & Birr (1927b)

Ref. 3: Walden, Ulich & Birr (1927c)

Note (a): equation obtained by least-squares analysis of the data given in Ref. 1

Note (b): equations obtained by least-squares analysis of the data given in Ref. 2

Note (c): equations obtained by least-squares analysis of the data given in Ref. 3

TABLE C3. Butylammonium picrate, $C_4H_9NH_3^+ \; C_6H_2(NO_2)_3O^-$

$\quad$ $T_F = 423$ K $\qquad\qquad\qquad\qquad\qquad\qquad\qquad\qquad\qquad\qquad$ Ref. 1

$\quad$ $\rho = 1.3023$ g cm^{-3} $\qquad$ at 423 K $\qquad\qquad\qquad\qquad\qquad$ Ref. 1

$\quad$ $\kappa = 0.005029$ S cm^{-1} $\qquad$ at 423 K $\qquad\qquad\qquad\qquad$ Ref. 2

$\quad$ $\Lambda = 1.167$ S cm^2 mol^{-1} $\qquad$ at 423 K $\qquad\qquad\qquad\qquad$ Ref. 2

$\quad$ $\eta = 0.02834$ Pa s $\qquad$ at 423 K $\qquad\qquad\qquad\qquad\qquad$ Ref. 3

$\quad$ $\Lambda\eta = 0.0331$ $\qquad\qquad$ at 423 K

Ref. 1: Walden, Ulich & Birr (1927a)

Ref. 2: Walden, Ulich & Birr (1927b)

Ref. 3: Walden, Ulich & Birr (1927c)

TABLE C4. *Iso*butylammonium picrate, $i.C_4H_9NH_3^+ \; C_6H_2(NO_2)_3O^-$

$\quad$ $T_F = 423$ K $\qquad\qquad\qquad\qquad\qquad\qquad\qquad\qquad\qquad\qquad$ Ref. 1

$\quad$ $\rho = 1.3104$ g cm^{-3} $\qquad$ at 423 K $\qquad\qquad\qquad\qquad\qquad$ Ref. 1

$\quad$ $\kappa = 0.00482$ S cm^{-1} $\qquad$ at 423 K $\qquad\qquad\qquad\qquad$ Ref. 2

$\quad$ $\Lambda = 1.111$ S cm^2 mol^{-1} $\qquad$ at 423 K $\qquad\qquad\qquad\qquad$ Ref. 2

$\quad$ $\eta = 0.0371$ Pa s $\qquad$ at 423 K $\qquad\qquad\qquad\qquad\qquad$ Ref. 3

$\quad$ $\Lambda\eta = 0.0412$ $\qquad\qquad$ at 423 K

Ref. 1: Walden, Ulich & Birr (1927a)

Ref. 2: Walden, Ulich & Birr (1927b)

Ref. 3: Walden, Ulich & Birr (1927c)

TABLE C5. Amylammonium picrate, $C_5H_{11}NH_3^+$ $C_6H_2(NO_2)_3O^-$

T_F= 417 K   Ref. 1

$\rho = 1.5271 - 6.4\ 10^{-4}T$ Note (a)

$\kappa = -0.04726 + 1.218\ 10^{-4}T$ Note (b)

$\eta = 0.51575 - 1.15\ 10^{-3}T$ Note (c)

T	ρ	κ	Λ	η	$\Lambda\eta$	
K	g cm^{-3}	S cm^{-1}	S cm^2 mol^{-1}	Pa s		ρ: Ref. 1 κ: Ref. 2 η: Ref. 3
418	1.2596	0.003643	0.9147	0.03505	0.0321	
423	1.2564	0.004252	1.070	0.02930	0.0314	

$\Lambda = 528409\ \exp\{-\dfrac{46107}{R\ T}\}$ Note (b)

$\eta = 9.1\ 10^{-9}\exp\{\dfrac{52684}{R\ T}\}$ Note (c)

Ref. 1: Walden, Ulich & Birr (1927a)

Ref. 2: Walden, Ulich & Birr (1927b)

Ref. 3: Walden, Ulich & Birr (1927c)

Note (a): equation obtained from the two only data given in Ref. 1

Note (b): equations obtained from the two only data given in Ref. 2

Note (c): equations obtained from the two only data given in Ref. 3

TABLE C6. *Iso*amylammonium picrate, $i.C_5H_{11}NH_3^+$ $C_6H_2(NO_2)_3O^-$

T_F= 406 K   Ref. 1

$\rho = 1.6228 - 8.46\ 10^{-4}T$ $r^2 = 0.9999$ Note (a)

$\kappa = 0.16319\ - 8.6383\ 10^{-4}T + 1.15\ 10^{-6}T^2$ Note (b)

$\eta = 13.1339 - 6.0993\ 10^{-2}T + 7.1\ 10^{-5}T^2$ Note (c)

T	ρ	κ	Λ	η	$\Lambda\eta$	
K	g cm^{-3}	S cm^{-1}	S cm^2 mol^{-1}	Pa s		ρ: Ref. 1 κ: Ref. 2 η: Ref. 3
408	1.2776	0.002180	0.5396	0.06761	0.0365	
413	1.2734	0.002583	0.6415	0.05389	0.0346	
418	1.2692	0.003040	0.7575	0.04424	0.0335	
423	1.2649	0.003558	0.8896	0.03762	0.0335	

$\Lambda = 715103\ \exp\{-\dfrac{47817}{R\ T}\}$ $r^2 = 0.9999$ Note (b)

$\eta = 4.3\ 10^{-9}\exp\{\dfrac{56173}{R\ T}\}$ $r^2 = 0.9960$ Note (c)

$\eta = \dfrac{6.6528\ 10^{-4}}{v - 0.7729}$ Note (c)

Ref.s 1, 2, 3: Walden, Ulich & Birr (1927 a, b, c)

Note (a): equation obtained by least-squares analysis of the data given in Ref. 1

Note (b): equations obtained by least-squares analysis of the data given in Ref. 2

Note (c): equations obtained by least-squares analysis of the data given in Ref. 3

TABLE C7. Heptylammonium picrate, $C_7H_{15}NH_3^+ \; C_6H_2(NO_2)_3O^-$

$T_F = 397$ K   Ref. 1

$\rho = 1.4765 - 6.548 \; 10^{-4}T$ $r^2 = 0.9999$ Note (a)

$\kappa = 0.095795 - 5.2332 \; 10^{-4}T + 7.1859 \; 10^{-7}T^2$ Note (b)

$\eta = 6.8350 - 3.1325 \; 10^{-2}T + 3.6023 \; 10^{-5}T^2$ Note (c)

T	ρ	κ	Λ	η	$\Lambda\eta$	ρ: Ref. 1
K	g cm^{-3}	S cm^{-1}	S cm^2 mol^{-1}	Pa s		κ: Ref. 2 η: Ref. 3
398	1.2160	0.001335	0.3780	0.07386	0.0279	
403	1.2126	0.001606	0.4560	0.06108	0.0279	
408	1.2093	0.001902	0.5415	0.05097	0.0276	
413	1.2061	0.002228	0.6360	0.04178	0.0266	
418	1.2028	0.002593	0.7422	0.03554	0.0264	
423	1.1996	0.003010	0.8639	0.02975	0.0257	

$\Lambda = 423323 \exp\{-\dfrac{46058}{R\,T}\}$ $r^2 = 0.9994$ Note (b)

$\eta = 1.51 \; 10^{-8} \exp\{\dfrac{50962}{R\,T}\}$ $r^2 = 0.9996$ Note (c)

$\eta = \dfrac{5.9134 \; 10^{-4}}{\upsilon \; -0.8147}$ Note (c)

Ref. 1: Walden, Ulich & Birr (1927a)

Ref. 2: Walden, Ulich & Birr (1927b)

Ref. 3: Walden, Ulich & Birr (1927c)

Note (a): equation obtained by least-squares analysis of the data given in Ref. 1

Note (b): equations obtained by least-squares analysis of the data given in Ref. 2

Note (c): equations obtained by least-squares analysis of the data given in Ref. 3

TABLE C8. Hexadecylammonium picrate, $C_{16}H_{33}NH_3^+ \; C_6H_2(NO_2)_3O^-$

$T_F = 388$ K   Ref. 1

$\rho = 1.3311 - 6.6475 \; 10^{-4}T$ $r^2 = 0.9998$ Note (a)

$\kappa = 0.0010733 - 2.4439 \; 10^{-5}T + 5.7498 \; 10^{-8}T^2$ Note (a)

$\eta = 12.1880 - 5.544 \; 10^{-2}T + 6.3187 \; 10^{-5}T^2$ Note (a)

T	ρ	κ	Λ	η	$\Lambda\eta$
K	g cm^{-3}	S cm^{-1}	S cm^2 mol^{-1}	Pa s	
393	1.0697	3.507 10^{-4}	0.1542	0.1637	0.0252
403	1.0632	5.098 10^{-4}	0.2256	0.1008	0.0227
413	1.0566	7.125 10^{-4}	0.3173	0.06792	0.0216
423	1.0498	9.645 10^{-4}	0.4323	0.04613	0.0199
433	1.0433	1.243 10^{-3}	0.5606	0.03375	0.0189
443	1.0368	1.589 10^{-3}	0.7211	0.02522	0.0182

$\Lambda = 150013 \exp\{-\dfrac{44975}{R\,T}\}$ $r^2 = 0.9983$ Note (a)

$\eta = 1.04 \; 10^{-8} \exp\{\dfrac{53995}{R\,T}\}$ $r^2 = 0.9967$ Note (a)

$$\eta = \frac{1.028 \ 10^{-3}}{v - 0.9290} \qquad \text{Note (a)}$$

Ref. 1: Walden & Birr (1932)   data in the table given according to Ref. 1

Note (a): equations obtained by least-squares analysis of the data given in Ref. 1

TABLE C9. Dimethylammonium picrate, $(CH_3)_2NH_2^+ \ C_6H_2(NO_2)_3O^-$

$T_F = 433$ K   Ref. 1

$\rho = 1.6645 - 6.8 \ 10^{-4}T$ $r^2 = 1.000$ Note (a)

$\kappa = 0.21533 - 1.2258 \ 10^{-3}T + 1.8 \ 10^{-6}T^2$ Note (b)

$\eta = 0.23740 - 7.786 \ 10^{-4}T + 6 \ 10^{-7}T^2$ Note (c)

T	ρ	κ	Λ	η	$\Lambda\eta$	
K	g cm$^-$	S cm$^-$	S cm mol$^-$	Pa s		ρ: Ref. 1 κ: Ref. 2 η: Ref. 3
433	1.3701	0.02204	4.410	0.01277	0.0563	
438	1.3667	0.02375	4.764	0.01149	0.0547	
443	1.3633	0.02555	5.138	0.01024	0.0526	

$$\Lambda = 3836.82 \ \exp\{-\frac{24367}{R \ T}\} \qquad r^2 = 0.9999 \qquad \text{Note (b)}$$

$$\eta = 7.244 \ 10^{-7} \ \exp\{\frac{35206}{R \ T}\} \qquad r^2 = 0.9990 \qquad \text{Note (c)}$$

$$\eta = \frac{1.323 \ 10^{-4}}{v - 0.7197} \qquad \text{Note (c)}$$

Ref. 1: Walden, Ulich & Birr (1927a)

Ref. 2: Walden, Ulich & Birr (1927b)

Ref. 3: Walden, Ulich & Birr (1927c)

Note (a): equation obtained by least-squares analysis of the data given in Ref. 1

Note (b): equations obtained by least-squares analysis of the data given in Ref. 2

Note (c): equations obtained by least-squares analysis of the data given in Ref. 3

TABLE C10. Methylethylammonium picrate, $(CH_3)(C_2H_5)NH_2^+ \ C_6H_2(NO_2)_3O^-$

$T_F = 370$ K   Ref. 1

$\rho = 1.6258 - 6.738 \ 10^{-4}T$ $r^2 = 0.9999$ Note (a)

$\kappa = 0.16812 - 1.0233 \ 10^{-3}T + 1.5666 \ 10^{-6}T^2$ Note (b)

$\eta = 2.6201 - 1.1988 \ 10^{-2}T + 1.3766 \ 10^{-5}T^2$ Note (c)

T	ρ	κ	Λ	η	$\Lambda\eta$
K	g cm^{-3}	S cm^{-1}	S cm^2 mol^{-1}	Pa s	
373	1.3748	4.388 10^{-3}	0.9198	0.06622	0.0609
383	1.3678	5.981 10^{-3}	1.260	0.04547	0.0573
393	1.3609	7.896 10^{-3}	1.672	0.03348	0.0560
403	1.3541	1.006 10^{-2}	2.141	0.02396	0.0513
413	1.3475	1.269 10^{-2}	2.714	0.01815	0.0493
423	1.3408	1.563 10^{-2}	3.359	0.01413	0.0475

| 433 | 1.3342 | 1.876 10^{-2} | 4.052 | 0.01126 | 0.0456 |
| 443 | 1.3275 | 2.217 10^{-2} | 4.813 | 0.009198 | 0.0443 |

$$\Lambda = 33539.4 \; \exp\{-\frac{32445}{R\,T}\} \qquad\qquad r^2 = 0.9976 \qquad \text{Note (b)}$$

$$\eta = 2.291 \; 10^{-7} \; \exp\{\frac{38859}{R\,T}\} \qquad\qquad r^2 = 0.9976 \qquad \text{Note (c)}$$

$$\eta = \frac{3.6051 \; 10^{-4}}{v - 0.7228} \qquad\qquad\qquad\qquad\qquad\qquad \text{Note (c)}$$

Ref. 1: Walden, Ulich & Birr (1927a)

Ref. 2: Walden, Ulich & Birr (1927b)

Ref. 3: Walden, Ulich & Birr (1927c)

Note (a): equation obtained by least-squares analysis of the data given in Ref. 1

Note (b): equations obtained by least-squares analysis of the data given in Ref. 2

Note (c): equations obtained by least-squares analysis of the data given in Ref. 3

TABLE C11. Diethylammonium picrate, $(C_2H_5)_2NH_2^+ \; C_6H_2(NO_2)_3O^-$

$$T_F = 348 \text{ K} \qquad\qquad\qquad\qquad\qquad\qquad\qquad\qquad \text{Ref. 1}$$

$$\rho = 1.5823 - 6.886 \; 10^{-4}T \qquad\qquad r^2 = 0.9997 \qquad \text{Note (a)}$$

$$\kappa = 0.027348 - 2.7934 \; 10^{-4}T + 5.8411 \; 10^{-7}T^2 \qquad \text{Note (a)}$$

$$\eta = 1.44797 - 6.5806 \; 10^{-3}T + 7.501 \; 10^{-6}T^2 \qquad \text{Note (b)}$$

T	ρ	κ	Λ	η	$\Lambda\eta$	ρ: Ref. 1
K	g cm^{-3}	S cm^{-1}	S cm^2 mol^{-1}	Pa s		κ: Ref. 1
						η: Ref. 2
353	1.3399	0.002053	0.4630	0.06671	0.0309	
363	1.3326	0.002982	0.6763	0.04519	0.0306	
373	1.3254	0.004207	0.9593	0.03203	0.0307	
383	1.3184	0.005625	1.289	0.02377	0.0307	
393	1.3113	0.007291	1.680	0.01831	0.0308	
403	1.3043	0.009250	2.143	0.01445	0.0310	
413	1.2975	0.01139	2.653	0.01163	0.0309	
423	1.2907	0.01405	3.290	0.009528	0.0313	
433	1.2839	0.01671	3.933	0.008005	0.0315	
443	1.2773	0.01969	4.659	0.006910	0.0322	
453	1.2706	0.02275	5.439	0.005937	0.0323	
463	1.2641	0.02599	6.214	0.005045	0.0313	

$$\Lambda = 26591 \; \exp\{-\frac{31806}{R\,T}\} \qquad\qquad r^2 = 0.9941 \qquad \text{Note (a)}$$

$$\eta = 1.379 \; 10^{-6} \; \exp\{\frac{31259}{R\,T}\} \qquad\qquad r^2 = 0.9928 \qquad \text{Note (b)}$$

$$\eta = \frac{3.1328 \; 10^{-4}}{v - 0.7435} \qquad\qquad\qquad\qquad\qquad\qquad \text{Note (b)}$$

Ref. 1: Walden & Birr (1932)

Ref. 2: Walden, Ulich & Birr (1927c)

Note (a): equations obtained by least-squares analysis of the data given in Ref. 1

Note (b): equations obtained by least-squares analysis of the data given in Ref. 2

TABLE C12. Dipropylammonium picrate, $(C_3H_7)_2NH_2^+ C_6H_2(NO_2)_3O^-$

T_F= 372 K		Ref. 1
$\rho = 1.4917 - 6.6815 \ 10^{-4}T$	$r^2 = 0.9880$	Note (a)
$\kappa = 0.112846 - 6.7248 \ 10^{-4}T + 1.0099 \ 10^{-6}T^2$		Note (b)
$\eta = 1.45447 - 6.4603 \ 10^{-3}T + 7.2031 \ 10^{-6}T^2$		Note (c)

T	ρ	κ	Λ	η	$\Lambda\eta$	
K	g cm^{-3}	S cm^{-1}	S cm^2 mol^{-1}	Pa s		ρ: Ref. 1 κ: Ref. 2 η: Ref. 3
373	1.2472	0.002540	0.6726	0.05001	0.0336	
383	1.2401	0.003454	0.9199	0.03502	0.0322	
393	1.2330	0.004515	1.209	0.02547	0.0308	
403	1.2261	0.005793	1.560	0.01962	0.0306	
413	1.2192	0.007356	1.992	0.01463	0.0291	
423	1.2124	0.009122	2.485	0.01173	0.0291	
433	1.2058	0.01106	3.029	0.009386	0.0284	
443	1.1990	0.01314	3.619	0.007685	0.0278	
453	1.1925	0.01547	4.284	0.006397	0.0274	
463	1.1860	0.01795	4.998	0.005480	0.0274	

$\Lambda = 21243.4 \exp\{-\dfrac{31969}{R\,T}\}$	$r^2 = 0.9976$	Note (b)
$\eta = 5.412 \ 10^{-7} \exp\{\dfrac{35239}{R\,T}\}$	$r^2 = 0.9966$	Note (c)
$\eta = \dfrac{3.874 \ 10^{-4}}{v - 0.7946}$		Note (c)

Ref. 1: Walden, Ulich & Birr (1927a)

Ref. 2: Walden, Ulich & Birr (1927b)

Ref. 3: Walden, Ulich & Birr (1927c)

Note (a): equation obtained by least-squares analysis of the data given in Ref. 1

Note (b): equations obtained by least-squares analysis of the data given in Ref. 2

Note (c): equations obtained by least-squares analysis of the data given in Ref. 3

TABLE C13. Dibutylammonium picrate, $(C_4H_9)_2NH_2^+ C_6H_2(NO_2)_3O^-$

T_F= 371 K		Ref. 1
$\rho = 1.4620 - 8.0889 \ 10^{-4}T$	$r^2 = 0.9919$	Note (a)
$\kappa = 0.093026 - 5.3213 \ 10^{-4}T + 7.6988 \ 10^{-7}T^2$		Note (a)
$\eta = 1.88613 - 8.4189 \ 10^{-3}T + 9.4244 \ 10^{-6}T^2$		Note (a)

T	ρ	κ	Λ	η	$\Lambda\eta$
K	g cm^{-3}	S cm^{-1}	S cm^2 mol^{-1}	Pa s	
373	1.1637	0.001585	0.4880	0.06139	0.0300
383	1.1547	0.002139	0.6638	0.04150	0.0275
393	1.1459	0.002879	0.9003	0.02989	0.0269
403	1.1371	0.003667	1.155	0.02187	0.0253

413	1.1284	0.004647	1.475	0.01635	0.0241
423	1.1199	0.005687	1.819	0.01275	0.0232
433	1.1116	0.006926	2.232	0.01011	0.0226
443	1.1034	0.008235	2.674	0.008212	0.0220
453	1.0952	0.009863	3.227	0.006793	0.0219
463	1.0872	0.01184	3.902	0.005720	0.0223

$\Lambda = 19629.9 \exp\{-\dfrac{32731}{R\,T}\}$ $r^2 = 0.9983$ Note (a)

$\eta = 2.934\ 10^{-7} \exp\{\dfrac{37731}{R\,T}\}$ $r^2 = 0.9959$ Note (a)

$\eta = \dfrac{4.760\ 10^{-4}}{v - 0.8533}$ Note (a)

Ref. 1: Walden & Birr (1932)   data in the table given according to Ref. 1

Note (a): equations obtained by least-squares analysis of the data given in Ref. 1

TABLE C14. Di*iso*amylammonium picrate, $(i.C_5H_{11})_2NH_2^+\ C_6H_2(NO_2)_3O^-$

$T_F = 368$ K   Ref. 1

$\rho = 1.4218 - 7.3193\ 10^{-4}T$ $r^2 = 0.9978$ Note (a)

$\kappa = 0.0088635 - 1.0231\ 10^{-4}T + 2.1528\ 10^{-7}T^2$ Note (b)

$\eta = 2.60418 - 1.18047\ 10^{-2}T + 1.3417\ 10^{-5}T^2$ Note (c)

T	ρ	κ	Λ	η	$\Lambda\eta$	
K	g cm^{-3}	S cm^{-1}	S cm^2 mol^{-1}	Pa s		ρ: Ref. 1 κ: Ref. 2 η: Ref. 3
373	1.1492	0.000917	0.3083	0.07178	0.0221	
383	1.1414	0.001310	0.4434	0.04799	0.0213	
393	1.1338	0.001793	0.6110	0.03432	0.0210	
403	1.1263	0.002419	0.8298	0.02416	0.0200	
413	1.1188	0.003144	1.085	0.01804	0.0196	
423	1.1115	0.003999	1.390	0.01385	0.0193	
433	1.1044	0.004954	1.733	0.01089	0.0189	
443	1.0971	0.006047	2.129	0.008744	0.0186	
453	1.0900	0.007281	2.581	0.007194	0.0186	
463	1.0831	0.008738	3.117	0.005939	0.0185	

$\Lambda = 44784.9 \exp\{-\dfrac{36652}{R\,T}\}$ $r^2 = 0.9972$ Note (b)

$\eta = 1.872\ 10^{-7} \exp\{\dfrac{39614}{R\,T}\}$ $r^2 = 0.9960$ Note (c)

$\eta = \dfrac{5.5571\ 10^{-4}}{v - 0.8633}$ Note (c)

Ref. 1: Walden, Ulich & Birr (1927a)

Ref. 2: Walden, Ulich & Birr (1927b)

Ref. 3: Walden, Ulich & Birr (1927c)

Note (a): equation obtained by least-squares analysis of the data given in Ref. 1

Note (b): equations obtained by least-squares analysis of the data given in Ref. 2

Note (c): equations obtained by least-squares analysis of the data given in Ref. 3

TABLE C15. Dihexadecylammonium picrate, $(C_{16}H_{33})_2NH_2^+ C_6H_2(NO_2)_3O^-$

$T_F = 328$ K   Ref. 1

$\rho = 1.1535 - 5.8886\ 10^{-4}T$ $r^2 = 0.9996$ Note (a)

$\kappa = 0.00701579 - 3.9780\ 10^{-5}T + 5.6502\ 10^{-8}T^2$ Note (a)

$\eta = 4.1626 - 2.01119\ 10^{-2}T + 2.4295\ 10^{-5}T^2$ Note (a)

T	ρ	κ	Λ	η	$\Lambda\eta$
K	g cm^{-3}	S cm^{-1}	S cm^2 mol^{-1}	Pa s	
333	0.9583	0.135 10^{-4}	0.00979	0.1864	0.00182
343	0.9519	0.228 10^{-4}	0.01664	0.1115	0.00186
353	0.9457	0.367 10^{-4}	0.02671	0.07349	0.00196
363	0.9396	0.543 10^{-4}	0.04016	0.04790	0.00192
373	0.9337	0.767 10^{-4}	0.05709	0.03393	0.00194
383	0.9276	1.054 10^{-4}	0.07897	0.02493	0.00197
393	0.9219	1.423 10^{-4}	0.1072	0.01870	0.00200
403	0.9260	1.865 10^{-4}	0.1415	0.01486	0.00210
413	0.9102	2.408 10^{-4}	0.1838	0.01180	0.00217
423	0.9046	3.070 10^{-4}	0.2356	0.009590	0.00226
433	0.8990	3.873 10^{-4}	0.2994	0.007930	0.00237
443	0.8834	4.749 10^{-4}	0.3694	0.006613	0.00244

$\Lambda = 19803.4\ \exp\{-\dfrac{39793}{R\ T}\}$ $r^2 = 0.9964$ Note (a)

$\eta = 2.71\ 10^{-7}\ \exp\{\dfrac{36729}{R\ T}\}$ $r^2 = 0.9922$ Note (a)

$\eta = \dfrac{8.003\ 10^{-4}}{v - 1.0411}$ Note (a)

Ref. 1: Walden & Birr (1932)   data in the table given according to Ref. 1

Note (a): equations obtained by least-squares analysis of the data given in Ref. 1

TABLE C16. Triethylammonium picrate, $(C_2H_5)_3NH^+ C_6H_2(NO_2)_3O^-$

$T_F = 448$ K   Ref. 1

$\rho = 1.4980 - 6.44\ 10^{-4}T$ $r^2 = 0.9999$ Note (a)

$\kappa = 0.09218 - 5.9028\ 10^{-4}T + 9.1428\ 10^{-7}T^2$ Note (b)

$\eta = 0.46225 - 1.9218\ 10^{-3}T + 2.0194\ 10^{-6}T^2$ Note (c)

T	ρ	κ	Λ	η	$\Lambda\eta$	ρ: Ref. 1
K	g cm^{-3}	S cm^{-1}	S cm^2 mol^{-1}	Pa s		κ: Ref. 2
						η: Ref. 3
448	1.2095	0.01125	3.072	0.006600	0.0203	
453	1.2063	0.01238	3.389	0.006039	0.0205	
458	1.2031	0.01358	3.728	0.005672	0.0211	
463	1.1999	0.01494	4.112	0.005374	0.0221	
468	1.1966	0.01615	4.457	0.005132	0.0229	

$$\Lambda = 19987 \exp\{- \frac{32700}{R\,T}\}$$
$r^2 = 0.9993$ Note (b)

$$\eta = 1.94975 \; 10^{-5} \exp\{\frac{21633}{R\,T}\}$$
$r^2 = 0.9859$ Note (c)

$$\eta = \frac{1.940 \; 10^{-3}}{v - 0.7971}$$
Note (c)

Ref. 1: Walden, Ulich & Birr (1927a)

Ref. 2: Walden, Ulich & Birr (1927b)

Ref. 3: Walden, Ulich & Birr (1927c)

Note (a): equation obtained by least-squares analysis of the data given in Ref. 1

Note (b): equations obtained by least-squares analysis of the data given in Ref. 2

Note (c): equations obtained by least-squares analysis of the data given in Ref. 3

TABLE C 17. Tripropylammonium picrate, $(C_3H_7)_3NH^+ \; C_6H_2(NO_2)_3O^-$

$T_F = 387$ K   Ref. 1

$\rho = 1.4442 - 7.22 \; 10^{-4}T$ $r^2 = 0.9998$ Note (a)

$\kappa = 0.10552 - 5.8005 \; 10^{-4}T + 8.0452 \; 10^{-7}T^2$ Note (b)

$\eta = 8.82985 - 4.2488 \; 10^{-2}T + 5.1096 \; 10^{-5}T^2$ Note (c)

T	ρ	κ	Λ	η	$\Lambda\eta$	
K	g cm^{-3}	S cm^{-1}	S cm^2 mol^{-1}	Pa s		ρ: Ref. 1 κ: Ref. 2 η: Ref. 3
388	1.1643	0.001563	0.4998	0.01849	0.00924	
393	1.1606	0.001820	0.5839	0.01570	0.00917	
398	1.1569	0.002105	0.6775	0.01366	0.00925	
403	1.1531	0.002412	0.7789	0.01206	0.00939	
408	1.1496	0.002793	0.9046	0.01059	0.00958	
413	1.1460	0.003221	1.046	0.009493	0.00993	
418	1.1423	0.003582	1.167	0.008576	0.0100	
423	1.1388	0.004137	1.352	0.007595	0.0103	
428	1.1352	0.004591	1.505	0.006952	0.0105	
433	1.1317	0.005187	1.706	0.006250	0.0107	
438	1.1282	0.005823	1.921	0.005757	0.0111	

$$\Lambda = 44229.9 \exp\{- \frac{36591}{R\,T}\}$$
$r^2 = 0.9912$ Note (b)

$$\eta = 7.136 \; 10^{-7} \exp\{\frac{32660}{R\,T}\}$$
$r^2 = 0.9976$ Note (c)

$$\eta = \frac{2.4045 \; 10^{-4}}{v - 0.8465}$$
Note (c)

Ref. 1: Walden, Ulich & Birr (1927a)

Ref. 2: Walden, Ulich & Birr (1927b)

Ref. 3: Walden, Ulich & Birr (1927c)

Note (a): equation obtained by least-squares analysis of the data given in Ref. 1

Note (b): equations obtained by least-squares analysis of the data given in Ref. 2

Note (c): equations obtained by least-squares anlysis of the data given in Ref. 3

TABLE C18. Tributylammonium picrate, $(C_4H_9)_3NH^+ C_6H_2(NO_2)_3O^-$

$T_F = 379$ K   Ref. 1
$\rho = 1.3812 - 6.7964\ 10^{-4}T$ $r^2 = 0.9999$ Note (a)
$\kappa = 0.0658275 - 3.6083\ 10^{-4}T + 4.9821\ 10^{-7}T^2$ Note (a)
$\eta = 0.58734 - 2.56914\ 10^{-3}T + 2.8325\ 10^{-6}T^2$ Note (a)

T / K	ρ / g cm^{-3}	κ / S cm^{-1}	Λ / S cm^2 mol^{-1}	η / Pa s	$\Lambda\eta$
383	1.1210	7.106 10^{-4}	0.2627	0.01892	0.00497
393	1.1141	9.668 10^{-4}	0.3596	0.01505	0.00541
403	1.1071	1.323 10^{-3}	0.4952	0.01198	0.00593
413	1.1004	1.781 10^{-3}	0.6708	0.009479	0.00636
423	1.0937	2.340 10^{-3}	0.8867	0.007387	0.00655
433	1.0870	2.997 10^{-3}	1.1427	0.006034	0.00690
443	1.0801	3.749 10^{-3}	1.4385	0.005042	0.00725

$\Lambda = 84119 \exp\{-\frac{40357}{R\,T}\}$ $r^2 = 0.9995$ Note (a)
$\eta = 9.271\ 10^{-7} \exp\{\frac{31655}{R\,T}\}$ $r^2 = 0.9990$ Note (a)
$\eta = \frac{2.4946\ 10^{-4}}{v - 0.8803}$ Note (a)

Ref. 1: Walden & Birr (1932)   data in the table given according to Ref. 1
Note (a): equations obtained by least-squares analysis of the data given in Ref. 1

TABLE C 19. Tri*iso*amylammonium picrate, $(i.C_5H_{11})_3NH^+ C_6H_2(NO_2)_3O^-$

$T_F = 398$ K   Ref. 1
$\rho = 1.3355 - 7.0307\ 10^{-4}T$ $r^2 = 0.9997$ Note (a)
$\kappa = 0.0530711 - 2.8043\ 10^{-4}T + 3.7324\ 10^{-7}T^2$ Note (b)
$\eta = 0.812623 - 3.5341\ 10^{-3}T + 3.866\ 10^{-6}T^2$ Note (c)

T / K	ρ / g cm^{-3}	κ / S cm^{-1}	Λ / S cm^2 mol^{-1}	η / Pa s	$\Lambda\eta$
398	1.0561	5.59 10^{-4}	0.2793	0.01876	0.00524
403	1.0524	6.73 10^{-4}	0.3374	0.01631	0.00550
408	1.0488	7.98 10^{-4}	0.4014	0.01392	0.00559
413	1.0450	9.34 10^{-4}	0.4716	0.01217	0.00574
418	1.0413	10.82 10^{-4}	0.5483	0.01066	0.00584
423	1.0379	12.42 10^{-4}	0.6314	0.009355	0.00591
428	1.0344	14.15 10^{-4}	0.7218	0.008281	0.00598
433	1.0310	16.15 10^{-4}	0.8265	0.007302	0.00604
438	1.0275	18.31 10^{-4}	0.9403	0.006519	0.00613
443	1.0241	20.69 10^{-4}	1.066	0.005874	0.00626
448	1.0206	23.36 10^{-4}	1.207	0.005381	0.00649

ρ: Ref. 1
κ: Ref. 2
η: Ref. 3

| 453 | 1.0172 | 26.28 10^{-4} | 1.363 | 0.004991 | 0.00680 |
| 458 | 1.0138 | 29.50 10^{-4} | 1.535 | 0.004652 | 0.00714 |

$$\Lambda = 107266 \ \exp\{-\frac{42417}{R\,T}\} \qquad\qquad r^2 = 0.9990 \qquad \text{Note (b)}$$

$$\eta = 2.75 \ 10^{-7} \ \exp\{\frac{36755}{R\,T}\} \qquad\qquad r^2 = 0.9976 \qquad \text{Note (c)}$$

$$\eta = \frac{2.5301 \ 10^{-4}}{v - 0.9348} \qquad\qquad \text{Note (c)}$$

Ref. 1: Walden, Ulich & Birr (1927a)

Ref. 2: Walden, Ulich & Birr (1927b)

Ref. 3: Walden, Ulich & Birr (1927c)

Note (a): equation obtained by least-squares analysis of the data given in Ref. 1

Note (b): equations obtained by least-squares analysis of the data given in Ref. 2

Note (c): equations obtained by least-squares analysis of the data given in Ref. 3

TABLE C20. Tetrapropylammonium picrate, $(C_3H_7)_4N^+ \ C_6H_2(NO_2)_3O^-$

T_F= 393 K			Ref. 1
$\rho = 1.3818 - 6.27 \ 10^{-4}T$		$r^2 = 0.9995$	Note (a)
$\kappa = 0.11696 - 7.5369 \ 10^{-4}T + 1.2061 \ 10^{-6}T^2$			Note (b)
$\eta = 0.456889 - 1.8956 \ 10^{-3}T + 1.9802 \ 10^{-6}T^2$			Note (c)

T	ρ	κ	Λ	η	$\Lambda\eta$	
K	g cm^{-3}	S cm^{-1}	S cm^2 mol^{-1}	Pa s		ρ: Ref. 1 κ: Ref. 2 η: Ref. 3 Note (d)
393	1.1361	0.007605	2.577	0.01875	0.0483	
403	1.1295	0.009031	3.313	0.01418	0.0470	
413	1.1228	0.01143	4.219	0.01095	0.0462	
423	1.1154	0.01401	5.201	0.008717	0.0453	
433	1.1102	0.01668	6.226	0.007111	0.0443	
443	1.1037	0.01978	7.427	0.005868	0.0436	
453	1.0972	0.02297	8.676	0.004998	0.0434	
463	1.0912	0.02608	9.905	0.004285	0.0424	
473	1.0853	0.02946	11.250	0.003714	0.0418	
483	1.0793	0.03274	12.623	0.003270	0.0413	
493	1.0735	0.03630	14.014	0.002953	0.0414	

$$\Lambda = 11260.8 \ \exp\{-\frac{27150}{R\,T}\} \qquad\qquad r^2 = 0.9938 \qquad \text{Note (b)}$$

$$\eta = 1.9338 \ 10^{-6} \ \exp\{\frac{29730}{R\,T}\} \qquad\qquad r^2 = 0.9938 \qquad \text{Note (c)}$$

$$\eta = \frac{1.979 \ 10^{-4}}{v - 0.8707} \qquad\qquad \text{Note c)}$$

Ref. 1: Walden, Ulich & Birr (1927a)

Ref. 2: Walden, Ulich & Birr (1927b)

Ref. 3: Walden, Ulich & Birr (1927c)

Note (a): equation obtained by least-squares analysis of the data given in Ref. 1

Note (b): equations obtained by least-squares analysis of the data given in Ref. 2

Note (c): equations obtained by least-squares analysis of the data given in Ref. 3

Note (d): recent measurements by Barreira & Barreira (1976) are in good agreement with Walden's data

TABLE C21. Tetrabutylammonium picrate, $(C_4H_9)_4N^+ \ C_6H_2(NO_2)_3O^-$

T_F= 363 K   Ref. 1

$\rho = 1.3182 - 5.9164 \ 10^{-4}T$ $r^2 = 0.9995$ Note (a)

$\kappa = 0.0717113 - 4.8156 \ 10^{-4}T + 7.9329 \ 10^{-7}T^2$ Note (a)

$\eta = 0.870719 - 3.7270 \ 10^{-3}T + 4.0014 \ 10^{-6}T^2$ Note (a)

T	ρ	κ	Λ	η	$\Lambda\eta$
K	g cm^{-3}	S cm^{-1}	S cm^2 mol^{-1}	Pa s	
373	1.0982	0.002686	1.150	0.04191	0.0482
383	1.0922	0.003676	1.583	0.02907	0.0460
393	1.0859	0.004898	2.122	0.02084	0.0442
403	1.0793	0.006256	2.727	0.01604	0.0437
413	1.0735	0.007964	3.490	0.01224	0.0427
423	1.0676	0.009849	4.341	0.009757	0.0424
433	1.0615	0.01191	5.279	0.007818	0.0413
443	1.0556	0.01407	6.272	0.006538	0.0410
453	1.0498	0.01648	7.346	0.005538	0.0407
463	1.0443	0.01910	8.606	0.004673	0.0402
473	1.0384	0.02163	9.801	0.004116	0.0403
483	1.0328	0.02422	11.035	0.003654	0.0403
493	1.0273	0.02676	12.257	0.003264	0.0400

$\Lambda = 20765.9 \ \exp\{-\dfrac{30048}{R \ T}\}$ $r^2 = 0.9932$ Note (a)

$\eta = 1.2507 \ 10^{-6} \ \exp\{\dfrac{31814}{R \ T}\}$ $r^2 = 0.9876$ Note (a)

$\eta = \dfrac{2.6500 \ 10^{-4}}{v - 0.9059}$ Note (a)

Ref. 1: Walden & Birr (1932), see Note (b)   data in the Table given according to Ref. 1

Note (a): equations obtained by least-squares analysis of the data given in Ref. 1

Note (b): recent measurements by Barreira & Barreira (1976) are in good agreement with Walden's data

TABLE C22. Tetra*iso*amylammonium picrate, $(i.C_5H_{11})_4N^+ \ C_6H_2(NO_2)_3O^-$

T_F= 360 K   Ref. 1

$\rho = 1.2624 - 5.895 \ 10^{-4}T$ $r^2 = 0.9995$ Note (a)

$\kappa = 0.080655 - 4.8352 \ 10^{-4}T + 7.256 \ 10^{-7}T^2$ Note (b)

$\eta = 2.55745 - 1.11897 \ 10^{-2}T + 1.2236 \ 10^{-5}T^2$ Note (c)

T	ρ	κ	Λ	η	$\Lambda\eta$	
K	g cm^{-3}	S cm^{-1}	S cm^2 mol^{-1}	Pa s		ρ: Ref. 1 κ: Ref. 2 η: Ref. 3 Note (d)
363	1.0493	8.352 10^{-4}	0.419	0.1304	0.0546	
373	1.0429	1.296 10^{-3}	0.654	0.08121	0.0531	
383	1.0368	1.878 10^{-3}	0.954	0.05321	0.0508	
393	1.0306	2.602 10^{-3}	1.329	0.03769	0.0501	
403	1.0244	3.456 10^{-3}	1.776	0.02705	0.0480	
413	1.0185	4.644 10^{-3}	2.445	0.01995	0.0488	
423	1.0126	5.962 10^{-3}	3.100	0.01517	0.0470	
433	1.0065	7.344 10^{-3}	3.842	0.01203	0.0462	
443	1.0007	8.942 10^{-3}	4.706	0.009623	0.0453	
453	0.9950	1.061 10^{-2}	5.616	0.007947	0.0446	
463	0.9893	1.240 10^{-2}	6.601	0.006680	0.0441	
473	0.9838	1.428 10^{-2}	7.644	0.005687	0.0435	
483	0.9781	1.636 10^{-2}	8.809	0.004844	0.0427	
493	0.9726	1.852 10^{-2}	10.028	0.004195	0.0421	

$\Lambda = 72566.5 \exp\{-\dfrac{35802}{R\,T}\}$ $\qquad$ $r^2 = 0.9896$ $\qquad$ Note (b)

$\eta = 2.807\ 10^{-7} \exp\{\dfrac{38747}{R\,T}\}$ $\qquad$ $r^2 = 0.9907$ $\qquad$ Note (c)

$\eta = \dfrac{4.9523\ 10^{-4}}{v - 0.9513}$ $\qquad$ Note (c)

Ref. 1: Walden, Ulich & Birr (1927a)

Ref. 2: Walden, Ulich & Birr (1927b)

Ref. 3: Walden, Ulich & Birr (1927c)

Note (a): equation obtained by least-squares analysis of the data given in Ref. 1

Note (b): equations obtained by least-squares analysis of the data given in Ref. 2

Note (c): equations obtained by least-squares analysis of the data given in Ref. 3

Note (d): recent measurements by Barreira & Barreira (1976) are in good agreement with Walden's data

TABLE C23. Dimethyldipropylammonium picrate, $(CH_3)_2(C_3H_7)_2N^+\ C_6H_2(NO_2)_3O^-$

$T_F = 366$ K $\qquad$ Ref. 1

$\rho = 1.4599 - 6.629\ 10^{-4}T$ $\qquad$ $r^2 = 0.9997$ $\qquad$ Note (a)

$\kappa = 0.109688 - 7.4466\ 10^{-4}T + 1.2462\ 10^{-6}T^2$ $\qquad$ Note (b)

$\eta = 0.878877 - 3.8403\ 10^{-3}T + 4.2154\ 10^{-6}T^2$ $\qquad$ Note (c)

T	ρ	κ	Λ	η	$\Lambda\eta$	
K	g cm^{-3}	S cm^{-1}	S cm^2 mol^{-1}	Pa s		ρ: Ref. 1 κ: Ref. 2 η: Ref. 3
373	1.2133	0.005477	1.617	0.03547	0.0574	
383	1.2063	0.007304	2.169	0.02505	0.0543	
393	1.1993	0.009442	2.821	0.01923	0.0542	
403	1.1926	0.01180	3.545	0.01445	0.0512	

T	ρ	κ	Λ	η	Λη
413	1.1858	0.01452	4.387	0.01138	0.0499
423	1.1792	0.01765	5.363	0.009018	0.0484
433	1.1726	0.02110	6.448	0.007420	0.0478
443	1.1660	0.02460	7.560	0.006225	0.0471
453	1.1597	0.02817	8.704	0.005302	0.0461
463	1.1533	0.03209	9.970	0.004617	0.0460
473	1.1469	0.03610	11.27	0.003973	0.0448

$\Lambda = 16368.5 \exp\{-\dfrac{28374}{R\,T}\}$ $r^2 = 0.9955$ Note (b)

$\eta = 1.1155\ 10^{-6} \exp\{\dfrac{31869}{R\,T}\}$ $r^2 = 0.9946$ Note (c)

$\eta = \dfrac{2.5195\ 10^{-4}}{v - 0.8184}$ Note (c)

Ref. 1: Walden, Ulich & Birr (1927a)

Ref. 2: Walden, Ulich & Birr (1927b)

Ref. 3: Walden, Ulich & Birr (1927c)

Note (a): equation obtained by least-squares analysis of the data given in Ref. 1

Note (b): equations obtained by least-squares analysis of the data given in Ref.2

Note (c): equations obtained by least-squares analysis of the data given in Ref.3

TABLE C24. Propyltriethylammonium picrate, $(C_3H_7)(C_2H_5)_3N^+\ C_6H_2(NO_2)_3 O^-$

$T_F = 417$ K   Ref. 1

$\rho = 1.4482 - 6.477\ 10^{-4}T$ $r^2 = 0.9953$ Note (a)

$\kappa = 0.00093601 - 2.3739\ 10^{-4}T + 6.4892\ 10^{-7}T^2$ Note (b)

$\eta = 0.30912 - 1.2426\ 10^{-3}T + 1.2641\ 10^{-6}T^2$ Note (c)

T	ρ	κ	Λ	η	Λη	
K	g cm⁻³	S cm⁻¹	S cm² mol⁻¹	Pa s		ρ: Ref. 1 κ: Ref. 2 η: Ref. 3
418	1.1786	0.01516	4.789	0.01067	0.0511	
423	1.1752	0.01651	5.230	0.009574	0.0501	
433	1.1686	0.01987	6.331	0.007958	0.0504	
443	1.1621	0.02316	7.421	0.006650	0.0493	
453	1.1550	0.02646	8.530	0.005599	0.0478	
463	1.1480	0.03014	9.833	0.004831	0.0475	
473	1.1414	0.03393	11.06	0.004223	0.0467	
483	1.1353	0.03761	12.33	0.003718	0.0458	

$\Lambda = 5744.9 \exp\{-\dfrac{24572}{R\,T}\}$ $r^2 = 0.9974$ Note (b)

$\eta = 4.0588\ 10^{-6} \exp\{\dfrac{27304}{R\,T}\}$ $r^2 = 0.9985$ Note (c)

$\eta = \dfrac{1.9227\ 10^{-4}}{v - 0.8309}$ Note (c)

Ref. 1: Walden, Ulich & Birr (1927a)

Ref. 2: Walden, Ulich & Birr (1927b)

Ref. 3: Walden, Ulich & Birr (1927c)

Note (a): equation obtained by least-squares analysis of the data given in Ref. 1

Note (b): equations obtained by least-squares analysis of the data given in Ref. 2

Note (c): equations obtained by least-squares analysis of the data given in Ref. 3

TABLE C25. Methyltripropylammonium picrate, $(CH_3)(C_3H_7)_3N^+ \ C_6H_2(NO_2)_3O^-$

T_F= 355 K   Ref. 1

$\rho = 1.4162 - 6.4083 \ 10^{-4}T$ $r^2 = 0.9997$ Note (a)

$\kappa = 0.118495 - 7.7246 \ 10^{-4}T + 1.2512 \ 10^{-6}T^2$ Note (b)

$\eta = 1.20203 - 5.3544 \ 10^{-3}T + 5.9792 \ 10^{-6}T^2$ Note (c)

T	ρ	κ	Λ	η	$\Lambda\eta$	
K	g cm^{-3}	S cm^{-1}	S cm^2 mol^{-1}	Pa s		ρ: Ref. 1 κ: Ref. 2 η: Ref. 3
363	1.1843	0.003078	1.004	0.05247	0.0527	
373	1.1774	0.004497	1.475	0.03410	0.0503	
383	1.1708	0.006138	2.025	0.02425	0.0491	
393	1.1643	0.008015	2.659	0.01830	0.0487	
403	1.1577	0.01020	3.404	0.01370	0.0466	
413	1.1511	0.01280	4.296	0.01095	0.0470	
423	1.1447	0.01567	5.289	0.008746	0.0463	
433	1.1384	0.01879	6.377	0.007257	0.0463	
443	1.1321	0.02200	7.508	0.006008	0.0451	
453	1.1260	0.02539	8.712	0.005234	0.0456	
463	1.1198	0.02907	10.030	0.004467	0.0448	
473	1.1138	0.03288	11.406	0.003837	0.0437	

$\Lambda = 33744.6 \ \exp\{-\dfrac{31041}{R \ T}\}$ $r^2 = 0.9923$ Note (b)

$\eta = 7.707 \ 10^{-7} \ \exp\{\dfrac{33092}{R \ T}\}$ $r^2 = 0.9911$ Note (c)

$\eta = \dfrac{2.6603 \ 10^{-4}}{v - 0.8408}$ Note (c)

Ref. 1: Walden, Ulich & Birr (1927a)

Ref. 2: Walden, Ulich & Birr (1927b)

Ref. 3: Walden, Ulich & Birr (1927c)

Note (a): equation obtained by least-squares analysis of the data given in Ref. 1

Note (b): equations obtained by least-squares analysis of the data given in Ref. 2

Note (c): equations obtained by least-squares analysis of the data given in Ref. 3

TABLE C26. Diethyldipropylammonium picrate, $(C_2H_5)_2(C_3H_7)_2N^+ \ C_6H_2(NO_2)_3O^-$

T_F= 353 K   Ref. 1

$\rho = 1.4453 - 6.4917 \ 10^{-4}T$ $r^2 = 0.9995$ Note (a)

$\kappa = 0.103322 - 7.0226 \ 10^{-4}T + 1.1781 \ 10^{-6}T^2$ Note (b)

$\eta = 1.400376 - 6.3132 \ 10^{-4}T + 7.1277 \ 10^{-6}T^2$ Note (c)

T	ρ	κ	Λ	η	$\Lambda\eta$	
K	g cm^{-3}	S cm^{-1}	S cm^2 mol^{-1}	Pa s		ρ: Ref. 1 κ: Ref. 2 η: Ref. 3
353	1.2170	0.002434	0.7728	0.07042	0.0544	
363	1.2101	0.003627	1.158	0.04428	0.0513	
373	1.2034	0.005237	1.681	0.03070	0.0516	
383	1.1966	0.007065	2.281	0.02257	0.0515	
393	1.1898	0.009118	2.961	0.01697	0.0502	
403	1.1830	0.01144	3.736	0.01326	0.0495	
413	1.1766	0.01418	4.656	0.01054	0.0491	
423	1.1701	0.01715	5.663	0.008478	0.0480	
433	1.1639	0.02095	6.755	0.007155	0.0483	
443	1.1578	0.02365	7.906	0.005973	0.0472	
453	1.1514	0.02706	9.081	0.005208	0.0473	
463	1.1451	0.03074	10.37	0.004440	0.0460	
473	1.1391	0.03446	11.68	0.003822	0.0444	

$$\Lambda = 33689.7 \ \exp\{-\frac{30815}{R\,T}\} \qquad\qquad r^2 = 0.9887 \qquad \text{Note (b)}$$

$$\eta = 8.425 \ 10^{-7} \ \exp\{\frac{32681}{R\,T}\} \qquad\qquad r^2 = 0.9903 \qquad \text{Note (c)}$$

$$\eta = \frac{2.989 \ 10^{-4}}{v - 0.8184} \qquad\qquad\qquad\qquad\qquad\quad \text{Note (c)}$$

Ref. 1: Walden, Ulich & Birr (1927a)

Ref. 2: Walden, Ulich & Birr (1927b)

Ref. 3: Walden, Ulich & Birr (1927c)

Note (a): equation obtained by least-squares analysis of the data given in Ref. 1

Note (b): equations obtained by least-squares analysis of the data given in Ref. 2

Note (c): equations obtained by least-squares analysis of the data given in Ref. 3

TABLE C27. Ethyltripropylammonium picrate, $(C_2H_5)(C_3H_7)_3N^+ \ C_6H_2(NO_2)_3 O^-$

$T_F = 380$ K   Ref. 1

$\rho = 1.4068 - 6.4488 \ 10^{-4}T \qquad\qquad r^2 = 0.9888$ Note (a)

$\kappa = 0.089690 - 6.2915 \ 10^{-4}T + 1.0706 \ 10^{-6}T^2$ Note (b)

$\eta = 0.768135 - 3.3559 \ 10^{-3}T + 3.6882 \ 10^{-6}T^2$ Note (c)

T	ρ	κ	Λ	η	$\Lambda\eta$	
K	g cm^{-3}	S cm^{-1}	S cm^2 mol^{-1}	Pa s		ρ: Ref. 1 κ: Ref. 2 η: Ref. 3
383	1.1613	0.005917	2.040	0.02470	0.0504	
393	1.1546	0.007782	2.698	0.01842	0.0497	
403	1.1477	0.009832	3.430	0.01388	0.0476	
413	1.1410	0.01231	4.320	0.01079	0.0466	
423	1.1346	0.01511	5.332	0.008608	0.0459	
433	1.1278	0.01813	6.437	0.007150	0.0461	
443	1.1213	0.02117	7.560	0.005893	0.0446	

453	1.1150	0.02449	8.795	0.005021	0.0442
463	1.1085	0.02787	10.067	0.004301	0.0433
473	1.1024	0.03153	11.452	0.003732	0.0427

$$\Lambda = 18322.4 \ \exp\{- \frac{28791}{R \ T}\} \qquad\qquad r^2 = 0.9955 \qquad \text{Note (b)}$$

$$\eta = 1.1674 \ 10^{-6} \ \exp\{\frac{31507}{R \ T}\} \qquad\qquad r^2 = 0.9956 \qquad \text{Note (c)}$$

$$\eta = \frac{2.2266 \ 10^{-4}}{v - 0.8533} \qquad\qquad \text{Note (c)}$$

Ref. 1: Walden, Ulich & Birr (1927a)

Ref. 2: Walden, Ulich & Birr (1927b)

Ref. 3: Walden, Ulich & Birr (1927c)

Note (a): equation obtained by least-squares analysis of the data given in Ref. 1

Note (b): equations obtained by least-squares analysis of the data given in Ref. 2

Note (c): equations obtained by least-squares analysis of the data given in Ref. 3

REFERENCES

ANGELL, C.A., *J. Phys. Chem.*, <u>68</u>, 1917 (1964)

ANGELL, C.A., *J. Chem. Phys.*, <u>46</u>, 4673 (1967)

BARANOWSKI, B., *"Nierownowagowa Termodynamika w Chemii Fizycznej"*, PAN, Warszawa, 1974

BARREIRA, M.L., and BARREIRA, F., *Electrochim. Acta*, <u>21</u>, 491 (1976)

BATCHINSKI, A.J., *Z. physik. Chem.*, <u>84</u>, 643 (1913)

COHEN, M.H., and TURNBULL, D., *J. Chem. Phys.*, <u>31</u>, 1164 (1959)

DE GROOT, R.S., and MAZUR, P., *"Non-Equilibrium Thermodynamics"*, North Holland Pub. Co.,
        Amsterdam, 1962

EYRING, H., *J. Chem. Phys.*, <u>4</u>, 283 (1936)

EYRING, H., and JHON, M.S., *"Significant Liquid Structures"*, Wiley, New York, 1969

FRENKEL, J., *"Kinetic Theory of Liquids"*, Oxford University Press, Oxford, 1946

GATNER, K., *Pol. J. Chem.*, in press (1978)

GATNER, K., and KISZA, A., *Z. physik. Chem.*, <u>241</u>, 1 (1969)

GORDON, J.E., *"Applications of Fused Salts in Organic Chemistry"* in *"Techniques and Methods of
        Organic and Organometallic Chemistry"*, ed. by Denney, D.B., M. Dekker, New York,
        1969, Vol. I, p. 51

GUZMAN, J., *An. Soc. Espan. Fis. y Quim.*, <u>9</u>, 353 (1913)

HARRAP, B.S., and HEYMAN, E., *Chem. Rev.*, <u>48</u>, 45 (1951)

HAZLEWOOD, F.J., RHODES, E., and UBBELOHDE, A.R., *Trans. Faraday Soc.*, <u>62</u>, 3101 (1966)

JANZ, G.J., DAMPIER, F.W., LAKSHMINARAYANAN, G.R., LORENZ, P.K., and TOMKINS, R.P.T., *"Molten
        Salts, Vol. 1, Electrical Conductivity, Density and Viscosity Data"*, NSRDS-NBS,
        1968

KISZA, A., *Z. physik. Chem.*, <u>237</u>, 97 (1968)

KISZA, A., and HAWRANEK, J., *Z. physik. Chem.*, <u>237</u>, 210 (1968)

KLEMM, A., *"Transport Properties of Molten Salts"* in *"Molten Salt Chemistry"*, ed. by Blander,
        M., Interscience Pub., New York, 1964

LAITY, R.W., *J. Chem. Phys.*, <u>30</u>, 682 (1959)

LAMPREIA, M.I., and BARREIRA, F., *Electrochim. Acta*, <u>21</u>, 485 (1976)

LEONESI, D., CINGOLANI, A., and BERCHIESI, G., *J. Chem. Eng. Data*, <u>18</u>, 391 (1973)

LIND, J.E., *"Molten Organic Salts - Physical Properties"* in *"Advances in Molten Salt Chemistry"*,
        Vol. 2, ed.s Braunstein, J., Mamantov, G., and Smith, G.P., Plenum Press, New York,
        1973

LIND, J.E., ABDEL-REHIM, H.A.A., and RUDICH, S.W., *J. Phys. Chem.*, <u>70</u>, 3610 (1966)

MACEDO, P.B., and LITOVITZ, T.A., *J. Chem. Phys.*, <u>42</u>, 245 (1965)

McLEOD, D.B., *Trans. Faraday Soc.*, <u>16</u>, 6 (1923/24)

MEISEL, T., HALMOS, Z., and PUNGOR, E., *Periodica Polytech. Chem. Eng. (Budapest)*, <u>17</u>, 89 (1973)

RICE, S.A., *Mol. Phys.*, <u>4</u>, 304 (1961)

RICE, S.A., *Trans. Faraday Soc.*, <u>58</u>, 499 (1962)

SUNDHEIM, B.R., *"Transport Properties of Liquid Electrolytes"* in *"Fused Salts"*, ed. by Sundheim,
        B.R., McGraw Hill, New York, 1964

WALDEN, P., *Z. physik. Chem.*, <u>A157</u>, 389 (1931)

WALDEN, P., and BIRR, E.J., *Z. physik. Chem.*, <u>A160</u>, 45, 57 (1932)

WALDEN, P., ULICH, H., and BIRR, E.J., *Z. physik. Chem.*, A130, 495 (1927)

WALDEN, P., ULICH, H., and BIRR, E.J., *Z. physik. Chem.*, A131, 1 (1927)

WALDEN, P., ULICH, H., and BIRR, E.J., *Z. physik. Chem.*, A131, 21 (1927)

WILLIAMS, M.L., LANDEL, R.F., and FERRY, J.D., *J. Amer. Chem. Soc.*, 77, 3701 (1955)

ADDENDUM

TABLE A27. Ethylammonium bromide, $C_2H_5NH_3^+Br^-$

$T_F = 432.5$ K   Ref. 1

$\rho = 2.1757 - 1.612 \ 10^{-3}T$, $r^2 = 0.9988$ Note (a)

$\kappa = -1.07918 + 3.8734 \ 10^{-3}T - 2.5804 \ 10^{-6}T^2$ Note (a)

$\eta = 0.102925 - 3.8321 \ 10^{-4}T + 3.6393 \ 10^{-7}T^2$ Note (a)

T	ρ	κ	Λ	η	$\Lambda\eta$	ρ, κ, η: Ref. 2
K	g cm^{-3}	S cm^{-1}	S cm^2 mol^{-1}	Pa s		
445	1.4583	0.1335	11.53	0.004473	0.05157	
450	1.4503	0.1413	12.28	0.004174	0.05126	
455	1.4422	0.1490	13.02	0.003901	0.05079	
460	1.4341	0.1566	13.76	0.003651	0.05024	
465	1.4261	0.1640	14.49	0.003422	0.04958	
470	1.4180	0.1713	15.22	0.003212	0.04889	
475	1.4100	0.1785	15.95	0.003019	0.04815	
480	1.4019	0.1855	16.67	0.002891	0.04819	
485	1.3938	0.1925	17.40	0.002677	0.04658	
490	1.3858	0.1992	18.81	0.002525	0.04750	

$\Lambda = 1891.23 \ \exp\{-\dfrac{18848}{R\,T}\}$ $r^2 = 0.9972$ Note (a)

$\eta = 8.8647 \ 10^{-6} \ \exp\{\dfrac{23027}{R\,T}\}$ $r^2 = 0.9995$ Note (a)

Ref. 1: Gordon (1969)

Ref. 2: Kisza & Zabinska, in press (1979)   } data in the table given according to Ref. 2

Note (a): equations obtained by least-squares analysis of the data given in Ref. 2

Part 2
SALT MIXTURES

Chapter 2.1

PHASE DIAGRAMS

P. Franzosini, P. Ferloni and M. Sanesi

SYSTEMS OF SALTS WITH ORGANIC ANION

In the last decades much work has been published, particularly by Soviet investigators, on phase diagrams of condensed salt systems containing at least one organic anion, both in view of practical applications (as non-aqueous ionic solvents, as reaction media, etc.) and for their theoretical significance.

In most cases, however, the methods employed allowed detection only of liquidus curves (or surfaces): this fact, along with the objective experimental difficulties already mentioned in Chapter 1.2 (many salts with organic anion are hygroscopic and/or exhibit metastable phases; their melts not infrequently show a marked tendency to undercooling or to glass-formation; etc.), often caused disagreement among different authors in stating the behaviour of a given system.

On the other hand, to attribute more reliability to one out of several diagrams referring to the same system seems rather arbitrary, because each of them usually looks formally as coherent; in a number of cases, however, the acquisition of improved investigation methods allowed some authors to revise, sometimes substantially, the topology of phase diagrams previously published by themselves.

An extreme example is offered by the system formed with potassium and sodium acetates, which was studied for the first time by Baskov in 1915 and subsequently by several other authors up to Storonkin, Vasil'kova & Tarasov in 1977. In spite of the repeated investigations (°), no unambiguous description of the phase relationships in this seemingly very simple system has yet been given. Such an ambiguity is of course also reflected on more complicated phase diagrams in which this acetate pair takes part.

An additional source of uncertainty is caused by the fact that many alkanoates can exist in a mesomorphic anisotropic liquid state. Therefore, in the phase diagrams of systems containing such salts the curves (or surfaces) on which the isotropic melt, when cooled, ceases to have a stable existence are not true liquidus but, at least in part, equilibrium curves (or surfaces)

(°) Baskov (1915): *continuous solid solutions.*
The fusion temperatures of potassium acetate (A) and sodium acetate (B) are T_A= 568.2 K and T_B= 593.2 K, respectively. Solid solutions are formed in all proportions and the lowest temperature at which the liquid can exist is about 506 K (or 496 K: there are discrepancies within the paper) at about 46 mol % B. Both the liquidus and solidus curves are given; the procedure employed, although not clearly described, is probably some kind of thermographic differential analysis.

Bergman & Evdokimova (1956): *simple eutectic.*
T_A= 575 K; T_B= 599 K. The system is characterized by a simple eutectic at 497 K and 45 mol % B. Both the liquidus and solidus curves are drawn from recorded heating traces.

Diogenov (1956 b, with the participation of Erlykov in the experimental part): *continuous solid solutions.*
T_A= 583.7 K; T_B= 610 K. Solid solutions are formed in all proportions and the lowest temperature at which the melt can exist is 501 K at 50 mol % (this figure is not reported in Table 1 because the same authors changed it in the immediately subsequent paper). The liquidus curve only is given, on the basis of visual-polythermal observations.

Diogenov & Erlykov (1958): *continuous solid solutions.*
All data are the same as in the preceding paper, but for the abscissa of the mimimum, now 45 mol % B.

Golubeva, Bergman & Grigor'eva (1958): *incongruently melting compound 2:1.*
T_A= 579 K; T_B= 599 K. The incongruently melting intermediate compound $2A.B$ is formed. Owing to insufficient information on the invariants, the relevant coordinates could not be included in Table 1. Only solid-liquid equilibria were detected by visual-polythermal method.

Sokolov & Pochtakova (1958 b): *congruently melting intermediate compound 3:2.*
T_A= 574 K; T_B= 604 K. The salts form the intermediate compound $3A.2B$, congruently melting at 514 K: the eutectics are at 513 K (and 38.5 mol % B) and at 508 K (and 46.5 mol % B), respectively. The visual-polythermal method was employed and the liquidus curve only is given.

between isotropic and anisotropic liquids: the situation looks as if a fog-bank prevents an observer from viewing a more or less important part of the "mountain-range outline" in a binary or of the "mountain landscape" in a more complicated system.

The following designations are usually employed:

binary, for a system formed either with one anionic and two cationic or with one cationic and two anionic species;

additive ternary, for a system formed with one anionic and three cationic or with one cationic and three anionic species;

reciprocal, for a system formed with more than one anionic and more than one cationic species;

reciprocal ternary, for a system formed with two anionic and two cationic species, which represents the simplest type of reciprocal system.

On account of the existence of several comprehensive monographs on phase diagrams (among which, e.g.: "Chemical Phase Theory", by Zernike, J., N.V. Uitgevers-Maatschappij A.E. Kluwer, Deventer, 1955; "Phase Diagrams of Fused Salts", by Ricci, J.E., a chapter in "Molten Salt Chemistry" ed. by Blander, M., Interscience Publishers, New York, 1964; "The Phase Rule and Heterogeneous Equilibrium", by Ricci, J.E., Dover Publications Inc., New York, 1966) basic information (on the nature of eutectics, distectics, peritectics; on the formation of intermediate compounds with either congruent or incongruent melting point; on the occurrence of either complete or partial miscibility in the solid state; etc.) will be considered as familiar: therefore, just a few details more closely related to the systems hereafter taken into account will be discussed briefly.

In the following examples only uni-univalent salts are considered; further it is assumed that both the components and the possible intermediate compounds are immiscible (or only partially miscible) in the solid state and do not undergo any solid-solid phase transition.

Let us first give a short comment on _binaries_. In the T vs x plane, the real liquidus curves may lie either below or above the ideal ones (negative and positive deviations from ideality,

---

(°) *(Continued)*

Il'yasov & Bergman (1960 a): *incongruently melting intermediate compound 2:1.*
T_A= 579 K; T_B= 601 K. The incongruently melting intermediate compound $2A.B$ is formed, with the peritectic at 528 K (and 35 mol % B) and the eutectic at 513 K (and 50 mol %).

Diogenov & Sarapulova (1964 b): *congruently melting intermediate compound 3:2.*
T_A= 583 K; T_B= 608 K. The authors abandonned the previous description by Diogenov and by Diogenov & Erlykov (continuous solid solutions) and agreed with that by Sokolov & Pochtakova (formation of the congruently melting compound $3A.2B$, with eutectics at 513 and 508 K). Once more the visual polythermal method is employed and the liquidus curve only is given.

Sokolov & Pochtakova (1967): *congruently melting intermediate compound 3:2.*
T_A= 575 K; T_B= 604 K. The existence of the congruently melting compound $3A.2B$ (already stated in the 1958 paper by the same authors) is confirmed. The visual-polythermal results are supplemented with microscopic and differential thermographic observations, permitting the authors to claim the existence of limited solid solutions rich in either component and the occurrence in the intermediate compound of a polymorphic transition at 463-471 K.

Diogenov & Chumakova (1975): *incongruently melting intermediate compound 3:2.*
T_A= 575 K; T_B= 599 K. The authors seem now inclined to consider the intermediate compound $3A.2B$ as melting incongruently instead of congruently.

Storonkin, Vasil'kova & Tarasov (1977): *eutectic.*
T_A= 584 K; T_B= 607 K. Undercooling phenomena in the range 40-50 mol % B preclude decision about the presence or absence of an intermediate compound merely on the basis of the thermographic procedure employed: additional information obtained through IR spectroscopy and contact-polythermal analysis, however, does not support the existence of any compound. Therefore, the system is classified as eutectic with limited solid solutions rich in either component, the coordinates of the invariant (511 K and 46 mol % B) having been determined by extrapolation. Among the quoted references, the 1967 paper by Sokolov & Pochtakova is unfortunately missing, although it could have offered the most relevant comparison.

respectively). In the latter case, increasing deviations cause a liquidus branch (usually that richer in the higher melting component) to assume a characteristic "S" shape progressively approaching flatness: the existence of a submerged (metastable) miscibility gap closer and closer to the liquidus curve is apparent. Should this tendency to demix become so remarkable as to produce a break in the liquidus (at the so-called monotectic temperature) the emersion of a stable miscibility gap in the liquid state would be observed.

Liquid layering has been detected in eleven binaries with common cation, characterized, *inter alia*, with a large difference in the anion sizes:

$K(C_9H_{17}O_2)$ $+$ $K(NO_2)$	$Na(C_2H_3O_2)$ $+$ $Na(C_{18}H_{35}O_2)$
$K(C_{10}H_{19}O_2)$ $+$ $K(NO_2)$	$Na(i.C_5H_9O_2)$ $+$ $Na(NO_2)$
$K(C_{18}H_{35}O_2)$ $+$ $K(NO_2)$	$Na(C_6H_{11}O_2)$ $+$ $Na(NO_2)$
$Na(CHO_2)$ $+$ $Na(C_6H_{11}O_2)$	$Na(C_6H_{11}O_2)$ $+$ $Na(NO_3)$
$Na(CHO_2)$ $+$ $Na(C_7H_5O_2)$	$Na(C_{18}H_{35}O_2)$ $+$ $Na(NO_3)$
$Na(CHO_2)$ $+$ $Na(C_{18}H_{35}O_2)$	

Such an occurrence is more frequent (see below) for the quasi-binaries representing stable diagonals in reciprocal ternaries.

Attention is now called on a few characteristic situations encountered with *additive ternaries*, for which the phase diagrams can be represented by means of prisms, having an equilateral triangle as the base. For the three binaries participating in the ternary the triangle sides are taken as composition axes and the prism edges as temperature axes. The composition of a given ternary mixture is defined by a point inside the triangle, while the temperatures at which this mixture undergoes phase changes are reported on the corresponding perpendicular.

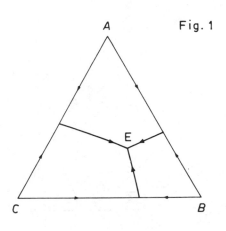

Fig. 1

In particular, when the binaries are all eutectic, three surfaces of primary crystallization can be drawn: in correspondence to each of them the proper component begins to separate(on cooling), while the system, trivariant in the liquid state, becomes bivariant. These surfaces cross along univariant curves, each starting from a binary eutectic and all joining at a ternary eutectic, the sole invariant point inside the prism.

The well known procedure of polythermal projection allows one to represent on the base triangle the solid-liquid equilibria of the system A, B, C, as shown in Fig. 1, where the arrows indicate the directions of decreasing temperature and E is the ternary eutectic at which equilibrium among one liquid and three solid phases occurs.

Should a binary intermediate compound (e.g., $A.C$) be formed and should it be able to exist (taking an expression by Zernike) "under all circumstances", the appearance shown in Fig. 2,a would be assumed by the polythermal projection. The latter is divided by the *triangulation line* $A.C-B$ in two *compatibility triangles*, corresponding to the subsystems $A, B, A.C$ (with eutectic E_1) and $A.C, B, C$ (with eutectic E_2), into which the system A, B, C can be resolved. Point s (*saddle* point) is a maximum along the curve joining the eutectics and in turn can be considered as a eutectic in the quasi-binary $A.C, B$.

In conditions of "not existence under all circumstances" for $A.C$, the univariant curve starting from one of the binary eutectics formed by $A.C$ with A and C, respectively, may cross the $A.C-B$ line before meeting the univariant curve coming from one of the two other eutectics, formed by A, B and B, C. In this case no saddle point can exist, and one of the ternary eutectics is transformed into the transition point P (Fig. 2,b). The $A.C-B$ line is still a triangulation line, even if $A.C, B$ no longer form a quasi-binary system.

Alternatively, the univariant curve starting from one of the binary eutectics formed by $A.C$

with A and C can reach that starting from the
other one, as shown in Fig. 2,c. This means
that $A.C$ decomposes on cooling into pure A and
C in correspondence with point R, here referred to as a *passage* point of the first type:
as a consequence the $A.C$–B line vanishes and
takes no more part in the triangulation.

Fig. 3 is self-explanatory and depicts three
cases (parallel to those just described) which
may occur when $A.C$ melts incongruently.

It must be added that a second type of R point

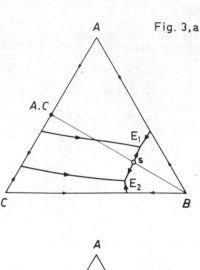

Fig. 3,a

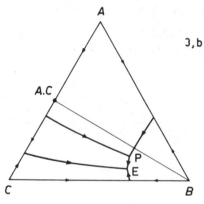

3,b

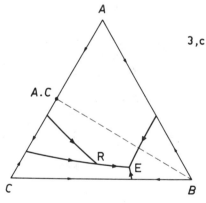

3,c

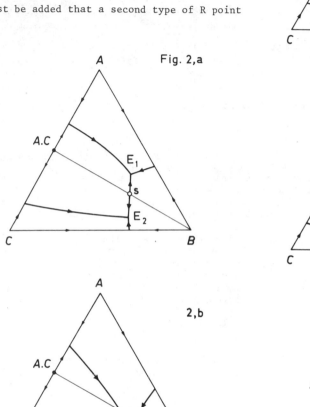

Fig. 2,a

2,b

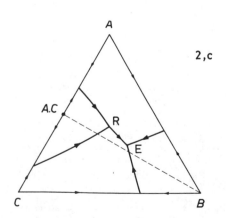

2,c

can exist as the meeting point of three univariant curves, one coming from higher and two going to lower temperatures: this is found in the
phase diagram when the crystallization field of
a binary compound (e.g., $A.C$ in Fig. 4) emerges
on cooling at the temperature of R. The compound
may then either remain stable (Fig. 4,a, where
$A.C$–B acts as a triangulation line), or vanish
in correspondence to an R point of the first
type (Fig. 4,b, where $A.C$–B is not a triangulation line and R_1, R_2 belong to the first and
second type, respectively).

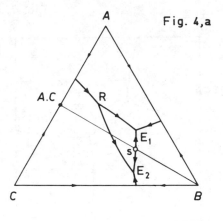

Fig. 4,a

Closed fields as those of Fig. 4 (which pertain to binary compounds) are obviously not to be confused with closed fields pertaining to ternary compounds, an example of which is offered by Fig. 5. The latter refers to the simple case of the congruently melting ternary compound *A.B.C*; the triangulation lines *A.B.C-A*, *A.B.C-B* and *A.B.C-C* subdivide the polythermal projection into three compatibility triangles corresponding to the subsystems *A,B,A.B.C*, *A.B.C,B,C* and *A.B.C,C,A*.

It can be realized easily that in any additive ternary, no matter how complex (provided that, as above said, both the components and the intermediate compounds be not completely miscible in the solid state), the number of compatibility triangles, the boundaries of which are either sides of the polythermal projection or (not crossing) triangulation lines, must be equal to the sum of the E + P points, inasmuch as no compatibility triangle must correspond to any R invariant.

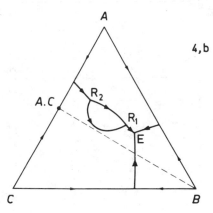

4,b

The nomenclature and symbolism concerning points P and R, however, is not unequivocal, as shown below.

(i) Zernike associates points R of the first type and points P, both meeting points of three univariant curves, two coming from higher and one going to lower temperatures, under the common designation of "peritectic of the first kind", indicated by the same symbol P. Moreover, he defines the above R points of the second type as "peritectic of the second kind", indicated by the symbol P[*].

(ii) Ricci designates points P as "transition or peritectic", and points R simply as "special triangular invariants", without any specific denomination.

(iii) Several Russian authors discriminate between *perekhodnaya tochka*, P, and *prokhodnaya tochka*, R, a distinction which we have tried to reproduce with the terms *transition point* and *passage point*. The latter was employed, e.g., in the English edition of a paper by Pochtakova (1965), and seems recommendable, although not currently employed.

(iv) Other Russian authors, however, sometimes mark also passage points with P: thus, in Diogenov and Sarapulova's paper (1964 b) on the additive ternary formed with potassium, sodium and rubidium acetates it is said, concerning the compound $3C_2H_3O_2K.2C_2H_3O_2Na$, that "its field tapers out rapidly" and ends in correspondence with an invariant at 483 K, which is apparently an R point, although indicated with the symbol P.

The possibility of misunderstanding is still increased by the fact that in some English editions expressions such as "transition point", "saddle point", "crossing point", etc., are found also when in the original Russian text the term *prokhodnaya tochka* is employed.

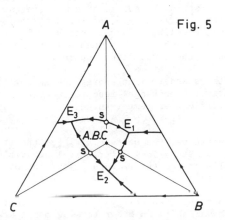

Fig. 5

For further details on the behaviour of additive ternaries, one is referred to specialized books. In any case, it is convenient to stress the fact that the obedience to triangulation rules is of basic importance for what concerns the internal consistence of a system. Thus, e. g., Pochtakova (1965) was induced to reinvesti-

gate the system formed with potassium, lithium
and sodium acetates previously studied by Diogenov (1956 b), just because on one hand the
latter author·"did not observe (...) triangulation", dividing "a system with four invariant
points (two peritectic and two eutectic) into
only two phase triangles", and on the other
hand she was unsuccessful when "attempted a triangulation using his results". Pochtakova was
indeed able to put into evidence, among other
facts, a ternary compound whose existence had
escaped Diogenov's attention, and then to perform
a more complex triangulation "into five phase
triangles", such as to fully justify the existence of the six (and not four) invariants found
by her, i.e., four transition points, one eutectic and one passage point.

As a final remark it can be noted that the
occurrence of continuous solid solutions may
be accompanied by a reduction in the number of
the ternary invariants: in the simplest case,
the polythermal projection shown in Fig. 1 may
change according either to Fig. 6,a (the eu=
tectic is not obliterated) or to Fig. 6,b (the
eutectic is obliterated).

A _reciprocal ternary system_ has already been defined as formed with two anionic (e.g., X⁻, Y⁻)

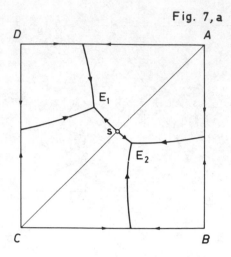

Fig. 7,a

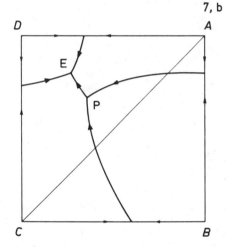

7,b

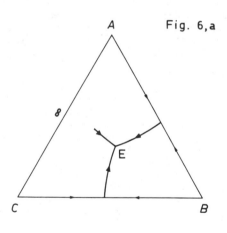

Fig. 6,a

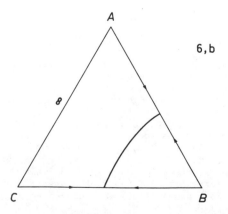

6,b

and two cationic (e.g., A^+, B^+) species, the combination of which can give rise to four different salts (AX = A, BX = B, BY = C, AY = D), so
that the pertinent phase diagram can be represented (as is well known) by means of a square
based prism.

In the simple case that complete miscibility does
not exist in the solid state and that no compound
of any kind is formed, pictures of the solid-
liquid equilibria like either Fig. 7,a or Fig.
7,b can be obtained through polythermal projection on the composition square.

If the standard free energy change for the meta-
thetic (or exchange) reaction occurring at the
solidification temperature

$$B + D \rightleftarrows A + C$$

is negative, the salt pair A,C is the stable
one and the A–C line (principal or stable diagonal) acts as the triangulation line of the
square.

As a rule, free energy values are unfortunately
not available: in order to single out stable

pairs it is therefore necessary to refer merely to enthalpy changes, $\Delta H°$, obtained from standard heats of formation at room temperature, $\Delta H_f°$(298 K).

To give an idea of the energy amounts involved in such exchange reactions, the pertinent figures found in the literature are listed hereafter: in each system the components underlined are those which form (or ought to form) the stable pair. The heats of formation employed by the different authors come either (a) from the Natl. Bur. Standards Circular 500, U.S. Dept. Commerce, Washington, 1952, or (b) from the compilation on "Heat Constants of Inorganic Compounds", ed. by Britske, E.V., & Kapustinskii, A.F., Izd. AN SSSR, Moscow–Leningrad, 1949 (in Russian), or (c) from unknown sources.

Component A	Component B	Component C	Component D	$-\Delta H°$ kJ equiv^{-1}	source of $\Delta H_f°$
$(CHO_2)K$	$\underline{(CHO_2)Na}$	$(Br)Na$	$\underline{(Br)K}$	19.8	a
Leonesi, Piantoni & Franzosini (1970)					
$(CHO_2)K$	$\underline{(CHO_2)Na}$	$(C_2H_3O_2)Na$	$\underline{(C_2H_3O_2)K}$	1.4	b
Sokolov (1965)					
$(CHO_2)K$	$\underline{(CHO_2)Na}$	$(Cl)Na$	$\underline{(Cl)K}$	12.4	a
Leonesi, Braghetti, Cingolani & Franzosini (1970)					
$(CHO_2)K$	$\underline{(CHO_2)Na}$	$(CNS)Na$	$\underline{(CNS)K}$	16.7	b
Sokolov & Pochtakova (1958 a)					
$(CHO_2)K$	$\underline{(CHO_2)Na}$	$(I)Na$	$\underline{(I)K}$	27.2	a
Leonesi, Piantoni & Franzosini (1970)					
$(CHO_2)K$	$\underline{(CHO_2)Na}$	$(NO_3)Na$	$\underline{(NO_3)K}$	14.1	b
Dmitrevskaya (1958 a)					
$(C_2H_3O_2)Cs$	$\underline{(C_2H_3O_2)K}$	$(NO_2)K$	$\underline{(NO_2)Cs}$	18.5	c
Diogenov & Morgen (1975 a)					
$(C_2H_3O_2)Cs$	$\underline{(C_2H_3O_2)K}$	$(NO_3)K$	$\underline{(NO_3)Cs}$	15.1	c
Diogenov, Erlykov & Gimel'shtein (1974)					
$(C_2H_3O_2)Cs$	$\underline{(C_2H_3O_2)Li}$	$(NO_3)Li$	$\underline{(NO_3)Cs}$	50.6	c
Diogenov & Gimel'shtein (1975)					
$(C_2H_3O_2)Cs$	$\underline{(C_2H_3O_2)Na}$	$(NO_2)Na$	$\underline{(NO_2)Cs}$	6.0	c
Diogenov & Morgen (1974 b)					
$(C_2H_3O_2)Cs$	$\underline{(C_2H_3O_2)Na}$	$(NO_3)Na$	$\underline{(NO_3)Cs}$	34.3	c
Gimel'shtein & Diogenov (1966)					
$(C_2H_3O_2)Cs$	$\underline{(C_2H_3O_2)Na}$	$(NO_3)Na$	$\underline{(NO_3)Cs}$	22.6	c
Diogenov, Erlykov & Gimel'shtein (1974)					
$(C_2H_3O_2)Cs$	$\underline{(C_2H_3O_2)Rb}$	$(NO_2)Rb$	$\underline{(NO_2)Cs}$	8.4	c
Diogenov & Morgen (1975 b)					
$(C_2H_3O_2)Cs$	$\underline{(C_2H_3O_2)Rb}$	$(NO_3)Rb$	$\underline{(NO_3)Cs}$	4.6	c
Diogenov, Erlykov & Gimel'shtein (1974)					
$(C_2H_3O_2)K$	$\underline{(C_2H_3O_2)Li}$	$(NO_3)Li$	$\underline{(NO_3)K}$	5.3	c
Diogenov, Nurminskii & Gimel'shtein (1957)					
$(C_2H_3O_2)K$	$\underline{(C_2H_3O_2)Li}$	$(NO_3)Li$	$\underline{(NO_3)K}$	52.9	c
Gimel'shtein & Diogenov (1958)					

$(C_2H_3O_2)K$	$\underline{(C_2H_3O_2)Li}$	$(NO_3)Li$	$\underline{(NO_3)K}$	33.9	c
Gimel'shtein (1970)					
$(C_2H_3O_2)K$	$\underline{(C_2H_3O_2)Li}$	$(NO_3)Li$	$\underline{(NO_3)K}$	35.6	c
Diogenov, Erlykov & Gimel'shtein (1974)					
$(C_2H_3O_2)K$	$\underline{(C_2H_3O_2)Na}$	$(Br)Na$	$\underline{(Br)K}$	17.6	c
Il'yasov & Bergman (1961)					
$(C_2H_3O_2)K$	$\underline{(C_2H_3O_2)Na}$	$(Cl)Na$	$\underline{(Cl)K}$	10.2	c
Il'yasov & Bergman (1960 a)					
$(C_2H_3O_2)K$	$\underline{(C_2H_3O_2)Na}$	$(CNS)Na$	$\underline{(CNS)K}$	15.5	b
Sokolov (1966)					
$(C_2H_3O_2)K$	$\underline{(C_2H_3O_2)Na}$	$(I)Na$	$\underline{(I)K}$	25.0	c
Diogenov & Erlykov (1958)					
$\underline{(C_2H_3O_2)K}$	$(C_2H_3O_2)Na$	$\underline{(NO_2)Na}$	$(NO_2)K$	3.1	b
Bergman & Evdokimova (1956)					
$(C_2H_3O_2)K$	$\underline{(C_2H_3O_2)Na}$	$(NO_3)Na$	$\underline{(NO_3)K}$	12.7	b
Bergman & Evdokimova (1956)					
$(C_2H_3O_2)K$	$\underline{(C_2H_3O_2)Na}$	$(NO_3)Na$	$\underline{(NO_3)K}$	12.6	c
Diogenov, Erlykov & Gimel'shtein (1974)					
$(C_2H_3O_2)_2K_2$	$\underline{(C_2H_3O_2)_2Na_2}$	$(S_2O_3)Na_2$	$\underline{(S_2O_3)K_2}$	11.2	b
Golubeva, Bergman & Grigor'eva (1958)					
$\underline{(C_2H_3O_2)K}$	$(C_2H_3O_2)Rb$	$\underline{(NO_3)Rb}$	$(NO_3)K$	10.5	c
Diogenov, Erlykov & Gimel'shtein (1974)					
$\underline{(C_2H_3O_2)Li}$	$(C_2H_3O_2)Na$	$\underline{(NO_3)Na}$	$(NO_3)Li$	3.3	c
Diogenov (1956 a)					
$\underline{(C_2H_3O_2)Li}$	$(C_2H_3O_2)Na$	$\underline{(NO_3)Na}$	$(NO_3)Li$	40.2	a
Gimel'shtein & Diogenov (1958)					
$\underline{(C_2H_3O_2)Li}$	$(C_2H_3O_2)Na$	$\underline{(NO_3)Na}$	$(NO_3)Li$	28.0	c
Diogenov, Erlykov & Gimel'shtein (1974)					
$\underline{(C_2H_3O_2)Li}$	$(C_2H_3O_2)Rb$	$\underline{(NO_3)Rb}$	$(NO_3)Li$	46.2	c
Diogenov, Erlykov & Gimel'shtein (1974)					
$\underline{(C_2H_3O_2)Na}$	$(C_2H_3O_2)Rb$	$\underline{(NO_3)Rb}$	$(NO_3)Na$	18.0	c
Gimel'shtein & Diogenov (1970)					

The substitution of enthalpy changes at room temperature for free energy changes at the solidification temperature obviously represents a crude approximation, and any inference therefrom may be erroneous, particularly when the $\Delta H°$ values are small. Help in identifying the stable pair in a given system can be obtained from the shape and relative size of the crystallization fields : on this ground, for example, Bergman & Evdokimova (1956) could reasonably assume the pair $(C_2H_3O_2)Na,(NO_2)K$ as the stable one in the reciprocal system formed with potassium and sodium acetates and nitrites, although the $\Delta H°$ value calculated by them (-3.1 kJ equiv^{-1}) would rather indicate the pair $(C_2H_3O_2)K,(NO_2)Na$. On the same ground, Golubeva, Bergman & Grigor'eva (1958) inferred the stable pair in the system formed with potassium and sodium acetates and thiosulfates to be $(C_2H_3O_2)_2K_2,(S_2O_3)Na_2$ instead of $(C_2H_3O_2)_2Na_2,(S_2O_3)K_2$. In other cases, e.g., for the systems formed with potassium and sodium formates and acetates and with potassium and rubidium acetates and nitrates, the particular features of the composi=

Fig. 8

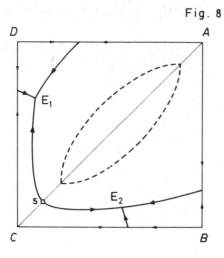

tion square led Sokolov (1965) and Diogenov & Gimel'shtein (1965), respectively, to designate them as "adiagonal", in spite of the actual $\Delta H°$ values.

In a general even if qualitative way, reciprocal ternaries are also classified by Russian authors in terms of "degree of irreversibility", although the latter quantity is not clearly defined from a thermodynamic point of view.

A system is thus indicated as "reversible", "irreversible" or "singular" when the pertinent value of $-\Delta G°$ (or $-\Delta H°$) does not exceed ≈ 10 kJ equiv^{-1}, or ranges between ≈ 10 and ≈ 40 kJ equiv^{-1}, or is still larger: these merely indicative figures may account for the increasing stability of the stable pair in the sequence reversible → irreversible → singular.

It is in the latter case that not infrequently liquid-liquid equilibria occur. The *stratification lens* (polythermal projection of the impinging curve of the miscibility gap on the liquidus surface) often looks like a more or less regular ellipse: the main axis as a rule coincides (or almost coincides) with the principal diagonal, so that the composition square assumes an appearance as shown in Fig. 8. The known liquid-liquid equilibria are listed hereafter: in a given system the stable pair is formed with the components underlined and usually behaves as a quasi-binary.

Component A	Component B	Component C	Component D	
$(C_2H_3O_2)Cs$	$\underline{(C_2H_3O_2)Li}$	$(NO_3)Li$	$\underline{(NO_3)Cs}$	
$(C_2H_3O_2)K$	$\underline{(C_2H_3O_2)Li}$	$(CNS)Li$	$\underline{(CNS)K}$	
$(C_2H_3O_2)K$	$\underline{(C_2H_3O_2)Li}$	$(NO_3)Li$	$\underline{(NO_3)K}$	
$\underline{(C_2H_3O_2)Li}$	$(C_2H_3O_2)Rb$	$\underline{(NO_3)Rb}$	$(NO_3)Li$	
$(C_3H_5O_2)K$	$\underline{(C_3H_5O_2)Li}$	$(CNS)Li$	$\underline{(CNS)K}$	
$(C_3H_5O_2)K$	$\underline{(C_3H_5O_2)Li}$	$(NO_3)Li$	$\underline{(NO_3)K}$	
$\underline{(C_3H_5O_2)Li}$	$(C_3H_5O_2)Na$	$\underline{(CNS)Na}$	$(CNS)Li$	
$\underline{(C_3H_5O_2)Li}$	$(C_3H_5O_2)Na$	$\underline{(NO_3)Na}$	$(NO_3)Li$	
$(C_4H_7O_2)K$	$\underline{(C_4H_7O_2)Li}$	$(CNS)Li$	$\underline{(CNS)K}$	
$(C_4H_7O_2)K$	$\underline{(C_4H_7O_2)Li}$	$(NO_3)Li$	$\underline{(NO_3)K}$	
$\underline{(C_4H_7O_2)Li}$	$(C_4H_7O_2)Na$	$\underline{(CNS)Na}$	$(CNS)Li$	
$\underline{(C_4H_7O_2)Li}$	$(C_4H_7O_2)Na$	$\underline{(NO_3)Na}$	$(NO_3)Li$	
$(i.C_4H_7O_2)K$	$(i.C_4H_7O_2)Li$	$(CNS)Li$	$(CNS)K$	stable pair not stated
$(i.C_4H_7O_2)K$	$(i.C_4H_7O_2)Li$	$(NO_3)Li$	$(NO_3)K$	stable pair not stated
$(i.C_4H_7O_2)Li$	$(i.C_4H_7O_2)Na$	$(CNS)Na$	$(CNS)Li$	stable pair not stated
$(i.C_4H_7O_2)Li$	$(i.C_4H_7O_2)Na$	$(NO_3)Na$	$(NO_3)Li$	stable pair not stated
$(C_5H_9O_2)K$	$\underline{(C_5H_9O_2)Li}$	$(CNS)Li$	$\underline{(CNS)K}$	
$(C_5H_9O_2)K$	$\underline{(C_5H_9O_2)Li}$	$(NO_3)Li$	$\underline{(NO_3)K}$	
$(C_5H_9O_2)K$	$\underline{(C_5H_9O_2)Na}$	$(NO_3)Na$	$\underline{(NO_3)K}$	
$\underline{(C_5H_9O_2)Li}$	$(C_5H_9O_2)Na$	$\underline{(CNS)Na}$	$(CNS)Li$	
$\underline{(C_5H_9O_2)Li}$	$(C_5H_9O_2)Na$	$\underline{(NO_3)Na}$	$(NO_3)Li$	
$(i.C_5H_9O_2)K$	$(i.C_5H_9O_2)Li$	$(CNS)Li$	$(CNS)K$	stable pair not stated
$(i.C_5H_9O_2)K$	$(i.C_5H_9O_2)Li$	$(NO_3)Li$	$(NO_3)K$	stable pair not stated
$\underline{(i.C_5H_9O_2)K}$	$(i.C_5H_9O_2)Na$	$\underline{(CNS)Na}$	$(CNS)K$	

$(i.C_5H_9O_2)K$	$(i.C_5H_9O_2)Na$	$(NO_3)Na$	$(NO_3)K$	stable pair not stated
$\underline{(i.C_5H_9O_2)Li}$	$(i.C_5H_9O_2)Na$	$\underline{(CNS)Na}$	$(CNS)Li$	
$(i.C_5H_9O_2)Li$	$(i.C_5H_9O_2)Na$	$(NO_3)Na$	$(NO_3)Li$	stable pair not stated
$(C_6H_{11}O_2)K$	$\underline{(C_6H_{11}O_2)Li}$	$(CNS)Li$	$\underline{(CNS)K}$	
$(C_6H_{11}O_2)K$	$\underline{(C_6H_{11}O_2)Li}$	$(NO_3)Li$	$\underline{(NO_3)K}$	
$(C_6H_{11}O_2)K$	$\underline{(C_6H_{11}O_2)Na}$	$(NO_3)Na$	$\underline{(NO_3)K}$	
$\underline{(C_6H_{11}O_2)Li}$	$(C_6H_{11}O_2)Na$	$\underline{(CNS)Na}$	$(CNS)Li$	
$\underline{(C_6H_{11}O_2)Li}$	$(C_6H_{11}O_2)Na$	$\underline{(NO_3)Na}$	$(NO_3)Li$	
$(C_7H_{13}O_2)K$	$\underline{(C_7H_{13}O_2)Na}$	$(CNS)Na$	$\underline{(CNS)Li}$	

Among the above systems a peculiar behaviour is offered by that of potassium and sodium *iso*-valerates and thiocyanates, where the *heteroionic* compound $(i.C_5H_9O_2)Na.(CNS)K$ "has limited miscibility with components" and gives rise to a stratification lens whose main axis is approximately parallel to the unstable (instead of the stable) diagonal, as reported by Sokolov, Tsindrik & Dmitrevskaya (1961).

Tables 1-5 summarize the main features of those systems on which sufficiently detailed data were found in the literature.

Compositions are given in mol per cent, unless otherwise specified.

In a continuous series of solid solutions "m" and "M" indicate a point of minimum and maximum, respectively.

The formulas of intermediate compounds (in italic) are underlined in the case of incongruent melting and bracketted when uncertain.

The solid-solid transition temperatures of pure components and the list of abbreviations are to be found in Chapter 1.2.

Invariants are classified as eutectic (marked with "E") and non-eutectic. Concerning binary systems, non-eutectic invariants are always peritectic (marked with "P"), whereas in the ternaries they may be either transition or passage points (marked with "P" and "$\underline{P}$", respectively). Among the latter are included all invariants corresponding to the definitions of either type of points R, independently of the fact that they were marked with "R" or with "P" in the original papers (°).

A further symbol, "$\underline{E}$", is employed in the following few special cases.

(i) Characteristic point at 389 K in the system of potassium and lithium formates and nitrates (Sokolov & Tsindrik, 1969): in the compatibility triangle $(HCO_2)Li,(NO_3)K,(NO_3)Li$ "a peritectic point P is formed (...) connected with the presence in this triangle of one more point {R_1 *in Fig. 4*, Editor's note} which from the drop in temperature is a eutectic, but in view of the fact that the field of the compound $KNO_3.LiNO_3$ tapers out is a peritectic {*passage point*} and does not participate in the triangulation".

(ii) Characteristic points at 396 K in the system of potassium and lithium acetates and nitrates (Sokolov & Tsindrik, 1969), at 399 K in the system of potassium and lithium propionates and nitrates (Sokolov & Tsindrik, 1969) and at 500 K in the system of potassium and sodium acetates and propionates (Sokolov & Pochtakova, 1958 b): the situations are analogous to that of (i).

(°) The symbol "P" is also used to designate a point of *double rise* occurring in the system of potassium and sodium formates and nitrates, the peculiar features of which are explained by Dmitrevskaya (1958 a) as follows: "The compound $2HCOOK.KNO_3$ formed on the $HCOOK-KNO_3$ side, tapers out in the reciprocal system and ceases to exist in the transition point {*passage point*, Editor's note} R at 164° {*437 K*}. The eutectics of the binary system $HCOOK-KNO_3$ have melting points {*423 and 419 K*} that are below either the transition point {*passage point*} or the melting point of the {*nearest*} ternary eutectic {*425 K*}. Point R in the reciprocal system is a point of double rise".

(iii) Characteristic point at 408 K in the system of cesium and lithium acetates and nitrates (Diogenov, Bruk & Nurminskii, 1965): in this case too the shape of the crystallization field suggests classification as a passage point in spite of the drop in temperature; in a revision of this system, in fact, a true passage point of the first type was found by Diogenov & Gimel' shtein (1975) at 425 K in the same region of the composition square, whereas the invariant at 408 K was not confirmed.

Binaries with common anion (Table 1) have been listed according to the increasing complexity of the anion and, for a given anion, following the alphabetical order of the cation. Binaries with common cation (Table 2) have been listed according to the alphabetical order of the cation and, for a given cation, following the increasing complexity of the component A anion, which may be either the only organic anion (when the other one is inorganic) or the simpler one, when both are organic. The fusion temperatures of components A and B are reported in the second and fourth column, respectively: those which actually are "clearing" temperatures (i.e., equilibrium temperatures between the isotropic and an anisotropic liquid phase, see Chapter 1.2) are in italic. Should such anisotropic phases be present, caution is to be taken in accepting some statements (reported in the Tables as given in the literature) on the topology of the concerned systems: e.g., the indication of "continuous solid solutions" might rather be interpreted as "complete miscibility between two mesomorphic liquids".

A number of temperatures have not been tabulated, inasmuch as it was doubtful whether they were to be attributed to peritectic invariants or to be intended as conjectural fusion temperatures of incongruently melting compounds.

Ternaries are arranged (Tables 3-5) in a way parallel to that chosen for binaries, with the additional specification that in a reciprocal ternary the salt formed with the only (or simpler) organic anion and with the cation coming first in the alphabetical order is taken as component A, component B being that obtained from A by changing the cation.

Characteristic points along the diagonals of a reciprocal ternary are designated as $(AC)_1$, $(AC)_2$, ..., $(BD)_1$, $(BD)_2$, ... and underlined when belonging to the diagonal recognized as the stable one.

Table 6 collects miscellaneously the information (less detailed than that summarized in Tables 1-5) available for some one hundred other binaries and reciprocal ternaries.

Among the pertinent papers, that by Baum, Demus & Sackmann (1970) deserves special attention inasmuch as it is the only one in the literature where a clear picture is given of the phase relationships occurring in systems formed with salts exhibiting liquid anisotropic phases. The authors subdivide the systems studied into four types, shown in Fig. 9 to which reference is made in Table 6. In order to better realize the features of such systems, the phase transition temperatures and designations of the pertinent pure components (see Chapter 1.2) are also to be taken into account.

The polythermal projections of a remarkable number of the systems listed in Tables 1-6 may be found in "Molten Salts with Organic Anions - An Atlas of Phase Diagrams" by Franzosini, P., Ferloni, P., and Spinolo, G., CNR - Institute of Physical Chemistry, University of Pavia (Italy), 1973.

Reciprocal systems more complicated than the ternaries have been studied only occasionally. Without entering into details, it should be mentioned here that some cross sections of the reciprocal quaternaries formed with cadmium, potassium and sodium acetates and bromides, and with potassium, lithium and sodium acetates and nitrates were investigated respectively by Il'yasov & Bergman (1960 b) and by Diogenov & Chumakova (1975).

The present tabulation covers the systems in which participate linear and branched alkanoate anions, or benzoate, whereas systems involving other (more unusual) organic anions, which anyway represent a negligible minority of cases, have been deliberately omitted.

References are listed in alphabetical order in the last section of this Chapter, while those pertinent to a given table are reported at the end of the table in abridged form.

TABLE 1. Binary systems with common anion

Component A	T/K	Component B	T/K	Characteristic point Type	Characteristic point T/K	Characteristic point $\% B$	Year	Ref.	Notes
$(CHO_2)_2Ba$	–	$(CHO_2)_2K_2$	441.9	P	465.4	62.7	1972	1	a
				E	435.8	92.6			
$(CHO_2)_2Ba$	–	$(CHO_2)_2Na_2$	530.7	P	515.2	48.2	1972	1	b
				E	498.0	64.6			
$(CHO_2)_2Ba$	–	$(CHO_2)_2Tl_2$	374.0	E	368.6	92.1	1972	1	c
$(CHO_2)_2Ca$	–	$(CHO_2)_2K_2$	441.9	E	436.4	94.3	1972	1	d
$(CHO_2)_2Ca$	–	$(CHO_2)_2Na_2$	530.7	E	506.6	75.7	1972	1	e
$(CHO_2)_2Ca$	–	$(CHO_2)_2Tl_2$	374.0	E	367.4	91.2	1972	1	f
$(CHO_2)K$	440	$(CHO_2)Li$	546	E_1	391	25	1969	2	
				$A.B$	–				
				E_2	413	60.5			
$(CHO_2)_2K_2$	441.9	$(CHO_2)_2Mg$	–	E	440.6	1.3	1972	1	g
$(CHO_2)_2K_2$	440	$(CHO_2)_2Mg$	–	–	–	–	1974	3	h
$(CHO_2)K$	440	$(CHO_2)Na$	531	E_1	440	4	1958	4	
				$3A.B$	455				
				E_2	441	50.5			
$(CHO_2)K$	441.9	$(CHO_2)Na$	530.7	E_1	436.7	4.3	1970	5	
				$3A.B$	453.2				
				E_2	438.2	49.5			
$(CHO_2)_2K_2$	441.9	$(CHO_2)_2Sr$	–	E	426.4	15.0	1972	1	i
				P	444.0	32.7			
$(CHO_2)Li$	546	$(CHO_2)Na$	531	m	443	50	1958	6	j
$(CHO_2)_2Mg$	–	$(CHO_2)_2Na_2$	531	E	525	79	1974	3	k

a: the system was investigated between 50 and 100 % B
b: the system was investigated between 45 and 100 % B
c: the system was investigated between 91 and 100 % B
d: the system was investigated between 89 and 100 % B
e: the system was investigated between 73 and 100 % B
f: the system was investigated between 90 and 100 % B
g: the system was investigated between 0 and 3 % B
h: the system was investigated between 0 and 25 % B; no eutectic was put into evidence (first addition of B: 5 %)
i: the system was investigated between 0 and 40 % B
j: the system was investigated between 20 and 100 % B; continuous solid solutions; the fusion temperature of component A was taken by extrapolation
k: the system was investigated between 70 and 100 % B

TABLE 1 (Continued)

$(CHO_2)_2Mg$	–	$(CHO_2)_2Tl_2$	374.0	m	370.2	97.0	1972	1	a
$(CHO_2)_2Na_2$	530.7	$(CHO_2)_2Sr$	–	E	508.6	24.6	1972	1	b
$(CHO_2)_2Sr$	–	$(CHO_2)_2Tl_2$	374.0	E	370.0	94.9	1972	1	c
$(C_2H_3O_2)_2Cd$	–	$(C_2H_3O_2)Cs$	454	E	437–440	85	1974	7	d
				2A.3B	528–530				
$(C_2H_3O_2)_2Cd$	–	$(C_2H_3O_2)K$	565	E_1	460	41	1954	8	e
				A.B	483				
				E_2	474	54			
				3A.4B	494				
				E_3	461	64			
				A.2B	475				
				E_4	456	73			
$(C_2H_3O_2)_2Cd$	–	$(C_2H_3O_2)_2K_2$	579	E	505	60	1962	9	f
$(C_2H_3O_2)_2Cd$	530–531	$(C_2H_3O_2)K$	579–581	E_1	503	18	1969	10	
				3A.B	515				
				E_2	423	42			
				A.B	461				
				E_3	421	58			
				A.2B	473				
				E_4	439	76			
$(C_2H_3O_2)_2Cd$	–	$(C_2H_3O_2)K$	579	E	461–469	54	1974	7	g
				<u>A.6B</u>					
$(C_2H_3O_2)_2Cd$	–	$(C_2H_3O_2)_2Na_2$	601	E	528	52	1962	9	h
$(C_2H_3O_2)_2Cd$	530–531	$(C_2H_3O_2)Na$	595	E_1	507	58	1969	10	i
				A.2B	527				
				E_2	496	75			
$(C_2H_3O_2)_2Cd$	–	$(C_2H_3O_2)Rb$	510	E	418–452	86	1974	7	j
				<u>A.2B</u>					

a: the system was investigated between 94 and 100 % B; solid solutions
b: the system was investigated between 0 and 30 % B
c: the system was investigated between 93 and 100 % B
d: the system was investigated between 40 and 100 % B; concerning the eutectic temperature the authors obtained 437 and 440 K by *cond* and *vis-pol*, respectively; concerning the fusion temperature of *2A.3B* the authors obtained 528 K by *DTA*, and 530 K by *cond* and *vis-pol*
e: the system was investigated between 30 and 100 % B
f: the system was investigated between 40 and 100 % B
g: the system was investigated between 25 and 100 % B; concerning the eutectic temperature the authors obtained 461 and 469 K by *cond* and *vis-pol*, respectively
h: the system was investigated between 35 and 100 % B
i: the system was investigated between 40 and 100 % B
j: the system was investigated between 40 and 100 % B; concerning the eutectic temperature the authors obtained 418, 442 and 452 K by *DTA*, *cond* and *vis-pol*, respectively

TABLE 1 (Continued)

A		B							
$(C_2H_3O_2)Cs$	453	$(C_2H_3O_2)K$	583	E	405	28.5	1960	11	
$(C_2H_3O_2)Cs$	453	$(C_2H_3O_2)K$	583	E	403	28.5	1965	12	
$(C_2H_3O_2)Cs$	460	$(C_2H_3O_2)K$	581	E	413	28.5	1975	13	
$(C_2H_3O_2)Cs$	467	$(C_2H_3O_2)K$	584	E	412	32	1977	14	
$(C_2H_3O_2)Cs$	458	$(C_2H_3O_2)Li$	563	E_1	408	24	1964	15	
				P	506	42.5			
				_A.B_	509				
				A.2B	566				
				E_2	513	88			
$(C_2H_3O_2)Cs$	458–459	$(C_2H_3O_2)Li$	557–561	E_1	420	23.0	1974	16	a
				A.2B	563–566				
				E_2	520	88.0			
$(C_2H_3O_2)Cs$	453	$(C_2H_3O_2)Na$	608	E	388	32	1964	17	
$(C_2H_3O_2)Cs$	467	$(C_2H_3O_2)Na$	607	E	392	36	1977	14	b
$(C_2H_3O_2)Cs$	453	$(C_2H_3O_2)Rb$	513	m	446	28	1964	17	c
$(C_2H_3O_2)Cs$	460	$(C_2H_3O_2)Rb$	511	m	445	–	1975	18	c
$(C_2H_3O_2)Cs$	–	$(C_2H_3O_2)_2Zn$	–	E_1	413	20	1972	19	d
				2A.B	463				
				E_2	377	45			
$(C_2H_3O_2)K$	583.7	$(C_2H_3O_2)Li$	564	E_1	454	36	1956	20	
				A.B	509				
				E_2	509	51.5			
				A.2B	548				
				E_3	495	84.5			
$(C_2H_3O_2)K$	575	$(C_2H_3O_2)Li$	557	E_1	470	37.6	1965	21	
				(A.2B)	(545)				
				E_2	511	87			
$(C_2H_3O_2)K$	574	$(C_2H_3O_2)Li$	557	E_1	470	38	1969	2	
				A.2B	–				
				E_2	507	87			

a: concerning the fusion temperatures of A, B and A.2B the lower values were obtained by DTA and the higher ones by _vis-pol_
b: owing to undercooling, the eutectic composition was singled out by extrapolation
c: continuous solid solutions
d: glassy region between 50 and 60 % B

TABLE 1 (Continued)

$(C_2H_3O_2)K$	–	$(C_2H_3O_2)Li$	–	E_1	470	37.5	1970	22	
				$A.2B$	548				
				E_2	507	87			
$(C_2H_3O_2)_2K_2$	575	$(C_2H_3O_2)_2Mg$	–	E	506–511	36.5	1974	3	a
$(C_2H_3O_2)K$	568.2	$(C_2H_3O_2)Na$	593.2	m	≈506	≈46	1915	23	b
$(C_2H_3O_2)K$	575	$(C_2H_3O_2)Na$	599	E	497	45	1956	24	
$(C_2H_3O_2)K$	583.7	$(C_2H_3O_2)Na$	610	m	501	45	1958	25	b
$(C_2H_3O_2)K$	579	$(C_2H_3O_2)Na$	599	$\underline{2A.B}$			1958	26	
$(C_2H_3O_2)K$	574–575	$(C_2H_3O_2)Na$	604	E_1	513	38.5	1958–1967	27, 28	
				$3A.2B$	514				
				E_2	508	46.5			
$(C_2H_3O_2)K$	579	$(C_2H_3O_2)Na$	601	$\underline{2A.B}$			1960	29	
				P	528	35			
				E	513	50			
$(C_2H_3O_2)K$	579	$(C_2H_3O_2)Na$	601	$\underline{2A.B}$			1960	30	
				P	511	36.5			
				E	505	50			
$(C_2H_3O_2)K$	583	$(C_2H_3O_2)Na$	608	E_1	513	–	1964	31	
				$3A.2B$	–				
				E_2	508	–			
$(C_2H_3O_2)K$	575	$(C_2H_3O_2)Na$	599	P	513	–	1975	32	
				$\underline{3A.2B}$					
				E	510	–			
$(C_2H_3O_2)K$	584	$(C_2H_3O_2)Na$	607	E	511	46	1977	14	c
$(C_2H_3O_2)K$	565	$(C_2H_3O_2)_2Pb$	477	E_1	448.1	28.3	1938	33	
				$2A.B$	(456)				
				E_2	442.7	47.4			
				$A.B$	(467)				
				E_3	(432)	62.5			
				$A.2B$	(442)				
				E_4	405.4	78.1			
$(C_2H_3O_2)K$	583	$(C_2H_3O_2)Rb$	509	m	508	85	1964	31	b

a: the system was investigated between 0 and 55 % B; concerning the eutectic temperature the author obtained 506 and 511 K by *DTA* and *vis-pol*, respectively
b: continuous solid solutions
c: owing to undercooling, the eutectic composition was singled out by extrapolation

TABLE 1 (Continued)

$(C_2H_3O_2)K$	579	$(C_2H_3O_2)_2Zn$	509	_8A.B_			1974	34	
				E	442	70			
$(C_2H_3O_2)Li$	564	$(C_2H_3O_2)Na$	610	E_1	499	18.5	1956	35, 20	
				_4A.B_	499–500				
				E_2	433	43			
$(C_2H_3O_2)Li$	557	$(C_2H_3O_2)Na$	604	E_1	492	31	1965	21	
				_3A.2B_	500				
				E_2	486	52.5			
$(C_2H_3O_2)Li$	564	$(C_2H_3O_2)Na$	599	E_1	492–494	–	1975	32	_a_
				_3A.B_	499				
				E_2	449–486				
$(C_2H_3O_2)Li$	563	$(C_2H_3O_2)Rb$	513	E_1	509	11.5	1964	15	
				_(3A.2B)_	582				
				_(2A.3B)_					
				E_2	449	74			
$(C_2H_3O_2)Li$	564	$(C_2H_3O_2)Rb$	509	E_1	509	12	1974	36	
				_2A.B_	573				
				E_2	460	74			
$(C_2H_3O_2)Li$	–	$(C_2H_3O_2)_2Zn$	–	_A.B_	538		1972	19	_b_
				_A.2B_					
				E	493	75			
$(C_2H_3O_2)_2Mg$	–	$(C_2H_3O_2)_2Na_2$	604	E_1	529–531	40.0	1974	3	_c_
				_A.B_	533				
				E_2	528	57.5			
$(C_2H_3O_2)Na$	600	$(C_2H_3O_2)Rb$	509	E_1	418	62	1958	37	
				_A.3B_	453				
				E_2	452–453	(76.5)			
$(C_2H_3O_2)Na$	601.5	$(C_2H_3O_2)_2Zn$	515.6	E_1	491–493	(28)	1939	38	
				_2A.B_	500.3				
				E_2	448.5–451.8	(54)			
$(C_2H_3O_2)Na$	–	$(C_2H_3O_2)_2Zn$	–	E_1	473	25	1972	19	
				_2A.B_	513				
				E_2	413	50			

_a_: concerning the eutectic temperatures the values 492, 486 and 494, 449 K were respectively taken from Fig. 2 and Fig. 4 of Ref. 32
_b_: glassy region between 15 and 30 % _B_
_c_: the system was investigated between 30 and 100 % _B_; concerning the fusion temperature of E_1 the author obtained 529 and 531 K by _vis-pol_ and _DTA_, respectively

TABLE 1 (Continued)

$(C_2H_3O_2)Na$	605	$(C_2H_3O_2)_2Zn$	509	2A.B				1974	34	a
				E	415–421	57				
$(C_2H_3O_2)_2Pb$	477	$(C_2H_3O_2)_2Zn$	517	E	433	16	wt	1914	39	b
$(C_2H_3O_2)Rb$	510	$(C_2H_3O_2)_2Zn$	509	E	432–436	60		1974	34	c
$(C_3H_5O_2)K$	638	$(C_3H_5O_2)Li$	602	E_1	564	32.5		1969	2	
				A.B	–					
				E_2	552	81				
$(C_3H_5O_2)_2K_2$	638	$(C_3H_5O_2)_2Mg$	577	E	579	36.5		1974	3	d
$(C_3H_5O_2)K$	638	$(C_3H_5O_2)Na$	571	E_1	585	34		1958	27	
				3A.2B	592					
				E_2	561	92				
$(C_3H_5O_2)K$	638	$(C_3H_5O_2)Na$	571	E_1	583	34		1958	40	
				3A.2B	592					
				E_2	560	92				
$(C_3H_5O_2)Li$	602	$(C_3H_5O_2)Na$	571	E	467	48		1958	41	
$(C_3H_5O_2)_2Mg$	577	$(C_3H_5O_2)_2Na_2$	571	E_1	508–512	35		1974	3	e
				A.B	517–521					
				E_2	515–517	57.5				
$(C_4H_7O_2)_2K_2$	677	$(C_4H_7O_2)_2Mg$	575	E_1	573–575	36.0		1974	3	f
				A.B	591					
				E_2	505–508	81.5				
$(C_4H_7O_2)K$	677	$(C_4H_7O_2)Na$	603	∞				1958	42	g
$(i.C_4H_7O_2)K$	633	$(i.C_4H_7O_2)Na$	535	E	521	92.5		1960	43	
$(C_4H_7O_2)Li$	602	$(C_4H_7O_2)Na$	603	E	495	50		1958	44	
$(C_4H_7O_2)_2Mg$	575	$(C_4H_7O_2)_2Na_2$	603	E_1	481–483	31		1974	3	h
				4A.3B	–					
				E_2	493	55				

a: concerning the eutectic temperature the authors obtained 415 and 421 K by vis-pol and cond, respectively
b: the system was investigated between 0 and 60 wt % B
c: concerning the eutectic temperature the authors obtained 432 and 436 K by vis-pol and cond, respectively
d: the system was investigated between 0 and 55 % B
e: concerning the fusion temperatures of E_1, A.B and E_2 the lower values were obtained by DTA and the higher ones by vis-pol
f: concerning the fusion temperatures of E_1 and E_2 the values 575, 505 and 573, 508 K were obtained by DTA and vis-pol, respectively
y: continuous solid solutions without any m or M point
h: concerning the fusion temperature of E_1 the author obtained 481 K by DTA and 483 K by vis-pol

TABLE 1 (Continued)

A		B							
$(C_5H_9O_2)K$	*717*	$(C_5H_9O_2)Na$	*630*	∞			1965	45	*a*
$(i.C_5H_9O_2)K$	*669*	$(i.C_5H_9O_2)Na$	*533*	∞			1963	46	*a*
$(i.C_5H_9O_2)K$	*669*	$(i.C_5H_9O_2)Na$	*535*	∞			1967	47	*a*
$(C_5H_9O_2)_2Mg$	537	$(C_5H_9O_2)_2Na_2$	*630*	P	497	32.5	1974	3	*b*
				A.3B					
				E	494	45			
$(C_6H_{11}O_2)K$	*717.7*	$(C_6H_{11}O_2)Na$	*638*	∞			1959	48	*a*

a: continuous solid solutions without any m or M point
b: the system was investigated between 27.5 and 100 % *B*; the fusion temperature of component *A* was taken by extrapolation

1. Berchiesi, Cingolani, Leonesi & Piantoni
2. Sokolov & Tsindrik (1969)
3. Pochtakova (1974)
4. Dmitrevskaya (1958 a)
5. Leonesi, Braghetti, Cingolani & Franzosini
6. Tsindrik
7. Nadirov & Bakeev (1974 b)
8. Lehrman & Schweitzer
9. Il'yasov
10. Pavlov & Golubkova (1969)
11. Nurminskii & Diogenov
12. Diogenov & Sergeeva
13. Diogenov & Morgen (1975 a)
14. Storonkin, Vasil'kova & Tarasov
15. Diogenov & Sarapulova (1964 a)
16. Sarapulova, Kashcheev & Diogenov
17. Diogenov & Sarapulova (1964 c)
18. Diogenov & Morgen (1975 b)
19. Pavlov & Golubkova (1972)
20. Diogenov (1956 b)
21. Pochtakova (1965)
22. Gimel'shtein (1970)
23. Baskov
24. Bergman & Evdokimova
25. Diogenov & Erlykov
26. Golubeva, Bergman & Grigor'eva
27. Sokolov & Pochtakova (1958 b)
28. Sokolov & Pochtakova (1967)
29. Il'yasov & Bergman (1960 b)
30. Nesterova & Bergman

31. Diogenov & Sarapulova (1964 b)
32. Diogenov & Chumakova
33. Lehrman & Leifer
34. Nadirov & Bakeev (1974 a)
35. Diogenov (1956 a)
36. Diogenov, Erlykov & Gimel'shtein
37. Gimel'shtein & Diogenov (1958)
38. Lehrman & Skell
39. Petersen
40. Dmitrevskaya & Sokolov (1958)
41. Tsindrik & Sokolov (1958 a)
42. Sokolov & Pochtakova (1958 c)
43. Sokolov & Pochtakova (1960 b)
44. Tsindrik & Sokolov (1958 b)
45. Dmitrevskaya & Sokolov (1965)
46. Pochtakova (1963)
47. Dmitrevskaya & Sokolov (1967)
48. Pochtakova (1959)

TABLE 2. Binary systems with common cation

Component A	T/K	Component B	T/K	Type	T/K	% B	Year	Ref.	Notes
				Characteristic point					

<table>
<thead>
<tr><th rowspan="2">Component A</th><th rowspan="2">T/K</th><th rowspan="2">Component B</th><th rowspan="2">T/K</th><th colspan="3">Characteristic point</th><th rowspan="2">Year</th><th rowspan="2">Ref.</th><th rowspan="2">Notes</th></tr>
<tr><th>Type</th><th>T/K</th><th>% B</th></tr>
</thead>
<tbody>
<tr><td>$Cs(C_2H_3O_2)$</td><td>460</td><td>$Cs(NO_2)$</td><td>678</td><td>E</td><td>398</td><td>36</td><td>1974</td><td>1</td><td></td></tr>
<tr><td>$Cs(C_2H_3O_2)$</td><td>453</td><td>$Cs(NO_3)$</td><td>680</td><td>E</td><td>415</td><td>25</td><td>1960</td><td>2</td><td>a</td></tr>
<tr><td>$Cs(C_2H_3O_2)$</td><td>455</td><td>$Cs(NO_3)$</td><td>680</td><td>E</td><td>429</td><td>25</td><td>1966</td><td>3</td><td></td></tr>
<tr><td>$K(CHO_2)$</td><td>441.9</td><td>$K(Br)$</td><td>–</td><td>E</td><td>434.5</td><td>5.3</td><td>1970</td><td>4</td><td>b</td></tr>
<tr><td>$K(CHO_2)$</td><td>440</td><td>$K(C_2H_3O_2)$</td><td>575</td><td>E</td><td>424</td><td>13</td><td>1965</td><td>5</td><td></td></tr>
<tr><td>$K(CHO_2)$</td><td>440</td><td>$K(C_3H_5O_2)$</td><td>638</td><td>E</td><td>433</td><td>5</td><td>1971</td><td>6</td><td></td></tr>
<tr><td>$K(CHO_2)$</td><td>440</td><td>$K(C_4H_7O_2)$</td><td>677</td><td>E</td><td>439</td><td>0.9</td><td>1974</td><td>7</td><td></td></tr>
<tr><td>$K(CHO_2)$</td><td>440</td><td>$K(i.C_4H_7O_2)$</td><td>629</td><td>E</td><td>437</td><td>1</td><td>1977</td><td>8</td><td></td></tr>
<tr><td>$K(CHO_2)$</td><td>441.9</td><td>$K(Cl)$</td><td>–</td><td>E</td><td>436.7</td><td>3.7</td><td>1970</td><td>4</td><td>c</td></tr>
<tr><td>$K(CHO_2)$</td><td>440</td><td>$K(CNS)$</td><td>450</td><td>E</td><td>356</td><td>47.5</td><td>1958</td><td>9</td><td></td></tr>
<tr><td>$K(CHO_2)$</td><td>441.9</td><td>$K(CNS)$</td><td>449.2</td><td>E</td><td>351.7</td><td>46</td><td>1971</td><td>10</td><td></td></tr>
<tr><td>$K(CHO_2)$</td><td>441.9</td><td>$K(I)$</td><td>–</td><td>E</td><td>429.5</td><td>8.7</td><td>1970</td><td>4</td><td>d</td></tr>
<tr><td>$K(CHO_2)$</td><td>441</td><td>$K(NO_2)$</td><td>709</td><td>E
P
<u>A.B</u></td><td>380
416</td><td>33.5
44</td><td>1961</td><td>11</td><td></td></tr>
<tr><td>$K(CHO_2)$</td><td>440</td><td>$K(NO_3)$</td><td>610</td><td>E$_1$
2A.B
E$_2$</td><td>423
(427)
419</td><td>32.5

44</td><td>1958</td><td>12</td><td></td></tr>
<tr><td>$K(CHO_2)$</td><td>441.9</td><td>$K(NO_3)$</td><td>–</td><td>(<u>4A.B</u>)
P
E</td><td>
399.7
387</td><td>
26.2
37.9</td><td>1970</td><td>13</td><td>e</td></tr>
<tr><td>$K(C_2H_3O_2)$</td><td>579</td><td>$K(Br)$</td><td>1013</td><td>E</td><td>563</td><td>10</td><td>1961</td><td>14</td><td>f</td></tr>
<tr><td>$K(C_2H_3O_2)$</td><td>578.7</td><td>$K(Br)$</td><td>–</td><td>E</td><td>561.1</td><td>10.5</td><td>1968</td><td>15</td><td>g</td></tr>
</tbody>
</table>

a: the system was investigated between 0 and 60 % B
b: the system was investigated between 0 and 14 % B
c: the system was investigated between 0 and 11 % B
d: the system was investigated between 0 and 17 % B
e: the system was investigated between 0 and 80 % B
f: the system was investigated between 0 and 22 % B
g: the system was investigated between 0 and 13 % B

TABLE 2 (Continued)

$K(C_2H_3O_2)$	574	$K(C_3H_5O_2)$	638	∞	–	–	1958	16	a
$K(C_2H_3O_2)$	574	$K(C_4H_7O_2)$	677	E	546	14.5	1960	17	
				P	623	79.5			
				(A.6B)					
$K(C_2H_3O_2)$	574	$K(i.C_4H_7O_2)$	633	E_1	564	13.5	1960	18	
				(3A.2B)	(578)				
				E_2	567	68			
$K(C_2H_3O_2)$	575	$K(C_5H_9O_2)$	717	E	553	12.5	1966	19	
				P	607	52.5			
				(2A.3B)					
$K(C_2H_3O_2)$	575	$K(i.C_5H_9O_2)$	669	(7A.B)			1963	20	
				P	543	18.5			
				E	542	50.0			
$K(C_2H_3O_2)$	574	$K(C_6H_{11}O_2)$	717.7	E	560	11.0	1959	21	
				P	592	39.0			
				(3A.2B)					
$K(C_2H_3O_2)$	579	$K(Cl)$	1045	E	566	10.5	1960	22	b
$K(C_2H_3O_2)$	578.7	$K(Cl)$	–	E	566.8	6.7	1968	15	c
$K(C_2H_3O_2)$	579	$K(CNS)$	449	E_1	405	57.5	1959	23	
				A.2B	407				
				E_2	403	73			
$K(C_2H_3O_2)$	575	$K(CNS)$	450	E_1	410	61	1966	24	
				A.2B	–				
				E_2	408	77.5			
$K(C_2H_3O_2)$	583.7	$K(I)$	956	E	550	(18)	1958	25	d
$K(C_2H_3O_2)$	578.7	$K(I)$	–	E	549.7	18.6	1968	15	e
$K(C_2H_3O_2)$	575	$K(NO_2)$	713	E	473	50	1956	26	
$K(C_2H_3O_2)$	575	$K(NO_2)$	709	P	483	45	1961	11	
				E	481	52			

a: continuous solid solutions without any m or M point
b: the system was investigated between 0 and 17.5 % B
c: the system was investigated between 0 and 8 % B
d: the system was investigated between 0 and 30 % B
e: the system was investigated between 0 and 20 % B

TABLE 2 (Continued)

$K(C_2H_3O_2)$	575	$K(NO_3)$	610	E_1	507	36	1956	26	
				$A.B$	511				
				E_2	495	61.5			
$K(C_2H_3O_2)$	583	$K(NO_3)$	610	E_1	493	39	1957	27	
				$A.B$	502				
				E_2	485	61			
$K(C_2H_3O_2)$	575	$K(NO_3)$	610	E_1	507	–	1975	28	
				$A.B$	511				
				E_2	497	–			
$K(C_3H_5O_2)$	638	$K(CNS)$	450	E	430	86	1966	24	
$K(C_3H_5O_2)$	639	$K(NO_2)$	709	*(3A.2B)*			1961	11	
				P	565	44			
				E	556	55			
$K(C_3H_5O_2)$	638	$K(NO_3)$	610	E	537	65	1958	29	
$K(C_4H_7O_2)$	*677*	$K(CNS)$	450	*6A.B*			1958	30	
				P	608	18			
				E	443	93.5			
$K(C_4H_7O_2)$	*677*	$K(NO_2)$	709	E_1	588	33.5	1961	11	
				5A.3B	(590)				
				E_2	579	45			
$K(C_4H_7O_2)$	*677*	$K(NO_3)$	610	E	556	58	1958	31	
$K(i.C_4H_7O_2)$	*638*	$K(NO_2)$	709	m	535	32.5	1961	11	a
$K(i.C_4H_7O_2)$	*629*	$K(NO_3)$	610	E	526	32.5	1960	32	
				P	529	47.5			
				A.B					
$K(C_5H_9O_2)$	*717*	$K(NO_2)$	709	E_1	594	37	1961	11	
				(4A.3B)–					
				E_2	596	47			
$K(C_5H_9O_2)$	*717*	$K(NO_3)$	610	E	583	49	1965	33	
$K(i.C_5H_9O_2)$	*669*	$K(NO_2)$	709	m	562	37.5	1961	11	a
$K(i.C_5H_9O_2)$	*669*	$K(NO_3)$	610	E_1	557	27.5	1967	34	b
				2A.B	–				
				E_2	549–553	46.0			

a: continuous solid solutions
b: concerning E_2, the lower and higher T values were taken by *DTA* and *vis-pol*, respectively

TABLE 2 (Continued)

$K(C_6H_{11}O_2)$	*717.7*	$K(NO_2)$	709	E_1	629	58	1961	11	
				(A.3B)	–				
				E_2	663	78.5			
$K(C_7H_{13}O_2)$	*725*	$K(NO_2)$	709	E_1	664	47.5	1961	11	
				2A.3B	(677)				
				E_2	662	74			
$K(C_8H_{15}O_2)$	*717*	$K(NO_2)$	709	E_1	593	26	1961	11	
				A.B	(642)				
				E_2	617	64.5			
$K(C_9H_{17}O_2)$	*694*	$K(NO_2)$	709	E	605	6.5	1961	11	*a*
$Li(CHO_2)$	546	$Li(C_2H_3O_2)$	557	E	513	62.5	1975	35	
$Li(CHO_2)$	546	$Li(CNS)$	539	E	429	48.5	1969	36	
$Li(CHO_2)$	546	$Li(NO_3)$	529	E	435	45	1958	37	*b*
$Li(C_2H_3O_2)$	557	$Li(CNS)$	539	E	439	51	1969	36	
$Li(C_2H_3O_2)$	564	$Li(NO_3)$	532	E	418	49	1956	38	
$Li(C_2H_3O_2)$	564	$Li(NO_3)$	532	E	449	51	1957	27	
$Li(C_2H_3O_2)$	557	$Li(NO_3)$	531	E	463	(51)	1969	39	
$Li(C_3H_5O_2)$	602	$Li(CNS)$	539	E_1	467	37.5	1969	36	
				(A.B)	(470)				
				E_2	466	60			
$Li(C_3H_5O_2)$	602	$Li(NO_3)$	529	E_1	517	30	1958	40	
				2A.3B	(526)				
				E_2	505	78.5			
$Li(C_4H_7O_2)$	602	$Li(CNS)$	539	E	481	50	1969	36	
$Li(C_4H_7O_2)$	602	$Li(NO_3)$	529	E	489	55	1958	41	
				P	505	85			
				(A.7B)					
$Na(CHO_2)$	530.7	$Na(Br)$	–	E	516.7	9.5	1970	4	*c*
$Na(CHO_2)$	531	$Na(C_2H_3O_2)$	604	E	515	10.5	1954	42	

a: liquid layering occurs at 643 K between 7.5 and 99 % *B*
b: the system was investigated between 20 and 100 % *B*; the fusion temperature of component *A* was taken by extrapolation
c: the system was investigated between 0 and 14 % *B*

TABLE 2 (Continued)

A	mp A	B	mp B	Type	T	%	Year	Ref	
Na(CHO$_2$)	531	Na(C$_3$H$_5$O$_2$)	571	E$_1$	528	6	1971	6	
				A.2B	–				
				E$_2$	566	98			
Na(CHO$_2$)	531	Na(C$_4$H$_7$O$_2$)	*603*	E$_1$	525	2.5	1954	42	
				A.B	614				
				E$_2$	581	89			
Na(CHO$_2$)	531	Na(i.C$_4$H$_7$O$_2$)	533	E$_1$	525	1.3	1954	42	
				A.B	603				
				E$_2$	523	96.5			
Na(CHO$_2$)	531	Na(i.C$_5$H$_9$O$_2$)	*535*	E$_1$	525	0.75	1954	42	
				3A.2B	593				
				E$_2$	518	94.5			
Na(CHO$_2$)	530.7	Na(Cl)	–	E	523.0	5.15	1970	4	*a*
Na(CHO$_2$)	531	Na(CNS)	584	E	460	36	1954	43	
Na(CHO$_2$)	531	Na(CNS)	584	E	462	38	1958	9	
Na(CHO$_2$)	530.7	Na(CNS)	580.7	<u>4A.B</u>			1971	44	*b*
				P	474.0	28.6			
				E	462.7	38.0			
Na(CHO$_2$)	528	Na(CNS)	581	E	443	36	1974	45	
Na(CHO$_2$)	530.7	Na(I)	–	E	500.9	17.25	1970	4	*c*
Na(CHO$_2$)	531	Na(NO$_2$)	557	E	442	48	1957	46	
Na(CHO$_2$)	531	Na(NO$_3$)	581	E	459	49	1954	43	
Na(CHO$_2$)	531	Na(NO$_3$)	581	E	460	48	1958	37	
Na(CHO$_2$)	530.7	Na(NO$_3$)	579.2	<u>3A.B</u>			1972	47	
				P	477	34.4			
				E	464	48.1			
Na(CHO$_2$)	528	Na(NO$_3$)	579	E	449	44	1974	45	
Na(C$_2$H$_3$O$_2$)	601	Na(Br)	1028	E	592	12.5	1961	14	*d*
Na(C$_2$H$_3$O$_2$)	601.3	Na(Br)	–	E	591.1	11.1	1968	15	*e*

a: the system was investigated between 0 and 8 % *B*
b: the system was investigated between 0 and 73 % *B*
c: the system was investigated between 0 and 25 % *B*
d: the system was investigated between 0 and 20 % *B*
e: the system was investigated between 0 and 12.5 % *B*

TABLE 2 *(Continued)*

$Na(C_2H_3O_2)$	604	$Na(C_3H_5O_2)$	571	E	564	95	1958	16	
$Na(C_2H_3O_2)$	604	$Na(C_4H_7O_2)$	*603*	E_1	539	33.5	1954,	42,	
				(3A.2B)	546		1960	17	
				E_2	523	69			
$Na(C_2H_3O_2)$	604	$Na(i.C_4H_7O_2)$	533–535	E	481	58	1954,	42,	
							1960	18	
$Na(C_2H_3O_2)$	604	$Na(C_5H_9O_2)$	*630*	E_1	537	31.5	1966	19	
				2A.B	541				
				E_2	526	54			
$Na(C_2H_3O_2)$	604	$Na(i.C_5H_9O_2)$	*535*	E	429	73	1954	42	
$Na(C_2H_3O_2)$	604	$Na(i.C_5H_9O_2)$	*533*	E	433	80	1963	20	
$Na(C_2H_3O_2)$	604	$Na(C_6H_{11}O_2)$	*638*	E_1	541	34.5	1954	42	
				5A.3B	(543)				
				E_2	533	49.5			
$Na(C_2H_3O_2)$	604	$Na(C_6H_{11}O_2)$	*638*	*4A.B*			1959	21	
				P	550	34			
				E	546	48.5			
$Na(C_2H_3O_2)$	604	$Na(C_7H_5O_2)$	738	E	588	2.6	1954	42	*a*
$Na(C_2H_3O_2)$	601	$Na(Cl)$	1073	E	601	10	1960	22	*b*
$Na(C_2H_3O_2)$	601.3	$Na(Cl)$	–	E	593.3	5.7	1968	15	*c*
$Na(C_2H_3O_2)$	604	$Na(CNS)$	584	E	517	54.5	1954	43	
$Na(C_2H_3O_2)$	599	$Na(CNS)$	581	E	509	55.5	1959	23	
$Na(C_2H_3O_2)$	601	$Na(CNS)$	581	E	507	55	1974	48	
$Na(C_2H_3O_2)$	610	$Na(I)$	943	E	583	23	1958	25	*d*
$Na(C_2H_3O_2)$	601.3	$Na(I)$	–	E	578.0	23.7	1968	15	*e*
$Na(C_2H_3O_2)$	599	$Na(NO_2)$	551	E	497	66	1956	26	
$Na(C_2H_3O_2)$	604	$Na(NO_2)$	557	E	499–500	65	1957,	46,	
							1970	49	
$Na(C_2H_3O_2)$	604	$Na(NO_3)$	581	E	497	58	1954	43	
$Na(C_2H_3O_2)$	599	$Na(NO_3)$	581	E	495	58	1956	26	

a: $C_7H_5O_2$ = benzoate; the system was investigated between 0 and 33 % *B*
b: the system was investigated between 0 and 17.5 % *B*
c: the system was investigated between 0 and 6.5 % *B*
d: the system was investigated between 0 and 32 % *B*
e: the system was investigated between 0 and 25 % *B*

TABLE 2 (Continued)

$Na(C_2H_3O_2)$	610	$Na(NO_3)$	581	*2A.B*			1956	38	
				P	539	(38.5)			
				E_1	498	57.5			
				A.4B	545				
				E_2	(541)	(82.5)			
$Na(C_2H_3O_2)$	600	$Na(NO_3)$	581	E	498	58	1966	3	
$Na(C_2H_3O_2)$	601	$Na(NO_3)$	579	E	491	56	1974	48	
$Na(C_2H_3O_2)$	599	$Na(NO_3)$	581	E	495	–	1975	28	
$Na(C_3H_5O_2)$	571	$Na(CNS)$	584	E	531	54	1954	43	
$Na(C_3H_5O_2)$	563	$Na(CNS)$	581	E	522	54	1974	45	
$Na(C_3H_5O_2)$	571	$Na(NO_2)$	557	E_1	566	1.4	1957	46	
				3A.B	588				
				E_2	527	80.5			
$Na(C_3H_5O_2)$	571	$Na(NO_3)$	581	E	528	56.5	1954	43	
$Na(C_4H_7O_2)$	*603*	$Na(i.C_4H_7O_2)$	533	m	494	72.5	1954	42	*a*
$Na(C_4H_7O_2)$	*603*	$Na(i.C_5H_9O_2)$	*535*	E	530	90.5	1954	42	
$Na(C_4H_7O_2)$	*603*	$Na(C_6H_{11}O_2)$	*638*	E_1	590	22.5	1954	42	
				(3A.B)	594				
				E_2	590	27.5			
$Na(C_4H_7O_2)$	*603*	$Na(C_7H_5O_2)$	736	(E)	603	0.13	1954	42	*b*
$Na(C_4H_7O_2)$	*603*	$Na(C_{18}H_{35}O_2)$	*581*	E	521	15	1954	42	
				3A.2B	663				
				P	582	96.5			
$Na(C_4H_7O_2)$	*603*	$Na(CNS)$	584	*3A.B*			1954	43	
				P	541	31.5			
				E	535	48.5			
$Na(C_4H_7O_2)$	*603*	$Na(NO_2)$	557	E_1	590	17.5	1957	46	
				3A.B	597				
				E_2	547	96			
$Na(C_4H_7O_2)$	*603*	$Na(NO_3)$	581	*(3A.B)*			1958	31	
				P	549	27			
				E	540	50			

a: continuous solid solutions
b: $C_7H_5O_2$ = benzoate; the system was investigated between 0 and 60 % B

TABLE 2 (Continued)

Na(i.C$_4$H$_7$O$_2$)	533	Na(i.C$_5$H$_9$O$_2$)	535	m	462	50	1954	42	a
Na(i.C$_4$H$_7$O$_2$)	533	Na(C$_6$H$_{11}$O$_2$)	638	E	433	23.5	1954	42	
Na(i.C$_4$H$_7$O$_2$)	533	Na(C$_7$H$_5$O$_2$)	736	E	501	3.5	1954	42	b
Na(i.C$_4$H$_7$O$_2$)	533	Na(C$_{18}$H$_{35}$O$_2$)	581	E	435	25.5	1954	42	
				2A.3B	(596)				
				P	585	94.5			
Na(i.C$_4$H$_7$O$_2$)	533	Na(CNS)	584	E	487	27.4	1954	43	
Na(i.C$_4$H$_7$O$_2$)	533	Na(NO$_2$)	557	E	520	8	1957	46	
Na(i.C$_4$H$_7$O$_2$)	533	Na(NO$_3$)	581	E	492	25.0	1954	43	
Na(i.C$_4$H$_7$O$_2$)	533	Na(NO$_3$)	581	E	493	25	1960	32	
Na(C$_5$H$_9$O$_2$)	629–630	Na(CNS)	584	E$_1$	562	46	1954,	43,	
				(A.B)	(564)		1972	50	
				E$_2$	560	56.5			
Na(C$_5$H$_9$O$_2$)	630	Na(NO$_2$)	557	E	555	99.96	1957	46	c
Na(C$_5$H$_9$O$_2$)	629–630	Na(NO$_3$)	581	E$_1$	564	40.5	1954,	43,	
				A.B	–		1972	50	
				E$_2$	554	58.5			
Na(i.C$_5$H$_9$O$_2$)	535	Na(C$_6$H$_{11}$O$_2$)	638	m	512	20	1954	42	a
Na(i.C$_5$H$_9$O$_2$)	535	Na(C$_7$H$_5$O$_2$)	736	E	534	3	1954	42	d
Na(i.C$_5$H$_9$O$_2$)	535	Na(C$_{18}$H$_{35}$O$_2$)	581	E	413	17.3	1954	42	
				A.2B	(596				
				P	582	93.5			
Na(i.C$_5$H$_9$O$_2$)	535	Na(CNS)	584	E	523	32	1954	43	
Na(i.C$_5$H$_9$O$_2$)	535	Na(NO$_2$)	557	E	542	21	1957	46	e
Na(i.C$_5$H$_9$O$_2$)	535	Na(NO$_3$)	581	E	527	31	1954	43	
Na(C$_6$H$_{11}$O$_2$)	638	Na(C$_7$H$_5$O$_2$)	736	P	644	13	1954	42	f

a: continuous solid solutions
b: C$_7$H$_5$O$_2$ = benzoate; the system was investigated between 0 and 75 % B
c: the system was investigated between 45 and 100 % B
d: C$_7$H$_5$O$_2$ = benzoate; the system was investigated between 0 and 70 % B
e: liquid layering occurs at 555 K between 66 and 98.4 % B
f: C$_7$H$_5$O$_2$ = benzoate; the system was investigated between 0 and 40 % B

TABLE 2 (Continued)

$Na(C_6H_{11}O_2)$	*638*	$Na(C_{18}H_{35}O_2)$	*581*	E	512	17.5	1954	42	
				2A.3B	(602)				
				P	587	94.5			
$Na(C_6H_{11}O_2)$	*638*	$Na(CNS)$	584	E	568	63	1954	43	
$Na(C_6H_{11}O_2)$	*638*	$Na(NO_3)$	581	E	560	56.5	1954	43	*a*
$Na(C_7H_5O_2)$	736	$Na(C_{18}H_{35}O_2)$	*581*	E	574	98.7	1954	42	*b*
$Rb(C_2H_3O_2)$	509	$Rb(NO_3)$	590	E_1	471	18.5	1958,	51,	*c*
				2A.B	475		1966	52	
				E_2	454-467	64.5-67			
$Rb(C_2H_3O_2)$	509	$Rb(NO_3)$	590	E_1	471	18	1974	53	
				A.B	476				
				E_2	467	66.5			

a: liquid layering occurs at 575 K between 60 and 99.84 % B
b: $C_7H_5O_2$ = benzoate; the system was investigated between 45 and 100 % B
c: concerning E_2 the lower and higher temperature and composition values are given in Ref. 51 and Ref. 52, respectively

1. Diogenov & Morgen (1974 b)
2. Nurminskii & Diogenov
3. Gimel'shtein & Diogenov (1966)
4. Leonesi, Braghetti, Cingolani & Franzosini
5. Sokolov (1965)
6. Sokolov & Minchenko (1971)
7. Sokolov & Minchenko (1974)
8. Sokolov & Minchenko (1977)
9. Sokolov & Pochtakova (1958 a)
10. Berchiesi, G., & Laffitte
11. Sokolov & Minich (1961)
12. Dmitrevskaya (1958 a)
13. Berchiesi, G., Cingolani & Leonesi
14. Il'yasov & Bergman (1961)
15. Piantoni, Leonesi, Braghetti & Franzosini
16. Sokolov & Pochtakova (1958 b)
17. Sokolov & Pochtakova (1960 a)
18. Sokolov & Pochtakova (1960 b)
19. Pochtakova (1966)
20. Pochtakova (1963)
21. Pochtakova (1959)
22. Il'yasov & Bergman (1960 a)
23. Golubeva, Aleshkina & Bergman
24. Sokolov (1966)
25. Diogenov & Erlykov
26. Bergman & Evdokimova
27. Diogenov, Nurminskii & Gimel'shtein
28. Diogenov & Chumakova
29. Dmitrevskaya & Sokolov (1958)
30. Sokolov & Pochtakova (1958 c)
31. Dmitrevskaya (1958 b)
32. Dmitrevskaya & Sokolov (1960)
33. Dmitrevskaya & Sokolov (1965)
34. Dmitrevskaya & Sokolov (1967)
35. Pochtakova (1975)
36. Sokolov & Dmitrevskaya
37. Tsindrik
38. Diogenov (1956 a)
39. Sokolov & Tsindrik (1969)
40. Tsindrik & Sokolov (1958 a)
41. Tsindrik & Sokolov (1958 b)
42. Sokolov (1954 b)
43. Sokolov (1954 a)
44. Cingolani, Berchiesi & Piantoni
45. Storonkin, Vasil'kova & Potemin (1974 a)
46. Sokolov (1957 b)
47. Berchiesi, M.A., Cingolani & Berchiesi
48. Storonkin, Vasil'kova & Potemin (1974 b)
49. Sokolov, Tsindrik & Khaitina (1970 a)
50. Sokolov & Khaitina (1972)
51. Gimel'shtein & Diogenov (1958)
52. Diogenov & Gimel'shtein (1966)
53. Diogenov, Erlykov & Gimel'shtein

TABLE 3. Ternary systems with common anion

Component A	Component B	Component C	Type	T/K	% A	% B	Year	Ref.	Notes
$(C_2H_3O_2)_2Cd$	$(C_2H_3O_2)_2K_2$	$(C_2H_3O_2)_2Na_2$	E	481	30	58	1962	1	a
			P	487	28	55			
$(C_2H_3O_2)Cs$	$(C_2H_3O_2)K$	$(C_2H_3O_2)Li$	E_1	346	52.5	27.5	1965	2	
			E_2	492	13	5			
			P_1	360	55	25			
			P_2	412	34	37			
			P_3	543	15	25			
$(C_2H_3O_2)Cs$	$(C_2H_3O_2)K$	$(C_2H_3O_2)Na$	E	363	50	26	1966	3	
			P	471	14	43			b
$(C_2H_3O_2)Cs$	$(C_2H_3O_2)K$	$(C_2H_3O_2)Na$	E	366	–	–	1977	4	c
			E	374	57	20			d
$(C_2H_3O_2)Cs$	$(C_2H_3O_2)K$	$(C_2H_3O_2)Rb$	–	–	–	–	1965	5	e
$(C_2H_3O_2)Cs$	$(C_2H_3O_2)Li$	$(C_2H_3O_2)Na$	E	353	59	17	1966	3	
			P_1	383	64	18			
			P_2	455	8	64			
			P_3	459	8	78			
$(C_2H_3O_2)Cs$	$(C_2H_3O_2)Li$	$(C_2H_3O_2)Rb$	E_1	398	62.0	21.0	1964	6	
			E_2	499	11.5	85.0			
			P_1	496	39.0	39.5			
			P_2	483	33.5	32.5			
			P_3	418	33.0	24.0			
$(C_2H_3O_2)Cs$	$(C_2H_3O_2)Na$	$(C_2H_3O_2)Rb$	E	368	55.5	30.5	1964	7	
			P	403	13.0	34.0			
$(C_2H_3O_2)K$	$(C_2H_3O_2)Li$	$(C_2H_3O_2)Na$	E_1	430	41.0	32.0	1956	8	
			E_2	430	9.0	57.0			
			P_1	431	44.0	31.0			
			P_2	478	13.0	78.0			
$(C_2H_3O_2)K$	$(C_2H_3O_2)Li$	$(C_2H_3O_2)Na$	E	435	38	32	1965	9	f
			P_1	465	49.5	15			

a: the field of component A was only partially investigated
b: the composition values are taken from Ref. 4
c: experimental value
d: calculated values
e: continuous solid solutions
f: the triple compound $4A.2B.C$ is formed by the reaction $3B.2C + A.2B + 3A = 4A.2B.C + 3B + C$

TABLE 3 (Continued)

				T/K	% A	% B	Year	Ref.	Notes
			P_2	445	45	25			
			P_3	460	55	31			
			P_4	453	20	44			
			P_5	484	21	56			
$(C_2H_3O_2)_2K_2$	$(C_2H_3O_2)_2Mg$	$(C_2H_3O_2)_2Na_2$	E_1	473	49	24	1976	10	*a*
			E_2	471	46	23			
			E_3	524	4	60			
			P_1	495	59	33			
			P_2	515	8	43			
$(C_2H_3O_2)K$	$(C_2H_3O_2)Na$	$(C_2H_3O_2)Rb$	E	417	8	33	1964	11	
			P_1	433	25	37			
			$\underline{P_2}$	483	46	43			
$(C_2H_3O_2)Li$	$(C_2H_3O_2)Na$	$(C_2H_3O_2)Rb$	E	411	21	25	1965	5	
			P_1	425	22	17			
			P_2	421	21.5	30			
			P_3	440	50	40			
			P_4	493	81	15			

a: the system was investigated up to 588 K; triple compound: *2A.5B.C*

1. Il'yasov	7. Diogenov & Sarapulova (1964 c)
2. Diogenov & Sergeeva	8. Diogenov (1956 b)
3. Diogenov & Shipitsyna	9. Pochtakova (1965)
4. Storonkin, Vasil'kova & Tarasov	10. Pochtakova (1976)
5. Diogenov, Borzova & Sarapulova	11. Diogenov & Sarapulova (1964 b)
6. Diogenov & Sarapulova (1964 a)	

TABLE 4. Ternary systems with common cation

Component *A*	Component *B*	Component *C*	-----Characteristic point-----				Year	Ref.	Notes
			Type	*T*/K	% *A*	% *B*			
$K(CHO_2)$	$K(Br)$	$K(CNS)$	E	348.2	55.2	2	1970	1	*a*
$K(CHO_2)$	$K(C_2H_3O_2)$	$K(CNS)$	-	-	-	-	1970	2	*b*

a: the system was investigated up to 513 K
b: no invariant points detected owing to glass formation

TABLE 4 (Continued)

$K(CHO_2)$	$K(Cl)$	$K(CNS)$	E	349.2	55	1.5	1970	1	a
$K(CHO_2)$	$K(CNS)$	$K(I)$	E	346.7	54.75	41.6	1970	1	a
$K(CHO_2)$	$K(CNS)$	$K(NO_3)$	E	328.7	48.3	34.9	1970	3	
			P	342.2	51.6	27.6			
$K(C_2H_3O_2)$	$K(CNS)$	$K(NO_3)$	E	379	19.5	54.5	1974	4	
$K_2(C_2H_3O_2)_2$	$K_2(CNS)_2$	$K_2(S_2O_3)$	E_1	383	21	62	1959	5	b
			E_2	389	31.5	52			
$Na(CHO_2)$	$Na(Br)$	$Na(CNS)$	E	458.2	60.7	3.7	1971	6	c
			P	468.2	68.7	4.6			
$Na(CHO_2)$	$Na(Cl)$	$Na(CNS)$	E	459.2	61.3	2.0	1971	6	c
			P	471.7	70.2	2.5			
$Na(CHO_2)$	$Na(CNS)$	$Na(I)$	E	453.2	58.6	32.8	1971	6	c
			P	463.7	66.7	23.1			
$Na(CHO_2)$	$Na(CNS)$	$Na(NO_3)$	E	421	52	26	1971	7	
$Na(CHO_2)$	$Na(CNS)$	$Na(NO_3)$	E	426.2	48.8	22.3	1972	8	
			P_1	440.7	56.5	21.9			
			P_2	430.7	50.2	24.5			
$Na(CHO_2)$	$Na(CNS)$	$Na(NO_3)$	E	411	42	26	1974	9	d
			E	430	42	26			e
$Na(CHO_2)$	$Na(NO_2)$	$Na(NO_3)$	E	415	37.5	33.5	1970	10	
			P	425	30.5	37.5			
$Na(C_2H_3O_2)$	$Na(CNS)$	$Na(NO_3)$	E	461	30	34	1971	7	
$Na(C_2H_3O_2)$	$Na(CNS)$	$Na(NO_3)$	E	456	32.5	30.5	1974	11	f
$Na_2(C_2H_3O_2)_2$	$Na_2(CNS)_2$	$Na_2(S_2O_3)$	E	495	32	40	1959	5	g
$Na(C_2H_3O_2)$	$Na(NO_2)$	$Na(NO_3)$	E	446	26	42	1970	10	
			P	451	20	47			
$Na(C_3H_5O_2)$	$Na(CNS)$	$Na(NO_3)$	E	487	18	39	1971	7	
$Na(C_3H_5O_2)$	$Na(CNS)$	$Na(NO_3)$	E	475	20.5	38.5	1974	9	d

a: the system was investigated up to 513 K
b: the system was investigated up to 523 K
c: the system was investigated up to 593 K
d: experimental values
e: calculated values
f: both calculated and experimental
g: the system was investigated up to 548 K

TABLE 4 (Continued)

			E	482	20.5	38.5			a
Na(C$_3$H$_5$O$_2$)	Na(NO$_2$)	Na(NO$_3$)	E$_1$	494	7	62	1970	12	
			E$_2$	490	8	50			
			E$_3$	523	38	7			
Na(C$_4$H$_7$O$_2$)	Na(CNS)	Na(NO$_2$)	E$_1$	487	2.5	40	1970	13	
			E$_2$	487	2	37			
Na(C$_4$H$_7$O$_2$)	Na(CNS)	Na(NO$_3$)	E	493	17.5	40	1970	13	
Na(C$_4$H$_7$O$_2$)	Na(NO$_2$)	Na(NO$_3$)	E$_1$	493	1	64	1970	12	
			E$_2$	493	1	54			
Na(i.C$_4$H$_7$O$_2$)	Na(NO$_2$)	Na(NO$_3$)	E	480	70	8	1970	12	
			P	497	2.5	53			
Na(C$_5$H$_9$O$_2$)	Na(CNS)	Na(NO$_3$)	E	501	0.5	45	1972	14	

a: calculated values

1. Cingolani, Gambugiati & Berchiesi
2. Sokolov & Minich (1970)
3. Berchiesi, G., Cingolani & Leonesi
4. Sokolov & Khaitina (1974)
5. Golubeva, Aleshkina & Bergman
6. Cingolani, Berchiesi & Piantoni
7. Sokolov & Khaitina (1971)
8. Berchiesi, M.A., Cingolani & Berchiesi
9. Storonkin, Vasil'kova & Potemin (1974 a)
10. Sokolov, Tsindrik & Khaitina (1970 a)
11. Storonkin, Vasil'kova & Potemin (1974 b)
12. Sokolov, Tsindrik & Khaitina (1970 b)
13. Sokolov & Khaitina (1970)
14. Sokolov & Khaitina (1972)

TABLE 5. Reciprocal ternary systems

Component A	Component B	Component C	Component D	Characteristic point						Year	Ref.	Notes
				Type	T/K	% A	% B	% C	% D			
$(CHO_2)K$	$(CHO_2)Li$	$(CNS)Li$	$(CNS)K$	E_1	395	–	10	44	46	1970	1	
				E_2	417	–	7	23	70			
				E_3	326	50	5	–	45			
				$\underline{P}$	337	52	12	–	36			
				$(AC)_1$	378	80						
				$(AC)_2$	390	67						
				$(AC)_3$	411	27.5						
				$(\underline{BD})$	425		14					
$(CHO_2)K$	$(CHO_2)Li$	$(NO_3)Li$	$(NO_3)K$	$\underline{E_1}$	389	–	10	44	46	1969	2	
				E_2	396	38.1	54.5	–	7.4			
				E_3	381	56.5	30.5	–	13			
				$\underline{P_1}$	425	–	51	42	7			
				$\underline{P_2}$	407	57.5	4.5	–	38			
				$(AC)_1$	397	78						
				$(AC)_2$	398	26						
				$(\underline{BD})$	479		81					
$(CHO_2)K$	$(CHO_2)Na$	$(Br)Na$	$(Br)K$	E_1	503	–	87.8	3.4	8.8	1970	3	a
				E_2	436	49.7	47.4	–	2.9			
				E_3	429	90.4	4.8	–	4.8			
				(AC)	428.9	95.2						
				$(\underline{BD})$	505.6		89.4					
$(CHO_2)K$	$(CHO_2)Na$	$(C_2H_3O_2)Na$	$(C_2H_3O_2)K$	E_1	419	51	38.5	10.5	–	1965	4	
				E_2	399	33.5	34	–	32.5			

a: the system was investigated up to 573 K

TAELE 5 (Continued)

System	Point	T					Year	Ref
(CHO₂)K — (CHO₂)Na — (C₃H₅O₂)Na — (C₃H₅O₂)K	E₃	414	83	4.5	–	12.5		
	P	459	25	–	37.5	37.5		
	(AC)₁	430	95.5					
	(AC)₂	404	68					
	(AC)₃	402	64.5					
	(BD)₁	503	89.5					
	(BD)₂	446	36				1971	5
(CHC₂)K — (CHO₂)Na — (C₄H₇O₂)Na — (C₄H₇O₂)K	E₁	434	50.5	47.5	–	2		
	E₂	558	–	2	87.5	10.5		
	E₃	437	68	24	–	8		
	E₄	421	93	3.5	–	3.5		
	P	539	–	10	52.5	37.5		
	(AC)₁	421	96					
	(AC)₂	478	62					
	(AC)₃	565	3					
	(BD)₁	520	94					
	(BD)₂	494	45				1974	6
(CHO₂)K — (CHO₂)Na — (i.C₄H₇O₂)Na — (i.C₄H₇O₂)K	E₁	435	46	53	1	–		
	E₂	436	72	25	3	–		
	E₃	432	97.5	1.5	–	1		
	(AC)₁	435	99					
	(AC)₂	489	60					
	(AC)₃	581	17					
	(BD)₁	525	97.5					
	(BD)₂	507	46.5					
	E₁	504	–	5	90	5		
	E₂	437	51	48.5	–	0.5		
	E₃	433	94	5	–	1	1977	7

TABLE 5 (Continued)

(CHO$_2$)K		(CHO$_2$)Na		Point	T					Year	Ref	α
(CHO$_2$)K	(Cl)Na	(CHO$_2$)Na	(Cl)K	P	443	89	10	—	1	1970	8	α
				(AC)$_1$	429	99						
				(AC)$_2$	487	66.5						
				(AC)$_3$	503	5						
				(BD)$_1$	525		99					
				(BD)$_2$	491		42					
				E$_1$	436	48.6	48.5	—	2.9			
				E$_2$	431.7	92.1	4.5	—	3.4			
				P	489	14.8	77.0	—	8.2			
				(AC)	432.6	96.6						
				(BD)	515.3		94.2					
(CHO$_2$)K	(CNS)Na	(CHO$_2$)Na	(CNS)K	E$_1$	386	—	21	26	53	1958	9	
				E$_2$	363	38.5	16	—	45.5			
				E$_3$	355	57.5	0.5	—	42			
				(AC)$_1$	429	96.4						
				(AC)$_2$	397	72						
				(AC)$_3$	423	34						
				(BD)	405		28.5					
(CHO$_2$)K	(I)Na	(CHO$_2$)Na	(I)K	E$_1$	499	—	81.7	14.8	3.5	1970	3	α
				E$_2$	436	49.6	46.7	—	3.7			
				E$_3$	426	88.4	3.7	—	7.9			
				(AC)$_1$	430	95.7						
				(AC)$_2$	436.7	92.0						
				(BD)	510.4		91.25					
(CHO$_2$)K	(NO$_3$)Na	(CHO$_2$)Na	(NO$_3$)K	E$_1$	429	—	38.5	34.5	27	1958	10	
				E$_2$	426	—	42	21.5	36.5			

α: the system was investigated up to 573 K

TABLE 5 (Continued)

Component A	Component B	Component C	Point	T (K)					Year	Ref
(CHO₂)Li	(C₂H₃O₂)Na	(C₂H₃O₂)Li	E₃	425	23.5	40	–	36.5	1975	11
			P₁	428	31.5	35	–	33.5		
			P₂	437	39	16	–	45		
			(AC)₁	431	65.5					
			(AC)₂	428	37					
			(AC)₃	437	31					
			(BD)	437		52				
(CHO₂)Na	(CNS)Na	(CNS)Li	E₁	435	46	40.5	13.5	–	1969	12
			E₂	463	23	–	51	26		
			P	481	12	–	36	52		
			(AC)	465	44					
			(BD)₁	498		85				
			(BD)₂	462		60				
			(BD)₃	503		16.5				
(CHO₂)Li	(NO₃)Na	(NO₃)Li	E₁	437	23.5	57.5	19	–	1958 [a]	13
			E₂	427	41	–	19	40		
			(AC)	477	74					
			(BD)₁	457		76				
			(BD)₂	491		23.5				
			E₁	419	40	–	14	46		
			E₂	427	49	37	14	–		
			(AC)	467	74					
			(BD)₁	448		76				
			(BD)₂	448		24				
(C₂H₃O₂)Cs	(C₂H₃O₂)K	(NO₂)Cs / (NO₂)K	E₁	375	–	39.5	20.0	40.5	1975	14
			E₂	382	29.5	35.0	–	35.5		

a: the field of component A was investigated up to 503 K

TABLE 5 (Continued)

Salt A	Salt B	Salt C	Salt D	Point	T	c1	c2	c3	c4	Year	Ref
$(C_2H_3O_2)Cs$	$(C_2H_3O_2)K$	$(NO_3)K$	$(NO_3)Cs$	$(AC)_1$	382	65.5			14.0	1960	15
				$(AC)_2$	378	39		44.5	26.5		
				(BD)	398		45	16.0	31.0		
$(C_2H_3O_2)Cs$	$(C_2H_3O_2)Li$	$(NO_3)Li$	$(NO_3)Cs$	E_1	394	57.5	28.5	-	14.0	1965	16
				E_2	438	-	29.0	44.5	26.5		
				P	458	-	53.0	16.0	31.0		
				$(AC)_1$	398	82					
				$(AC)_2$	453	40					
				$(AC)_3$	441	28					
				(BD)	444	-	57				
$(C_2H_3O_2)Cs$	$(C_2H_3O_2)Li$	$(NO_3)Li$	$(NO_3)Cs$	E_1	399	75.5	24.5	-	4.0		[a]
				E_2	411	16.5	75.0	-	8.5		
				E_3	408	-	11.0	50.0	39.0		
				P_1	483	65.0	23.0	-	12.0		
				P_2	430	-	50.5	46.0	3.5		
				$(AC)_1$	408	85.5					
				$(AC)_2$	420	22					
				(BD)	518		90.5				
$(C_2H_3O_2)Cs$	$(C_2H_3O_2)Li$	$(NO_3)Li$	$(NO_3)Cs$	E_1	400	69.5	26.5	-	4.0	1975	17
				E_2	509	11.5	83.5	-	5.0		
				E_3	409	-	51.0	43.5	5.5		
				P	425	-	6.5	58.0	35.5		
				$(AC)_1$	407	85					
				$(AC)_2$	421	20					
				(BD)	518		90				
$(C_2H_3O_2)Cs$	$(C_2H_3O_2)Na$	$(NO_2)Na$	$(NO_2)Cs$	E_1	395	-	5	46.5	48.5	1974	18

a: liquid layering occurs along the *AC* diagonal between 41 and 59.5 % *A*, and along the *BD* diagonal between 30 and 62.5 % *B* at 577 K

TABLE 5 (Continued)

Component A	Component B	Point	t					Year	Ref	Note
$(C_2H_3O_2)Cs$	$(NO_3)Na$	E_2	367	57.0	21.5	–	21.5	1966	19	a
		$(AC)_1$	367	78						
		$(AC)_2$	473	19						
		$(\underline{BD})$	–							
$(C_2H_3O_2)Na$	$(NO_3)Cs$	E_1	442	–	17	50	33			
		E_2	373	67	25	–	8			
		$(AC)_1$	388							
		$(AC)_2$	419							
		$(AC)_3$	464							a
		$(\underline{BD})$	–							
$(C_2H_3O_2)Rb$	$(NO_2)Cs$	E	389	32.5	32.5	–	35.0	1975	20	b
		$\underline{P}$	391	52.0	–	48.0	–			
$(C_2H_3O_2)Rb$	$(NO_3)Cs$	E_1	401	55.0	22.0	–	23.0	1966	21	
		E_2	438	–	44.0	26.0	30.0			
		P	433	15.0	53.5	–	31.5			
		(AC)	408	76						
		$(\underline{BD})_1$	448		78					
		$(\underline{BD})_2$	441		56					c
$(C_2H_3O_2)K$	$(CNS)Li$	E_1	411	–	3.5	53	43.5	1970	1	d
		E_2	433	–	2.5	22.5	75			
		E_3	395	26.5	2.5	–	71			
		E_4	390	38	2.5	–	59.5			
		$\underline{P}$	447	58	29	–	13			
		$(AC)_1$	435	78						
		$(AC)_2$	415	25						
		$(\underline{BD})$	437		5					

a: solid solutions
b: three and two characteristic points occur along the AC and BD diagonals, respectively (co-ordinates not clearly reported)
c: the BD diagonal is to be identified as the stable one according to Ref. 40
d: liquid layering occurs along the AC diagonal between 40 and 57.5 % A, and along the BD diagonal between 18 and 85 % B

TABLE 5 (Continued)

Point	T	$(C_2H_3O_2)K$	$(C_2H_3O_2)Li$	$(NO_3)Li$	$(NO_3)K$	Year	Ref
E	463	60	29	–	11	1957	22
P_1	467	50.5	27.5	–	22		
P_2	488	54	29.5	–	16.5		
P_3	498	15	82.5	–	2.5		
P_4	448	–	48.5	49	2.5		
$(AC)_1$	489	82.5					
$(AC)_2$	471	74					
$(AC)_3$	422	23.5					
$\underline{(BD)}$	513		87				*a*
E_1	464	72	18	10	–	1969	2
E_2	503	20.5	76	3.5	–		
$\underline{E_3}$	396	–	3	46	51		
P_1	483	71	8	21	–		
P_2	447	–	43	53	4		
$(AC)_1$	498	85					
$(AC)_2$	484	74					
$(AC)_3$	425	22.5					
$\underline{(BD)}$	527		94				*a*
E_1	463	60.5	27.5	–	12.0	1970	23
E_2	503	11.0	79.0	–	10.0		
$\underline{E_3}$	393	–	5.0	44.0	51.0		
$\underline{P_1}$	471	54.0	29.5	–	16.5		
P_2	445	–	51.5	46.0	2.5		
$(AC)_1$	493	84					
$(AC)_2$	475	74					
$(AC)_3$	428	23					
$\underline{(BD)}$	513		88				*b*

a: liquid layering occurs along the *AC* diagonal between 32 and 57 % *A*, and along the *BD* diagonal between 20 and 80 % *B*

b: liquid layering occurs along the *AC* diagonal between 35 and 56 % *A*, and along the *BD* diagonal between 22 and 73 % *B* at $\simeq$565 K

TABLE 5 (Continued)

System 1: $(C_2H_3O_2)K$ – $(C_2H_3O_2)Na$ – $(Br)Na$ – $(Br)K$ — 1961, [24], a

Point	T	$(C_2H_3O_2)K$	$(C_2H_3O_2)Na$	$(Br)Na$	$(Br)K$
E_1	498	48.5	48.5	-	3
E_2	567	-	88	1.5	10.5
P	518	62	34.5	-	3.5
(AC)	548	91			
(BD)	581		88		

System 2: $(C_2H_3O_2)K$ – $(C_2H_3O_2)Na$ – $(C_3H_5O_2)Na$ – $(C_3H_5O_2)K$ — 1958, [26]

Point	T	$(C_2H_3O_2)K$	$(C_2H_3O_2)Na$	$(C_3H_5O_2)Na$	$(C_3H_5O_2)K$
E_1	505	-	50.5	7.5	42
E_2	500	42	47	-	11
P	517	-	33.5	21.5	45
$(AC)_1$	513	43			
$(AC)_2$	519	36			
$(AC)_3$	558	6			
(BD)	509		57		

System 3: $(C_2H_3O_2)Na$ – $(C_2H_3O_2)K$ – $(C_4H_7O_2)K$ – $(C_4H_7O_2)Na$ — 1960, [27]

Point	T	$(C_2H_3O_2)Na$	$(C_2H_3O_2)K$	$(C_4H_7O_2)K$	$(C_4H_7O_2)Na$
E	453	54	26	20	-
P_1	472	47.5	9	43.5	-
P_2	458	51	27	22	-
P_3	464	55	29	16	-
$(AC)_1$	519	84.5			
$(AC)_2$	490	52			
$(BD)_1$	493		71		
$(BD)_2$	473		55.5		

System 4: $(C_2H_3O_2)K$ – $(C_2H_3O_2)Na$ – $(i \cdot C_4H_7O_2)Na$ – $(i \cdot C_4H_7O_2)K$ — 1960, [28], b

Point	T	$(C_2H_3O_2)K$	$(C_2H_3O_2)Na$	$(i \cdot C_4H_7O_2)Na$	$(i \cdot C_4H_7O_2)K$
E_1	469	13	23	64	-
E_2	455	48.5	18	33.5	-
E_3	467	52.5	29.5	18	-
P_1	482	15.5	7.5	77	-
P_2	465	49.5	4.5	46	-
P_3	478	56	30	14	-

a: the composition square was investigated up to 723 K; according to Ref. 25 the characteristic points along the diagonals occur at 549.2 K and 91.4 % A, and at 575.8 K and 88.2 % B

b: Probable heteroionic compound: A.2C

TABLE 5 (Continued)

Salts	Point						Year	No.	
	(AC)₁	549	89						
	(AC)₂	466	51						
	(AC)₃	498	14						
	(BD)₁	464		63.5					
	(BD)₂	475		52					
(C₂H₃O₂)K (C₂H₃O₂)Na (C₅H₉O₂)Na (C₅H₉O₂)K	E	463	52.5	28.5	19	–	1966	29	α
	P₁	471	44.0	22.5	33.5	–			
	P₂	465	25.0	52	–	23			
	P₃	465	52.5	29.5	18	–			
	(AC)₁	531	85.5						
	(AC)₂	503	62						
	(BD)₁	500		71					
	(BD)₂	482		62.5					
	E₁	420	51.5	13.5	35	–			
	E₂	423	53	15.5	31.5	–			
(C₂H₃O₂)K (C₂H₃O₂)Na (i.C₅H₉O₂)Na (i.C₅H₉O₂)K	P₁	436	18.5	19	62.5	–	1963	30	
	P₂	437	48	20	32	–			
	P₃	428	53.5	20.5	26	–			
	P₄	442	57.5	20	22.5	–			
	P₅	475	54.5	32	13.5	–			
	(AC)₁	509	82.5						
	(AC)₂	441	57						
	(BD)₁	443		61					
	(BD)₂	425		55.5					
(C₂H₃O₂)K (C₂H₃O₂)Na (C₆H₁₁O₂)Na (C₆H₁₁O₂)K	E₁	469	46	24.5	29.5	–	1959	31	
	E₂	466	49.5	28.0	22.5	–			
	P₁	470	52.5	29	18.5	–			

α: heteroionic compounds: A.B.2C and 3A.2B.C

TABLE 5 (Continued)

System: $(C_2H_3O_2)K$ — $(C_2H_3O_2)Na$ / $(Cl)Na$ — $(Cl)K$

Point	T (K)	$(C_2H_3O_2)K$	$(C_2H_3O_2)Na$	$(Cl)Na$	$(Cl)K$	Year	Ref
P_2	475	55.5	30.5	14	–	1960	32 a
$(AC)_1$	542	88.5					
$(AC)_2$	521	65.0					
$(BD)_1$	504		72.0				
$(BD)_2$	497		65.5				

System: $(C_2H_3O_2)Na$ / $(Cl)K$

Point	T (K)	a	b	c	d	Year	Ref
E	503	55	40	5	–	1966	33
P_1	516	66	29	5	–		
P_2	559	22	70.5	7.5	–		
(AC)	553	94					
(BD)	577		91				

System: $(C_2H_3O_2)K$ — $(C_2H_3O_2)Na$ / $(CNS)K$ — $(CNS)Na$

Point	T (K)	a	b	c	d	Year	Ref
E_1	396	–	5	25	70		
E_2	397	26	9	–	65		
E_3	398	34.5	9.5	–	56		
P	475	48	35	–	17		
$(AC)_1$	455	73					
$(AC)_2$	483	26					
(BD)	422		14				

System: $(C_2H_3O_2)K$ — $(C_2H_3O_2)Na$ / $(I)K$ — $(I)Na$

Point	T (K)	a	b	c	d	Year	Ref
E	493	53.5	38.5	8	–	1958	34 b
P	574	13	74	13	–		
(AC)	526	86.5					
(BD)	573		87				

System: $(C_2H_3O_2)K$ — $(C_2H_3O_2)Na$ / $(NO_2)K$ — $(NO_2)Na$

Point	T (K)	a	b	c	d	Year	Ref
E_1	447	–	13.5	54.5	32.0	1956	35
E_2	431	38.0	26.0	–	36.0		
$(AC)_1$	444	70					
$(AC)_2$	467	22					
(BD)	486		45.0				

a: according to Ref. 25 the characteristic points along the diagonals occur at 557.0 K and 93.6 % A, and at 583.7 K and 93.6 % B

b: according to Ref. 25 the characteristic points along the diagonals occur at 530.9 K and 86.6 % A, and at 574.5 K and 87.6 % B

TABLE 5 (Continued)

Components				Point	T	A	B	C	D		
A = $(C_2H_3O_2)K$	B = $(C_2H_3O_2)Na$	C = $(NO_3)K$	D = $(NO_3)Na$	E_1	458	-	20.0	47.5	32.5	1956	35 *a*
				E_2	457	-	19.5	39.0	41.5		
				E_3	461	30.0	29.5	-	40.5		
				E_4	461	48.0	35.0	-	17.0		
				$(AC)_1$	481	77					
				$(AC)_2$	467	66					
				$(AC)_3$	463	26					
				(BD)	485		39.5				
A = $(C_2H_3O_2)_2K_2$	B = $(C_2H_3O_2)_2Na_2$	C = $(S_2O_3)K_2$	D = $(S_2O_3)Na_2$	E	496	55	37	8	-	1958	37 *b*
				P	502	57.5	31.5	11	-		
				(AC)	553	82.5					
A = $(C_2H_3O_2)K$	B = $(C_2H_3O_2)Rb$	C = $(NO_3)K$	D = $(NO_3)Rb$	$(AC)_1$	490	68				1965	38 *c*
				$(AC)_2$	459	26					
				$(BD)_1$	476		82				
				$(BD)_2$	479		30				
A = $(C_2H_3O_2)Li$	B = $(C_2H_3O_2)Na$	C = $(CNS)Li$	D = $(CNS)Na$	E_1	459	47.0	35.0	18	-	1969	12
				E_2	465	62.5	19.5	18.0	-		
				E_3	437	36.5	-	16	47.5		
				(AC)	475	77.5					
				$(BD)_1$	473		72.5				
				$(BD)_2$	483		20				
A = $(C_2H_3O_2)Li$	B = $(C_2H_3O_2)Na$	C = $(NO_3)Li$	D = $(NO_3)Na$	E_1	410	49.5	-	8.0	42.5	1956	39 *d*
				E_2	421	51.0	35.3	13.5	-		
				P_1	461	70.5	20.0	9.5	-		

a: according to Ref. 36 (1975) the system is characterized by two eutectics only (approximately corresponding to E_3, E_4), and by a P point occurring at 47.5 % *A*, 40 % *B* and 12.5 % *D*

b: the *BD* diagonal was not investigated; components *C* and *D* decompose before melting

c: owing to the existence of solid solutions between *A,B* and *C,D* no invariant points occur

d: according to Ref.s 40 (1974) and 36 (1975) the system is characterized by two eutectics and a P point only

TABLE 5 (Continued)

System	Point	T (K)					Year	Ref	
(C₂H₃O₂)Li (C₂H₃O₂)Rb (NO₃)Li (NO₃)Rb	P_2	484	13.5	47.0	39.5	–	1974	40	a
	P_3	493	19.0	37.5	43.5	–			
	(AC)	470	85						
	$(BD)_1$	473		71.5					
	$(BD)_2$	451		23.5					
(C₂H₃O₂)Rb (NO₃)Li	E_1	429	16.0	71.0	13.0	–			
	E_2	503	85.5	11.0	3.5	–			
	E_3	401	52.0	–	5.5	42.5			
	P_1	451	19.5	64.0	16.5	–			
	P_2	416	8.0	–	39.0	53.0			
	(AC)	516	92						
	$(BD)_1$	445		85.5					
	$(BD)_2$	453		81					
	$(BD)_3$	411		22.5					
(C₂H₃O₂)Na (C₂H₃O₂)Rb (NO₃)Na (NO₃)Rb	E_1	414	31.0	60.0	9.0	–	1958	41	
	E_2	425	15.5	35.0	49.5	–			
	E_3	418	10.0	–	51.5	38.5			
	P_1	434	24.0	69.0	7.0	–			
	P_2	487	32.0	–	8.0	60.0			
	P_3	492	42.0	–	5.5	52.5			
	(AC)	471	25						
	$(BD)_1$	447		82					
	$(BD)_2$	453		71					
	$(BD)_3$	473		25					
(C₂H₃O₂)Na (C₂H₃O₂)Rb (NO₃)Na (NO₃)Rb	E_1	414	31.0	60.0	9.0	–	1970	42	b
	E_2	425	15.5	35.0	49.5	–			

a: liquid layering occurs along the AC diagonal between 12.5 and 78 % A at 526 K, and along the BD diagonal between 32.5 and 70 % B

b: as for the characteristic points along the diagonals, the temperatures fully agree with Ref. 41 whereas the pertinent compositions are not reported

TABLE 5 (Continued)

Component A	Component B	Component C	Point	T (K)					Year	Ref
$(C_3H_5O_2)K$	$(C_3H_5O_2)Li$	$(NO_3)K$	E_3	423	12.0	–	61.0	27.0	1969	2 [a]
			E_4	418	11.0	–	48.0	41.0		
			P	434	24.0	69.0	7.0	–		
$(C_3H_5O_2)Li$	$(NO_3)Li$	$(NO_3)K$	E_1	503	–	65	32.5	2.5		
			E_2	541	18.5	77	–	4.5		
			E_3	523	51.5	13.5	–	35		
			E_4	399	–	1	44	55		
			P	501	–	29	63	8		
			$(AC)_1$	537	80					
			$(AC)_2$	551	71					
			$(AC)_3$	485	16.5					
			(BD)	579		92.5				
$(C_3H_5O_2)K$	$(C_3H_5O_2)Na$	$(CNS)K$ $(CNS)Na$	E_1	401	–	0.5	26	73.5	1966	33
			E_2	423	11.5	6	–	82.5		
			P	503	31	51	–	18		
			$(AC)_1$	490	64					
			$(AC)_2$	513	25					
			(BD)	432		8				
$(C_3H_5O_2)K$	$(C_3H_5O_2)Na$	$(NO_3)K$ $(NO_3)Na$	E_1	479	–	11.5	41	47.5	1958	43
			E_2	511	49.5	13.5	–	37		
			P	523	35.5	40	–	15.5		
			$(AC)_1$	515	70					
			$(AC)_2$	501	22.5					
			(BD)	–		–				
$(C_3H_5O_2)Li$	$(C_3H_5O_2)Na$	$(CNS)Li$ $(CNS)Na$	E_1	463	44.5	39	16.5	–	1969	12 [b]
			E_2	465	58.5	–	8	33.5		

a: liquid layering occurs along the AC diagonal between 25 and 65 % A, and along the BD diagonal between 9 and 90 % B

b: continuous solid solutions with a minimum at 519 K and 52.5 % B

TABLE 5 (Continued)

A	B	C	D	Point	T, K	% A	% B	% C	% D	Year	Ref	Note
$(C_3H_5O_2)Li$	$(C_3H_5O_2)Na$	$(NO_3)Na$	$(NO_3)Li$	E_3	463	43	–	7	50			a
				(AC)	531	80						
				$(BD)_1$	509		76					
				$(BD)_2$	496		15					
$(C_3H_5O_2)Li$	$(C_3H_5O_2)Na$	$(NO_3)Na$	$(NO_3)Li$	E_1	463	10	–	32.5	57.5	1958	44	b
				E_2	483	47.5	–	14	38.5			
				E_3	457	47.5	45	7.5	–			
				(AC)	537	78						
				$(BD)_1$	505		78					
				$(BD)_2$	497		29					
				$(BD)_3$	489		14					
$(C_4H_7O_2)K$	$(C_4H_7O_2)Na$	$(CNS)Na$	$(CNS)K$	E_1	398	–	0.4	26	73.6	1958	45	
				E_2	437	4	2.5	–	93.5			
				P_1	514	–	30	47	23			
				P_2	497	35	12.5	–	52.5			
				(AC)	512	26						
				(BD)	438		4					
$(C_4H_7O_2)K$	$(C_4H_7O_2)Na$	$(NO_3)Na$	$(NO_3)K$	E	548	50.5	13	–	36.5	1958	46	
				P	522	–	29.5	51	19.5			
				$(AC)_1$	573	72.0						
				$(AC)_2$	519	22.5						
				(BD)	552		15					
$(C_4H_7O_2)Li$	$(C_4H_7O_2)Na$	$(CNS)Na$	$(CNS)Li$	E_1	491	47.5	37.0	15.5	–	1969	12	c
				E_2	479	49	–	5	46			
				P	531	7.5	52.2	40	–			

a: liquid layering occurs between 5 and 28 % A

b: liquid layering occurs along the *AC* diagonal between 15 and 71 % A at 557 K, and along the *BD* diagonal between 38 and 62.5 % B

c: liquid layering occurs along the *AC* diagonal between 10 and 77.5 % A, and along the *BD* diagonal between 35 and 62.5 % B

TABLE 5 (Continued)

System a: $(C_4H_7O_2)Li$ — $(C_4H_7O_2)Na$ — $(NO_3)Na$ — $(NO_3)Li$ — 1958, 47, *a*

Point	t				
(AC)	545	82.5			
$(BD)_1$	517		75		
$(BD)_2$	493		14		
E_1	479	-	40	2.5	57.5
E_2	489	52	44	4	-
P_1	491	18.5	-	14	67.5
P_2	529	12	58	30	-
(AC)	571	91			
$(BD)_1$	527		78		
$(BD)_2$	489		16		

System b: $(i \cdot C_4H_7O_2)K$ — $(i \cdot C_4H_7O_2)Na$ — $(NO_3)Na$ — $(NO_3)K$ — 1960, 48, *b*

Point	t				
E_1	487	-	18.5	39.5	42
E_2	490	-	58.5	21	20.5
E_3	515	17.5	62.5	-	20
E_4	521	47.5	14.5	-	38
P	527	44	11	-	45
$(AC)_1$	533	71			
$(AC)_2$	499	31			
$(BD)_1$	517		77.5		
$(BD)_2$	551		16.5		

System c: $(C_5H_9O_2)K$ — $(C_5H_9O_2)Na$ — $(NO_3)Na$ — $(NO_3)K$ — 1965, 49, *c*

Point	t				
E_1	493	-	2	50.5	47.5
E_2	577	28.0	15.0	-	57.0
P	553	-	34.0	59.0	7.0
$(AC)_1$	613	70			
$(AC)_2$	543	17.5			
(BD)	583		10		

a: liquid layering occurs along the *AC* diagonal between 2.5 and 90 % *A*, and along the *BD* diagonal between 25 and 65 % *B*
b: probable heteroionic compound: *B.D*
c: liquid layering occurs along the *AC* diagonal between 25 and 67.5 % *A*, and along the *BD* diagonal between 17.5 and 70 % *B*

TABLE 5 (Continued)

	(i·C5H9O2)K	(i·C5H9O2)Na	(NO3)Na	(NO3)K		1967	50	a
E_1	495	1.5	–	49.0	49.5			
E_2	513	12.5	51.0	36.5	–			
E_3	546	45.5	15.5	–	39.0			
P	548	44.5	9.0	–	46.5			
$(AC)_1$	557	69.5						
$(AC)_2$	543	15.0						
$(BD)_1$	539	82.0						
$(BD)_2$	571	7.5						

a: liquid layering occurs along the AC diagonal between 20.0 and 62.5 % A; heteroionic compound: $B.3D$

1. Sokolov & Tsindrik (1970)
2. Sokolov & Tsindrik (1969)
3. Leonesi, Piantoni & Franzosini
4. Sokolov (1965)
5. Sokolov & Minchenko (1971)
6. Sokolov & Minchenko (1974)
7. Sokolov & Minchenko (1977)
8. Leonesi, Braghetti, Cingolani & Franzosini
9. Sokolov & Pochtakova (1958 a)
10. Dmitrevskaya (1958 a)
11. Pochtakova (1975)
12. Sokolov & Dmitrevskaya
13. Tsindrik
14. Diogenov & Morgen (1975 a)
15. Nurminskii & Diogenov
16. Diogenov, Bruk & Nurminskii
17. Diogenov & Gimel'shtein (1975)
18. Diogenov & Morgen (1974 b)
19. Gimel'shtein & Diogenov (1966)
20. Diogenov & Morgen (1975 b)
21. Diogenov & Gimel'shtein (1966)
22. Diogenov, Nurminskii & Gimel'shtein
23. Gimel'shtein (1970)
24. Il'yasov & Bergman (1961)
25. Piantoni, Leonesi, Braghetti & Franzosini
26. Sokolov & Pochtakova (1958 b)
27. Sokolov & Pochtakova (1960 a)
28. Sokolov & Pochtakova (1960 b)
29. Pochtakova (1966)
30. Pochtakova (1963)
31. Pochtakova (1959)
32. Il'yasov & Bergman (1960 a)
33. Sokolov (1966)
34. Diogenov & Erlykov
35. Bergman & Evdokimova
36. Diogenov & Chumakova
37. Golubeva, Bergman & Grigor'eva
38. Diogenov & Gimel'shtein (1965)
39. Diogenov (1956 a)
40. Diogenov, Erlykov & Gimel'shtein (1958)
41. Gimel'shtein & Diogenov (1958)
42. Gimel'shtein & Diogenov (1970)
43. Dmitrevskaya & Sokolov (1958)
44. Tsindrik & Sokolov (1958 a)
45. Sokolov & Pochtakova (1958 c)
46. Dmitrevskaya (1958 b)
47. Tsindrik & Sokolov (1958 b)
48. Dmitrevskaya & Sokolov (1960)
49. Dmitrevskaya & Sokolov (1965)
50. Dmitrevskaya & Sokolov (1967)

TABLE 6. Miscellanea

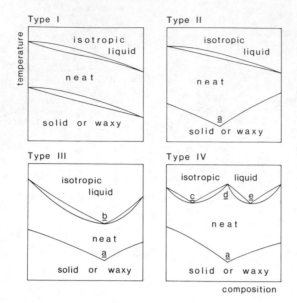

Fig. 9. "Das Mischbarkeitsver-
halten der neat–Modifikationen"
in binary systems according to
Baum, Demus & Sackmann (Ref. 1)

Miscellaneous binary systems with common anion

Component A	Component B	--------- Characteristic features ---------	Year	Ref.	Notes
$(C_{14}H_{27}O_2)K$	$(C_{14}H_{27}O_2)Na$	Type II in Fig. 9 with a at 508 K	1970	1	
$(C_{18}H_{35}O_2)Cs$	$(C_{18}H_{35}O_2)Rb$	Type II in Fig. 9 with a at 538 K	1970	1	
$(C_{18}H_{35}O_2)K$	$(C_{18}H_{35}O_2)Na$	Type II in Fig. 9 with a at 523 K	1970	1	
$(C_{18}H_{35}O_2)K$	$(C_{18}H_{35}O_2)Rb$	Type II in Fig. 9 with a at 528 K	1970	1	
$(C_{18}H_{35}O_2)Na$	$(C_{18}H_{35}O_2)Tl$	Type II in Fig. 9 with a at 387 K	1970	1	

1. Baum, Demus & Sackmann

Miscellaneous binary systems with common cation

Component A	Component B	--------- Characteristic features ---------	Year	Ref.	Notes
$Cs(C_{12}H_{23}O_2)$	$Cs(C_{18}H_{35}O_2)$	Type II in Fig. 9 with a at 533 K	1970	1	
$K(CHO_2)$	$K(C_5H_9O_2)$	eutectic system	1957	2	
$K(CHO_2)$	$K(i.C_5H_9O_2)$	eutectic system	1957	2	
$K(CHO_2)$	$K(C_6H_{11}O_2)$	eutectic system	1957	2	
$K(C_3H_5O_2)$	$K(C_4H_7O_2)$	continuous solid solutions with a m point	1957	2	
$K(C_3H_5O_2)$	$K(i.C_4H_7O_2)$	continuous solid solutions with a m point	1957	2	
$K(C_3H_5O_2)$	$K(C_5H_9O_2)$	eutectic system	1957	2	
$K(C_3H_5O_2)$	$K(i.C_5H_9O_2)$	continuous solid solutions with a m point	1957	2	
$K(C_3H_5O_2)$	$K(C_6H_{11}O_2)$	eutectic system	1957	2	
$K(C_4H_7O_2)$	$K(i.C_4H_7O_2)$	continuous solid solutions with a m point	1957	2	

Miscellaneous binary systems with common cation (Continued)

$K(C_4H_7O_2)$	$K(C_5H_9O_2)$	system with intermediate compounds	1957	2	
$K(C_4H_7O_2)$	$K(i.C_5H_9O_2)$	peritectic system	1957	2	
$K(C_4H_7O_2)$	$K(C_6H_{11}O_2)$	system with intermediate compounds	1957	2	
$K(i.C_4H_7O_2)$	$K(C_5H_9O_2)$	continuous solid solutions with a m point	1957	2	
$K(i.C_4H_7O_2)$	$K(i.C_5H_9O_2)$	eutectic system	1957	2	
$K(i.C_4H_7O_2)$	$K(C_6H_{11}O_2)$	eutectic system	1957	2	
$K(C_5H_9O_2)$	$K(i.C_5H_9O_2)$	eutectic system	1957	2	
$K(C_5H_9O_2)$	$K(C_6H_{11}O_2)$	continuous solid solutions with a m point	1957	2	
$K(C_5H_9O_2)$	$K(C_{18}H_{35}O_2)$	Type III in Fig. 9 with <u>a</u> at 523 K and <u>b</u> at 618 K	1970	1	
$K(i.C_5H_9O_2)$	$K(C_6H_{11}O_2)$	eutectic system	1957	2	
$K(C_{10}H_{19}O_2)$	$K(NO_2)$	system with liquid layering	1961	3	
$K(C_{18}H_{35}O_2)$	$K(NO_2)$	system with liquid layering	1961	3	
$Na(CHO_2)$	$Na(C_6H_{11}O_2)$	system with liquid layering	1954	4	
$Na(CHO_2)$	$Na(C_7H_5O_2)$	system with liquid layering	1954	4	*a*
$Na(CHO_2)$	$Na(C_{18}H_{35}O_2)$	system with liquid layering	1954	4	
$Na(C_2H_3O_2)$	$Na(C_{18}H_{35}O_2)$	system with liquid layering	1954	4	
$Na(C_6H_{11}O_2)$	$Na(C_{18}H_{35}O_2)$	Type III in Fig. 9 with <u>a</u> at 488 K and <u>b</u> at 553 K	1970	1	
$Na(C_6H_{11}O_2)$	$Na(NO_2)$	system with liquid layering	1957	5	
$Na(C_{12}H_{23}O_2)$	$Na(C_{13}H_{25}O_2)$	complete miscibility for all phases	1968	6	
$Na(C_{12}H_{23}O_2)$	$Na(C_{16}H_{31}O_2)$	complete miscibility for several phases	1941	7	
$Na(C_{12}H_{23}O_2)$	$Na(C_{18}H_{35}O_2)$	complete miscibility for several phases	1941	7	
$Na(C_{12}H_{23}O_2)$	$Na(C_{18}H_{35}O_2)$	complete miscibility for several phases	1968	6	
$Na(C_{13}H_{25}O_2)$	$Na(C_{16}H_{31}O_2)$	complete miscibility for several phases	1968	6	
$Na(C_{13}H_{25}O_2)$	$Na(C_{18}H_{35}O_2)$	complete miscibility for several phases	1968	6	
$Na(C_{16}H_{31}O_2)$	$Na(C_{18}H_{35}O_2)$	complete miscibility for all phases	1941	7	
$Na(C_{16}H_{31}O_2)$	$Na(C_{22}H_{43}O_2)$	complete miscibility for all phases	1941	7	
$Na(C_{18}H_{33}O_2)$	$Na(C_{18}H_{35}O_2)$	complete miscibility for all phases	1941	7	*b*
$Na(C_{18}H_{35}O_2)$	$Na(NO_3)$	system with liquid layering	1954	8	
$Rb(C_{12}H_{23}O_2)$	$Rb(C_{18}H_{35}O_2)$	Type II in Fig. 9 with <u>a</u> at 538 K	1970	1	
$Tl(C_6H_{11}O_2)$	$Tl(C_{12}H_{23}O_2)$	Type IV in Fig. 9 with <u>a</u> at 391, <u>c</u> at 474, <u>d</u> at 485 and <u>e</u> at 468 K	1970	1	
$Tl(C_6H_{11}O_2)$	$Tl(C_{18}H_{35}O_2)$	Type IV in Fig. 9 with <u>a</u> at 390, <u>c</u> at 445, <u>d</u> at 479 and <u>e</u> at 438 K	1970	1	
$Tl(C_{10}H_{19}O_2)$	$Tl(C_{18}H_{35}O_2)$	Type III in Fig. 9 with <u>a</u> at 379 K and <u>b</u> at 435 K	1970	1	
$Tl(C_{12}H_{23}O_2)$	$Tl(C_{16}H_{31}O_2)$	Type I in Fig. 9	1970	1	
$Tl(C_{12}H_{23}O_2)$	$Tl(C_{18}H_{35}O_2)$	Type I in Fig. 9	1970	1	
$Tl(C_{16}H_{31}O_2)$	$Tl(C_{18}H_{35}O_2)$	Type I in Fig. 9	1970	1	

a: $C_7H_5O_2$ = benzoate   *b*: $C_{18}H_{33}O_2$ = oleate

1. Baum, Demus & Sackmann   4. Sokolov (1954 b)   7. Vold, M.J. (1941)
2. Sokolov (1957 a)   5. Sokolov (1957 b)   8. Sokolov (1954 a)
3. Sokolov & Minich (1961)   6. Pacor & Spier

Miscellaneous reciprocal ternary systems

Component A	Component B	Component C	Component D	Characteristic point Type T/K	% A	% B	Year	Ref.	Notes
$(CHO_2)Cs$	$(CHO_2)K$	$(Br)K$	$(Br)Cs$	(BD) 425.7		93.65	1970	1	a
$(CHO_2)Cs$	$(CHO_2)K$	$(Cl)K$	$(Cl)Cs$	(BD) 431.5		96.1	1970	1	b
$(CHO_2)Cs$	$(CHO_2)K$	$(I)K$	$(I)Cs$	(BD) 425.4		93.05	1970	1	c
$(CHO_2)Cs$	$(CHO_2)Na$	$(Br)Na$	$(Br)Cs$	(BD) 512.8		91.75	1970	1	d
$(CHO_2)Cs$	$(CHO_2)Na$	$(Cl)Na$	$(Cl)Cs$	(BD) 515.0		93.7	1970	1	d
$(CHO_2)Cs$	$(CHO_2)Na$	$(I)Na$	$(I)Cs$	(BD) 523.2		96.9	1970	1	e
$(CHO_2)K$	$(CHO_2)Li$	$(Br)Li$	$(Br)K$	(AC) 427.9	95.55		1970	1	f
$(CHO_2)K$	$(CHO_2)Li$	$(Cl)Li$	$(Cl)K$	(AC) 431.6	96.7		1970	1	g
$(CHO_2)K$	$(CHO_2)Rb$	$(Br)Rb$	$(Br)K$	(AC) 432.0	94.5		1970	1	f
$(CHO_2)K$	$(CHO_2)Rb$	$(Cl)Rb$	$(Cl)K$	(AC) 434.5	96.2		1970	1	g
$(CHO_2)K$	$(CHO_2)Rb$	$(I)Rb$	$(I)K$	(AC) 425.6	91.0		1970	1	h
$(CHO_2)Li$	$(CHO_2)Na$	$(Br)Na$	$(Br)Li$	(BD) 506.7		92.4	1970	1	i
$(CHO_2)Li$	$(CHO_2)Na$	$(Cl)Na$	$(Cl)Li$	(BD) 517.2		95.5	1970	1	j
$(CHO_2)Na$	$(CHO_2)Rb$	$(Br)Rb$	$(Br)Na$	(AC) 507.8	89.65		1970	1	k
$(CHO_2)Na$	$(CHO_2)Rb$	$(Cl)Rb$	$(Cl)Na$	(AC) 514.8	93.75		1970	1	l
$(CHO_2)Na$	$(CHO_2)Rb$	$(I)Rb$	$(I)Na$	(AC) 517.4	93.95		1970	1	l
$(C_2H_3O_2)Cs$	$(C_2H_3O_2)K$	$(Br)K$	$(Br)Cs$	(BD) 541.0		87.0	1968	2	m
$(C_2H_3O_2)Cs$	$(C_2H_3O_2)K$	$(Cl)K$	$(Cl)Cs$	(BD) 556.0		92.8	1968	2	d
$(C_2H_3O_2)Cs$	$(C_2H_3O_2)K$	$(I)K$	$(I)Cs$	(BD) 531.5		81.3	1968	2	n
$(C_2H_3O_2)Cs$	$(C_2H_3O_2)Na$	$(Br)Na$	$(Br)Cs$	(BD) 577.5		88.9	1968	2	o
$(C_2H_3O_2)Cs$	$(C_2H_3O_2)Na$	$(Cl)Na$	$(Cl)Cs$	(BD) 581.4		93.1	1968	2	c
$(C_2H_3O_2)Cs$	$(C_2H_3O_2)Na$	$(I)Na$	$(I)Cs$	(BD) 588.1		94.3	1968	2	c
$(C_2H_3O_2)K$	$(C_2H_3O_2)Li$	$(Br)Li$	$(Br)K$	(AC) 547.1	92.4		1968	2	p
$(C_2H_3O_2)K$	$(C_2H_3O_2)Li$	$(Cl)Li$	$(Cl)K$	(AC) 557.0	94.4		1968	2	l
$(C_2H_3O_2)K$	$(C_2H_3O_2)Rb$	$(Br)Rb$	$(Br)K$	(AC) 552.6	89.2		1968	2	q
$(C_2H_3O_2)K$	$(C_2H_3O_2)Rb$	$(Cl)Rb$	$(Cl)K$	(AC) 560.5	92.9		1968	2	p
$(C_2H_3O_2)K$	$(C_2H_3O_2)Rb$	$(I)Rb$	$(I)K$	(AC) 536.6	81.1		1968	2	r
$(C_2H_3O_2)Li$	$(C_2H_3O_2)Na$	$(Br)Na$	$(Br)Li$	(BD) 576.2		91.3	1968	2	s
$(C_2H_3O_2)Li$	$(C_2H_3O_2)Na$	$(Cl)Na$	$(Cl)Li$	(BD) 586.0		95.4	1968	2	j

a: the system was investigated between 93 and 100 % B
b: the system was investigated between 95 and 100 % B
c: the system was investigated between 92 and 100 % B
d: the system was investigated between 90 and 100 % B
e: the system was investigated between 96 and 100 % B
f: the system was investigated between 94 and 100 % A
g: the system was investigated between 96 and 100 % A
h: the system was investigated between 90 and 100 % A
i: the system was investigated between 91 and 100 % B
j: the system was investigated between 94 and 100 % B
k: the system was investigated between 88 and 100 % A
l: the system was investigated between 93 and 100 % A
m: the system was investigated between 84 and 100 % B
n: the system was investigated between 80 and 100 % B
o: the system was investigated between 86 and 100 % B
p: the system was investigated between 91 and 100 % B
q: the system was investigated between 87 and 100 % A
r: the system was investigated between 78 and 100 % A
s: the system was investigated between 88 and 100 % B

Miscellaneous reciprocal ternary systems (Continued)

$(C_2H_3O_2)Na$	$(C_2H_3O_2)Rb$	$(Br)Rb$	$(Br)Na$	*(AC)* 574.8 88.0	1968	2	a
$(C_2H_3O_2)Na$	$(C_2H_3O_2)Rb$	$(Cl)Rb$	$(Cl)Na$	*(AC)* 582.5 93.6	1968	2	b
$(C_2H_3O_2)Na$	$(C_2H_3O_2)Rb$	$(I)Rb$	$(I)Na$	*(AC)* 580.7 90.6	1968	2	c
$(C_3H_5O_2)K$	$(C_3H_5O_2)Li$	$(CNS)Li$	$(CNS)K$	liquid layering	1961	3	d
$(C_4H_7O_2)K$	$(C_4H_7O_2)Li$	$(CNS)Li$	$(CNS)K$	liquid layering	1961	3	d
$(C_4H_7O_2)K$	$(C_4H_7O_2)Li$	$(NO_3)Li$	$(NO_3)K$	liquid layering	1961	3	d
$(i.C_4H_7O_2)K$	$(i.C_4H_7O_2)Li$	$(CNS)Li$	$(CNS)K$	liquid layering	1961	3	
$(i.C_4H_7O_2)K$	$(i.C_4H_7O_2)Li$	$(NO_3)Li$	$(NO_3)K$	liquid layering	1961	3	
$(i.C_4H_7O_2)Li$	$(i.C_4H_7O_2)Na$	$(CNS)Na$	$(CNS)Li$	liquid layering	1961	3	
$(i.C_4H_7O_2)Li$	$(i.C_4H_7O_2)Na$	$(NO_3)Na$	$(NO_3)Li$	liquid layering	1961	3	
$(C_5H_9O_2)K$	$(C_5H_9O_2)Li$	$(CNS)Li$	$(CNS)K$	liquid layering	1961	3	d
$(C_5H_9O_2)K$	$(C_5H_9O_2)Li$	$(NO_3)Li$	$(NO_3)K$	liquid layering	1961	3	d
$(C_5H_9O_2)K$	$(C_5H_9O_2)Na$	$(CNS)Na$	$(CNS)K$		1961	3	d
$(C_5H_9O_2)Li$	$(C_5H_9O_2)Na$	$(CNS)Na$	$(CNS)Li$	liquid layering	1961	3	e
$(C_5H_9O_2)Li$	$(C_5H_9O_2)Na$	$(NO_3)Na$	$(NO_3)Li$	liquid layering	1961	3	e
$(i.C_5H_9O_2)K$	$(i.C_5H_9O_2)Li$	$(CNS)Li$	$(CNS)K$	liquid layering	1961	3	
$(i.C_5H_9O_2)K$	$(i.C_5H_9O_2)Li$	$(NO_3)Li$	$(NO_3)K$	liquid layering	1961	3	
$(i.C_5H_9O_2)K$	$(i.C_5H_9O_2)Na$	$(CNS)Na$	$(CNS)K$	heteroionic compound $B.D$ only partially mis= cible with the compo= nent salts A, B, C, D	1961	3	e
$(i.C_5H_9O_2)Li$	$(i.C_5H_9O_2)Na$	$(CNS)Na$	$(CNS)Li$	liquid layering	1961	3	
$(i.C_5H_9O_2)Li$	$(i.C_5H_9O_2)Na$	$(NO_3)Na$	$(NO_3)Li$	liquid layering	1961	3	
$(C_6H_{11}O_2)K$	$(C_6H_{11}O_2)Li$	$(CNS)Li$	$(CNS)K$	liquid layering	1961	3	d
$(C_6H_{11}O_2)K$	$(C_6H_{11}O_2)Li$	$(NO_3)Li$	$(NO_3)K$	liquid layering	1961	3	d
$(C_6H_{11}O_2)K$	$(C_6H_{11}O_2)Na$	$(CNS)Na$	$(CNS)K$	liquid layering	1961	3	d
$(C_6H_{11}O_2)K$	$(C_6H_{11}O_2)Na$	$(NO_3)Na$	$(NO_3)K$	liquid layering	1961	3	d
$(C_6H_{11}O_2)Li$	$(C_6H_{11}O_2)Na$	$(CNS)Na$	$(CNS)Li$	liquid layering	1961	3	e
$(C_6H_{11}O_2)Li$	$(C_6H_{11}O_2)Na$	$(NO_3)Na$	$(NO_3)Li$	liquid layering	1961	3	e
$(C_7H_{13}O_2)K$	$(C_7H_{13}O_2)Na$	$(CNS)Na$	$(CNS)K$	liquid layering	1961	3	d
$(C_9H_{17}O_2)Na$	$(C_9H_{17}O_2)Tl$	$(C_{14}H_{27}O_2)Tl$	$(C_{14}H_{27}O_2)Na$	*(BD)*: Type II in Fig. 9 with $\underline{a}$ at 378 K	1970	4	
$(C_{10}H_{19}O_2)Na$	$(C_{10}H_{19}O_2)Tl$	$(C_{18}H_{35}O_2)Tl$	$(C_{18}H_{35}O_2)Na$	*(BD)*: Type III in Fig. 9 with $\underline{a}$ at 375 K and $\underline{b}$ at 478 K	1970	4	
$(C_{12}H_{23}O_2)K$	$(C_{12}H_{23}O_2)Na$	$(C_{18}H_{35}O_2)Na$	$(C_{18}H_{35}O_2)K$	*(BD)*: Type II in Fig. 9 with $\underline{a}$ at 503 K	1970	4	
$(C_{12}H_{23}O_2)Na$	$(C_{12}H_{23}O_2)Tl$	$(C_{18}H_{35}O_2)Tl$	$(C_{18}H_{35}O_2)Na$	*(BD)*: Type II in Fig. 9 with $\underline{a}$ at 378 K	1970	4	
$(C_{16}H_{33}O_2)Na$	$(C_{16}H_{33}O_2)Tl$	$(C_{18}H_{35}O_2)Tl$	$(C_{18}H_{35}O_2)Na$	*(BD)*: Type III in Fig 9 with $\underline{a}$ at 413 K and $\underline{b}$ at 450 K	1970	4	

a: the system was investigated between 86 and 100 % A d: *(BD)* is the stable diagonal
b: the system was investigated between 92 and 100 % A e: *(AC)* is the stable diagonal
c: the system was investigated between 88 and 100 % A

1. Leonesi, Braghetti, Cingolani & Franzosini   3. Sokolov, Tsindrik & Dmitrevskaya
2. Piantoni, Leonesi, Braghetti & Franzosini   4. Baum, Demus & Sackmann

SYSTEMS OF SALTS WITH ORGANIC CATION

Only a few authors dealt so far with systems of this type.

Worthy of a particular mention is a 1965 paper by Gordon who, employing the Kofler contact method, obtained "qualitative and semiquantitative descriptions" of most of the fusion diagrams summarized in Table 7.

Also to be cited is the phase equilibrium study carried out by Easteal & Angell (1970) on the binary pyridinium chloride + zinc chloride, where the existence of four congruently melting compounds (4:1; 2:1; 1:1; 1:2) was observed: this system, however, is not included in the table inasmuch as the pertinent results were given by the authors only graphically.

TABLE 7. Systems with organic cation

a) Binaries with common anion

Anion	Cation in A	Cation in B	Characteristic point Type	T/K	% B	Year	Ref.	Notes
Br	$(C_4H_9)_4N$	N_2H_5	–	–	–	1962	1	a
Br	$(C_5H_{11})_4N$	$(C_7H_{15})_4N$	E	348.2	61.5	1965	2	
Br	$(C_6H_{13})_4N$	$(C_6H_{13})_3(C_7H_{15})N$	–	–	–	1965	2	b
Cl	$(C_5H_{11})NH_3$	$(C_5H_{11})NH_3$	m	450	25	1939	3	b, c
ClO$_4$	$(C_5H_{11})_4N$	$(C_6H_{13})_3(C_7H_{15})N$	E	367.0	–	1965	2	
ClO$_4$	$(C_6H_{13})_4N$	$(C_6H_{13})_3(C_7H_{15})N$	m	371.9	–	1965	2	b
I	$(C_4H_9)_4N$	$(C_5H_{11})_4N$	E	384.3	–	1965	2	
I	$(C_5H_{11})_4N$	$(C_6H_{13})_4N$	E	364	–	1965	2	
I	$(C_5H_{11})_4N$	$(C_6H_{13})_3(C_7H_{15})N$	E	363.7	–	1965	2	
I	$(C_5H_{11})_4N$	$(C_6H_{13})_2(C_7H_{15})_2N$	E	362	–	1965	2	
I	$(C_5H_{11})_4N$	$(C_7H_{15})_4N$	E	377	–	1965	2	
I	$(C_6H_{13})_4N$	$(C_6H_{13})_3(C_7H_{15})N$	m	365	–	1965	2	b
NO$_3$	$(C_5H_{11})_4N$	Ag	E	314.9 ±0.5	42.4±0.5	1973	4	
NO$_3$	$(C_6H_5)NH_3$	NH$_4$	E	429.9	39	1945	5	d

a: liquid layering occurs; N_2H_5 = hydrazinium
b: continuous solid solutions
c: (C_5H_{11}) = 2-methyl-butyl in component A, and 3-methyl-butyl in component B
d: $(C_6H_5)NH_3$ = anilinium

b) Binaries with common cation

Cation	Anion in A	Anion in B	Characteristic point Type	T/K	% B	Year	Ref.	Notes
$(C_5H_{11})_4N$	Br	ClO$_4$	m	370.2	–	1965	2	a
$(C_5H_{11})_4N$	Br	CNS	m	313.5	–	1965	2	a, b
$(C_5H_{11})_4N$	Br	I	–	–	–	1965	2	a
$(C_5H_{11})_4N$	Br	NO$_2$	m	365.5	–	1965	2	a
$(C_5H_{11})_4N$	Br	NO$_3$	m	373.6	23	1969	6	a
$(C_5H_{11})_4N$	$C_6H_2(NO_2)_3O$	Br	E	331.7	31	1969	6	c

TABLE 7 (Continued)

$(C_5H_{11})_4N$	$C_6H_2(NO_2)_3O$	CNS	E	308.2	–	1965	2	*b, c*
$(C_5H_{11})_4N$	$C_6H_2(NO_2)_3O$	I	E	345.0	–	1965	2	*c*
$(C_5H_{11})_4N$	$C_6H_2(NO_2)_3O$	NO_3	E	333.5	–	1965	2	*c*
$(C_5H_{11})_4N$	ClO_4	CNS	m	321.2	–	1965	2	*a, b*
$(C_5H_{11})_4N$	ClO_4	CNS	m	316.8	–	1965	2	*a, d*
$(C_5H_{11})_4N$	ClO_4	I	–	–	–	1965	2	*a*
$(C_5H_{11})_4N$	ClO_4	NO_2	m	367	–	1965	2	*a*
$(C_5H_{11})_4N$	I	NO_2	–	–	–	1965	2	*a*
$(C_5H_{11})_4N$	I	NO_3	m	386.0	–	1965	2	*a*
$(C_6H_{13})_4N$	ClO_4	I	E	371	–	1965	2	
$(C_6H_{13})_4N$	I	NO_3	m	341	–	1965	2	*a*
$(C_6H_{13})_3(C_7H_{15})N$	I	NO_3	m	340	–	1965	2	*a*
$(C_7H_{15})_4N$	Br	I	m	361.7	–	1965	2	*a*
$(C_7H_{15})_4N$	ClO_4	I	m	393	–	1965	2	*a*

a: continuous solid solutions
b: component *B* present as polymorph (I)
c: $C_6H_2(NO_2)_3O$ = picrate
d: component *B* present as polymorph (II)

1. Seward	4. Gordon & Varughese
2. Gordon (1965)	5. Klug & Pardee
3. Seigle & Hass	6. Gordon (1969)

Systems that can be considered as diagonals of reciprocal ternaries have also been occasionally investigated. As examples, the following may be mentioned:

(i) methylammonium chloride + ammonium nitrate (Klug & Pardee, 1945): the system exhibits an intermediate compound 1:1 congruently melting at 461 K, and two eutectics at 15 and 79 mol % NH_4NO_3, and at 377.5 and 427 K, respectively;

(ii) anilinium chloride + ammonium nitrate (Klug & Pardee, 1945): the system exhibits an intermediate compound 1:4 congruently melting at 428 K, and two eutectics at 10.1 and 29.2 mol % NH_4NO_3, and at 420.0 and 423.3 K, respectively;

(iii) N-ethylpyridinium bromide + aluminum chloride (Hurley & Wier, 1951): the system exhibits an intermediate compound 1:1 congruently melting at 361 K, and two eutectics at 33 and 67 mol % $AlCl_3$, and at 318 and 233 K, respectively.

A few more systems containing, e.g., amidine salts have been deliberately neglected.

GLASS-FORMING SYSTEMS

The glass-forming ability of pure organic salts has been briefly discussed in Chapter 1.2. Some information is now to be added on glassy materials which can be obtained on quenching a number of molten mixtures containing either short chain organic anions or organic cations.

Bartholomew & Lewek (1970) were able to get glasses from an equimolar mixture of $(CHO_2)K$ and $(CHO_2)Na$ with the addition of a small percentage of $(CHO_2)_2Ca$: a more extensive investigation on formate glasses was however prevented by the poor stability of these salts in the molten state.

More details are instead available in the case of acetate glasses.

Independent investigations by Bartholomew & Holland and by Duffy & Ingram led in 1969 to par-allel conclusions, i.e., respectively that "melts containing at least one univalent and one divalent acetate salts give excellent glasses on quenching in a manner similar to that used for making the nitrate glasses" and that glasses can be "obtained readily from molten acetate mixtures".

The extent of glass formation in a wide variety of binaries {$(C_2H_3O_2)K$ mixed with Ba, Ca, Cd, Pb, Sr and Zn acetate; $(C_2H_3O_2)Li$ mixed with Ba, Ca, Cd, K, Na, NH_4, Pb and Sr acetate; $(C_2H_3O_2)Na$ mixed with Ba, Ca, Cd, Mg, Pb, Sr and Zn acetate; etc.} and ternaries {formed with $(C_2H_3O_2)K$, $(C_2H_3O_2)Na$ and one among the alkali-earth acetates} was investigated by Bartholomew & Lewek (1970) who, by the way, could also confirm the validity of the known statement that in glass-forming binaries the most stable glassy material is produced when mixtures, whose compo-sitions lie not far from those corresponding to eutectics, are quenched (see Fig. 10).

Bartholomew also patented in 1972 the production of glassy materials from acetate melts, and listed for them a number of applications.

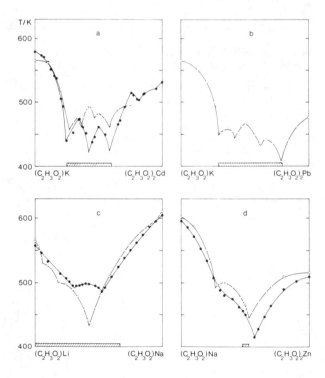

Fig. 10. a) fusion diagram of K and Cd acetates (open circles: Lehrman & Schweitzer, 1954; filled circles: Pavlov & Golubkova, 1969); b) fusion diagram of K and Pb acet-ates (Lehrman & Leifer, 1938); c) fusion diagram of Li and Na acetates (open circles: Diogenov, 1956 a; filled circles: Pochtakova, 1965); d) fusion diagram of Na and Zn acetates (open circles: Lehrman & Skell, 1939; filled circles: Nadirov & Bakeev, 1974 a). The hatched strips indicate the extent of the glass-forming regions.

Furthermore, the ability of several trivalent metal acetates and trifluoroacetates to give glasses when mixed with the corresponding salts of a number of uni- and divalent metals was proved by Van Uitert, Bonner & Grodkiewicz (1971).

The characteristic parameter T_g (glass transition temperature, see Chapter 1.2) has been determined for a series of binary and ternary acetate mixtures by *DTA* or *DSC* (at the heating rate of 10 K min^{-1}): a list of pertinent data is given hereafter.

------------- Composition of the mixtures in mol % ----------------

$(C_2H_3O_2)_2Ca$	$(C_2H_3O_2)K$	$(C_2H_3O_2)Li$	$(C_2H_3O_2)Na$	$(C_2H_3O_2)_2Pb$	T_g/K	Ref.
33.3	66.7	–	–	–	360	1
33.3	66.7	–	–	–	373	2
40	60	–	–	–	373	2
50	50	–	–	–	383	2
–	25	–	–	75	315	2
–	50	–	–	50	326.5	2
–	60	–	–	40	321	2
33.3	–	–	66.7	–	383	1
–	–	57	43	–	350	2
–	–	66.7	33.3	–	357	2
–	–	75	25	–	365	2
–	–	50	–	50	323	2
–	–	66.7	–	33.3	326	2
50	20	–	30	–	388	1
50	30	–	20	–	373	1
50	40	–	10	–	373	1

1. Bartholomew (1970)   2. Ingram, Lewis & Duffy (1972)

Salts with organic cation when mixed either with one another or with a proper inorganic salt often exhibit a tendency to glass formation: detailed investigation, however, has been so far performed only in a few cases.

Several properties of the system formed with pyridinium and zinc chlorides have been studied by Easteal & Angell (1970): of interest here is the fact that glasses could be obtained on quenching mixtures containing 39–45 and 55–100 mol % $ZnCl_2$, and also on slowly cooling melts of composition ranging between 60 and 90 mol %.

Moreover, glass-forming mixtures which contained salts such as α-, β- or γ-picolinium chloride, quinolinium chloride, etc., were submitted to investigation with advanced techniques by Abkemeier (1972), Hodge (1974) and Angell, Hodge & Cheeseman (1976).

The ability of different families of salt melts to form glassy solids was discussed in a comprehensive paper by Angell & Tucker (1974).

REFERENCES

ABKEMEIER, M.C., *Ph.D. Thesis*, Purdue University (1972)

ANGELL, C.A., HODGE, I.M., and CHEESEMAN, P.A., *"Molten Salts" Proc. Int. Conf. Molten Salts*, Ed. Pemsler, J.P., The Electrochemical Soc. Inc., 138 (1976)

ANGELL, C.A., and TUCKER, J.C., *"Physical Chemistry of Process Metallurgy - The Richardson Conference"*, Ed. Jeffes, J.H.E., and Tait, R.J., Inst. Mining Metallurgy Publ., 207 (1974)

BARTHOLOMEW, R.F., *J. Phys. Chem.*, $\underline{74}$, 2507 (1970)

BARTHOLOMEW, R.F., *U.S. Patent 3,649,551* (1972)

BARTHOLOMEW, R.F., and HOLLAND, H.J., *J. Amer. Ceram. Soc.*, $\underline{52}$, 402 (1969)

BARTHOLOMEW, R.F., and LEWEK, S.S., *J. Amer. Ceram. Soc.*, $\underline{53}$, 445 (1970)

BASKOV, A., *Zh. Russk. Fiz. Khim. Obshch.*, $\underline{47}$, 1533 (1915)

BAUM, E., DEMUS, D., and SACKMANN, H., *Wiss. Z. Univ. Halle XIX '70*, 37

BERCHIESI, G., CINGOLANI, A., and LEONESI, D., *Z. Naturforsch.*, $\underline{25a}$, 1766 (1970)

BERCHIESI, G., CINGOLANI, A., LEONESI, D., and PIANTONI, G., *Can. J. Chem.*, $\underline{50}$, 1972 (1972)

BERCHIESI, G., and LAFFITTE, M., *J. Chim. Phys.*, $\underline{1971}$, 877

BERCHIESI, M.A., CINGOLANI, A., and BERCHIESI, G., *J. Chem. Eng. Data*, $\underline{17}$, 61 (1972)

BERGMAN, A.G., and EVDOKIMOVA, K.A., *Izv. Sekt. Fiz.-Khim. Anal., Inst. Obshch. i Neorg. Khim. Akad. Nauk SSSR*, $\underline{27}$, 296 (1956)

CINGOLANI, A., BERCHIESI, G., and PIANTONI, G., *J. Chem. Eng. Data*, $\underline{16}$, 464 (1971)

CINGOLANI, A., GAMBUGIATI, M.A., and BERCHIESI, G., *Z. Naturforsch.*, $\underline{25a}$, 1519 (1970)

DIOGENOV, G.G., *Russ. J. Inorg. Chem.*, $\underline{1}$(4), 199 (1956); translated from *Zh. Neorg. Khim.*, $\underline{1}$, 799 (1956)

DIOGENOV, G.G., *Russ. J. Inorg. Chem.*, $\underline{1}$(11), 122 (1956); translated from *Zh. Neorg. Khim.*, $\underline{1}$, 2551 (1956)

DIOGENOV, G.G., BORZOVA, L.L., and SARAPULOVA, I.F., *Russ. J. Inorg. Chem.*, $\underline{10}$, 948 (1965); translated from *Zh. Neorg. Khim.*, $\underline{10}$, 1738 (1965)

DIOGENOV, G.G., BRUK, T.I., and NURMINSKII, N.N., *Russ. J. Inorg. Chem.*, $\underline{10}$, 814 (1965); translated from *Zh. Neorg. Khim.*, $\underline{10}$, 1496 (1965)

DIOGENOV, G.G., and CHUMAKOVA, V.P., *Fiz.-Khim. Issled. Rasplavov Solei*, $\underline{1975}$, 7

DIOGENOV, G.G., and ERLYKOV, A.M., *Nauch. Doklady Vysshei Shkoly, Khim. i Khim. Tekhnol.*, $\underline{1958}$, 3, 413

DIOGENOV, G.G., ERLYKOV, A.M., and GIMEL'SHTEIN, V.G., *Russ. J. Inorg. Chem.*, $\underline{19}$, 1069 (1974); translated from *Zh. Neorg. Khim.*, $\underline{19}$, 1955 (1974)

DIOGENOV, G.G., and GIMEL'SHTEIN, V.G., *Russ. J. Inorg. Chem.*, $\underline{10}$, 1395 (1965); translated from *Zh. Neorg. Khim.*, $\underline{10}$, 2567 (1965)

DIOGENOV, G.G., and GIMEL'SHTEIN, V.G., *Russ. J. Inorg. Chem.*, $\underline{11}$, 113 (1966); translated from *Zh. Neorg. Khim.*, $\underline{11}$, 207 (1966)

DIOGENOV, G.G., and GIMEL'SHTEIN, V.G., *Fiz.-Khim. Issled. Rasplavov Solei*, $\underline{1975}$, 13

DIOGENOV, G.G., and MORGEN, L.T., *Nekotorye Vopr. Khimii Rasplavlen. Solei i Produktov Destruktsii Sapropelitov*, $\underline{1974}$, 32

DIOGENOV, G.G., and MORGEN, L.T., *Fiz.-Khim. Issled. Rasplavov Solei*, $\underline{1975}$, 59

DIOGENOV, G.G., and MORGEN, L.T., *Fiz.-Khim. Issled. Rasplavov Solei*, $\underline{1975}$, 62

DIOGENOV, G.G., NURMINSKII, N.N., and GIMEL'SHTEIN, V.G., *Russ. J. Inorg. Chem.*, 2(7), 237, (1957); translated from *Zh. Neorg. Khim.*, 2, 1596 (1957)

DIOGENOV, G.G., and SARAPULOVA, I.F., *Russ. J. Inorg. Chem.*, 9(2), 265 (1964); translated from *Zh. Neorg. Khim.*, 9, 482 (1964)

DIOGENOV, G.G., and SARAPULOVA, I.F., *Russ. J. Inorg. Chem.*, 9(5), 704 (1964); translated from *Zh. Neorg. Khim.*, 9, 1292 (1964)

DIOGENOV, G.G., and SARAPULOVA, I.F., *Russ. J. Inorg. Chem.*, 9(6), 814 (1964); translated from *Zh. Neorg. Khim.*, 9, 1499 (1964)

DIOGENOV, G.G., and SERGEEVA, G.S., *Russ. J. Inorg. Chem.*, 10, 153 (1965); translated from *Zh. Neorg. Khim.*, 10, 292 (1965)

DIOGENOV, G.G., and SHIPITSYNA, V.K., *Tr. Irkutsk. Politekh. Inst.*, 27, 60 (1966)

DMITREVSKAYA, O.I., *J. Gen. Chem. USSR*, 28, 295 (1958); translated from *Zh. Obshch. Khim.*, 28, 299 (1958)

DMITREVSKAYA, O.I., *J. Gen. Chem. USSR*, 28, 2046 (1958); translated from *Zh. Obshch. Khim.*, 28, 2007 (1958)

DMITREVSKAYA, O.I., and SOKOLOV, N.M., *J. Gen. Chem. USSR*, 28, 2949 (1958); translated from *Zh. Obshch. Khim.*, 28, 2920 (1958)

DMITREVSKAYA, O.I., and SOKOLOV, N.M., *J. Gen. Chem. USSR*, 30, 19 (1960); translated from *Zh. Obshch. Khim.*, 30, 20 (1960)

DMITREVSKAYA, O.I., and SOKOLOV, N.M., *Zh. Obshch. Khim.*, 35, 1905 (1965)

DMITREVSKAYA, O.I., and SOKOLOV, N.M., *J. Gen. Chem. USSR*, 37, 2050 (1967); translated from *Zh. Obshch. Khim.*, 37, 2160 (1967)

DUFFY, J.A., and INGRAM, M.D., *J. Amer. Ceram. Soc.*, 52, 224 (1969)

EASTEAL, A.J., and ANGELL, C.A., *J. Phys. Chem.*, 74, 3987 (1970)

GIMEL'SHTEIN, V.G., *Symposium, "Fiziko-Khimicheskii Analiz Solevykh Sistem", Izd. Irkutsk. Politekh. Inst. , Irkutsk*, 1970, 39

GIMEL'SHTEIN, V.G., and DIOGENOV, G.G., *Russ. J. Inorg. Chem.*, 3(7), 230 (1958); translated from *Zh. Neorg. Khim.*, 3, 1644 (1958)

GIMEL'SHTEIN, V.G., and DIOGENOV, G.G., *Tr. Irkutsk. Politeckh. Inst.*, 27, Ser. Khim., 69 (1966)

GIMEL'SHTEIN, V.G., and DIOGENOV, G.G., *Tr. Irkutsk. Politeckh. Inst.*, 44, Ser. Khim., 124 (1970)

GOLUBEVA, M.S., ALESHKINA, N.N., and BERGMAN, A.G., *Russ. J. Inorg. Chem.*, 4, 1201 (1959); translated from *Zh. Neorg. Khim.*, 4, 2606 (1959)

GOLUBEVA, M.S., BERGMAN, A.G., and GRIGOR'EVA, E.A., *Uch. Zap. Rostovsk.-na-Donu Gos. Univ.*, 41, 145 (1958)

GORDON, J.E., *J. Amer. Chem. Soc.*, 87, 4347 (1965)

GORDON, J.E., *"Applications of Fused Salts in Organic Chemistry" in "Techniques and Methods of Organic and Organometallic Chemistry"*, Ed. by Denney, D.B., M. Dekker, New York, 1969, Vol. I, 51

GORDON, J.E., and VARUGHESE, P., *J. Org. Chem.*, 38, 3726 (1973)

HODGE, I.M., *Ph.D. Thesis*, Purdue University (1974)

HURLEY, F.H., and WIER, Jr, T.P., *J. Electrochem. Soc.*, 98, 203 (1951)

IL'YASOV, I.I., *Zh. Obshch. Khim.*, 32, 347 (1962)

IL'YASOV, I.I., and BERGMAN, A.G., *Zh. Obshch. Khim.*, 30, 355 (1960)

IL'YASOV, I.I., and BERGMAN, A.G., *J. Gen. Chem. USSR*, 30, 1091 (1960); translated from *Zh. Obshch. Khim.*, 30, 1075 (1960)

IL'YASOV, I.I., and BERGMAN, A.G., *Zh. Obshch. Khim.*, 31, 368 (1961)

INGRAM, M.D., LEWIS, G.G., and DUFFY, J.A., *J. Phys. Chem.*, 76, 1035 (1972)

KLUG, H.L., and PARDEE, A.M., *Proc. S. Dakota Acad. Sci.*, 25, 48 (1945); taken from GORDON, J. E., 1969

LEHRMAN, A., and LEIFER, E., *J. Amer. Chem. Soc.*, 60, 142 (1938)

LEHRMAN, A., and SCHWEITZER, D., *J. Phys. Chem.*, 58, 383 (1954)

LEHRMAN, A., and SKELL, P., *J. Amer. Chem. Soc.*, 61, 3340 (1939)

LEONESI, D., BRAGHETTI, M., CINGOLANI, A., and FRANZOSINI, P., *Z. Naturforsch.*, 25a, 52 (1970)

LEONESI, D., PIANTONI, G., and FRANZOSINI, P., *Z. Naturforsch.*, 25a, 56 (1970)

NADIROV, E.G., and BAKEEV, M.I., *Tr. Khim.-Metall. Inst. Akad. Nauk Kaz. SSR*, 25, 115 (1974)

NADIROV, E.G., and BAKEEV, M.I., *Tr. Khim.-Metall. Inst. Akad. Nauk Kaz. SSR*, 25, 129 (1974)

NURMINSKII, N.N., and DIOGENOV, G.G., *Russ. J. Inorg. Chem.*, 5, 1011 (1960); translated from *Zh. Neorg. Khim.*, 5, 2084 (1960)

PACOR, P., and SPIER, H.L., *J. Amer. Oil Chem. Soc.*, 45, 338 (1968)

PAVLOV, V.L., and GOLUBKOVA, V.V., *Vestn. Kiev. Politekh. Inst. Ser. Khim. Mashinostr. Tekhnol.*, 1969, 6, 76

PAVLOV, V.L., and GOLUBKOVA, V.V., *Visn. Kiiv. Univ. Ser. Khim.*, 1972, 13, 28

PETERSEN, J., *Z. Elektrochem.*, 20, 328 (1914)

PIANTONI, G., LEONESI, D., BRAGHETTI, M., and FRANZOSINI, P., *Ric. Sci.*, 38, 127 (1968)

POCHTAKOVA, E.I., *J. Gen. Chem. USSR*, 29, 3149 (1959); translated from *Zh. Obshch. Khim.*, 29, 3183 (1959)

POCHTAKOVA, E.I., *Zh. Obshch. Khim.*, 33, 342 (1963)

POCHTAKOVA, E.I., *Russ. J. Inorg. Chem.*, 10, 1268 (1965); translated from *Zh. Neorg. Khim.*, 10, 2333 (1965)

POCHTAKOVA, E.I., *Zh. Obshch. Khim.*, 36, 3 (1966)

POCHTAKOVA, E.I., *Zh. Obshch. Khim.*, 44, 241 (1974)

POCHTAKOVA, E.I., *Zh. Obshch. Khim.*, 45, 503 (1975)

POCHTAKOVA, E.I., *Zh. Obshch. Khim.*, 46, 6 (1976)

SARAPULOVA, I.F., KASHCHEEV, G.N., and DIOGENOV, G.G., *Nekotorye Vopr. Khimii Rasplavlen. Solei i Produktov Destruktsii Sapropelitov*, 1974, 3

SEIGLE, L.W., and HASS, H.B., *Ind. Eng. Chem.*, 31, 648 (1939)

SEWARD, R.P., *J. Phys. Chem.*, 66, 1125 (1962)

SOKOLOV, N.M., *Zh. Obshch. Khim.*, 24, 1150 (1954)

SOKOLOV, N.M., *Zh. Obshch. Khim.*, 24, 1581 (1954)

SOKOLOV, N.M., *Tezisy Dokl. XII Nauch. Konf. S.M.I.*, 1957, 100

SOKOLOV, N.M., *J. Gen. Chem. USSR*, 27, 917 (1957); translated from *Zh. Obshch. Khim.*, 27, 840 (1957)

SOKOLOV, N.M., *Zh. Obshch. Khim.*, 35, 1897 (1965)

SOKOLOV, N.M., *Zh. Obshch. Khim.*, 36, 577 (1966)

SOKOLOV, N.M., and DMITREVSKAYA, O.I., *Russ. J. Inorg. Chem.*, 14, 148 (1969); translated from *Zh. Neorg. Khim.*, 14, 286 (1969)

SOKOLOV, N.M., and KHAITINA, M.V., *Russ. J. Inorg. Chem.*, 15, 1482 (1970); translated from *Zh. Neorg. Khim.*, 15, 2850 (1970)

SOKOLOV, N.M., and KHAITINA, M.V., *Zh. Obshch. Khim.*, 41, 1417 (1971)

SOKOLOV, N.M., and KHAITINA, M.V., *Zh. Obshch. Khim.*, 42, 2121 (1972)

SOKOLOV, N.M., and KHAITINA, M.V., *Zh. Obshch. Khim.*, 44, 2113 (1974)

SOKOLOV, N.M., and MINCHENKO, S.P., *Zh. Obshch. Khim.*, 41, 1656 (1971)

SOKOLOV, N.M., and MINCHENKO, S.P., *Zh. Obshch. Khim.*, 44, 1429 (1974)

SOKOLOV, N.M., and MINCHENKO, S.P., *Zh. Obshch. Khim.*, 47, 740 (1977)

SOKOLOV, N.M., and MINICH, M.A., *Russ. J. Inorg. Chem.*, 6, 1293 (1961); translated from *Zh. Neorg. Khim.*, 6, 2258 (1961)

SOKOLOV, N.M., and MINICH, M.A., *Russ. J. Inorg. Chem.*, 15, 735 (1970); translated from *Zh. Neorg. Khim.*, 15, 1433 (1970)

SOKOLOV, N.M., and POCHTAKOVA, E.I., *J. Gen. Chem. USSR*, 28, 1449 (1958); translated from *Zh. Obshch. Khim.*, 28, 1391 (1958)

SOKOLOV, N.M., and POCHTAKOVA, E.I., *Zh. Obshch. Khim.*, 28, 1397 (1958)

SOKOLOV, N.M., and POCHTAKOVA, E.I., *J. Gen. Chem. USSR*, 28, 1741 (1958); translated from *Zh. Obshch. Khim.*, 28, 1693 (1958)

SOKOLOV, N.M., and POCHTAKOVA, E.I., *J. Gen. Chem. USSR*, 30, 1429 (1960); translated from *Zh. Obshch. Khim.*, 30, 1401 (1960)

SOKOLOV, N.M., and POCHTAKOVA, E.I., *J. Gen. Chem. USSR*, 30, 1433 (1960); translated from *Zh. Obshch. Khim.*, 30, 1405 (1960)

SOKOLOV, N.M., and POCHTAKOVA, E.I., *Zh. Obshch. Khim.*, 37, 1420 (1967)

SOKOLOV, N.M., and TSINDRIK, N.M., *Russ. J. Inorg. Chem.*, 14, 302 (1969); translated from *Zh. Neorg. Khim.*, 14, 584 (1969)

SOKOLOV, N.M., and TSINDRIK, N.M., *Zh. Obshch. Khim.*, 41, 951 (1970)

SOKOLOV, N.M., TSINDRIK, N.M., and DMITREVSKAYA, O.I., *J. Gen. Chem. USSR*, 31, 971 (1961); translated from *Zh. Obshch. Khim.*, 31, 1051 (1961)

SOKOLOV, N.M., TSINDRIK, N.M., and KHAITINA, M.V., *Russ. J. Inorg. Chem.*, 15, 433 (1970); translated from *Zh. Neorg. Khim.*, 15, 433 (1970)

SOKOLOV, N.M., TSINDRIK, N.M., and KHAITINA, M.V., *Russ. J. Inorg. Chem.*, 15, 721 (1970); translated from *Zh. Neorg. Khim.*, 15, 1405 (1970)

STORONKIN, A.V., VASIL'KOVA, I.V., and POTEMIN, S.S., *Vestn. Leningr. Univ., Fiz., Khim.*, 1974, 10, 84

STORONKIN, A.V., VASIL'KOVA, I.V., and POTEMIN, S.S., *Vestn. Leningr. Univ., Fiz., Khim.*, 1974, 16, 73

STORONKIN, A.V., VASIL'KOVA, I.V., and TARASOV, A.A. *Vestn. Leningr. Univ., Fiz., Khim.*, 1977, 4, 80

TSINDRIK, N.M., *Zh. Obshch. Khim.*, 28, 830 (1958)

TSINDRIK, N.M., and SOKOLOV, N.M., *J. Gen. Chem. USSR*, 28, 1462 (1958); translated from *Zh. Obshch. Khim.*, 28, 1404 (1958)

TSINDRIK, N.M., and SOKOLOV, N.M., *J. Gen. Chem. USSR*, 28, 1775 (1958); translated from *Zh. Obshch. Khim.*, 28, 1728 (1958)

VAN UITERT, L.G., BONNER, W.A., and GRODKIEWICZ, W.H., *Mat. Res. Bull.*, 6, 513 (1971)

VOLD, M.J., *J. Amer. Chem. Soc.*, 63, 160 (1941)

ADDENDA

Addendum *to*: TABLE 1. Binary systems with common anion

Component A	T/K	Component B	T/K	Characteristic point Type	T/K	% B	Year	Ref.	Notes
$(C_2H_3O_2)K$	583–585	$(C_2H_3O_2)Li$	564–565	E_1	470	37.5	1971	1	
				A.2B	548				
				E_2	507–510	87.0			
$(C_2H_3O_2)Na$	600–601	$(C_2H_3O_2)Rb$	508–509	E_1	419	61.5	1971	1	
				A.3B	452				
				E_2	451	76.5			
$(C_2H_3O_2)Na$	599–600	$(C_2H_3O_2)Rb$	511	E	419	61.0	1974	2	

1. GIMEL'SHTEIN, V.G., *Tr. Irkutsk. Politekh. Inst.*, **66**, 80 (1971)
2. DIOGENOV, G.G., and MORGEN, L.T., *Nekotorye Vopr. Khimii Rasplavlen. Solei i Produktov Destruktsii Sapropelitov*, <u>1974</u>, 27

Addendum *to*: TABLE 2. Binary systems with common cation

Component A	T/K	Component B	T/K	Characteristic point Type	T/K	% B	Year	Ref.	Notes
$K(C_2H_3O_2)$	583–585	$K(NO_3)$	610–611	E_1	507	35.5	1971	1	
				A.B	511				
				E_2	497	62.5			
$Na(C_2H_3O_2)$	599–600	$Na(NO_2)$	551	E	497	66	1974	2	
$Rb(C_2H_3O_2)$	511	$Rb(NO_2)$	692	E_1	419	44.5	1974	2	
				A.B	425				
				E_2	422	53.5			
$Rb(C_2H_3O_2)$	508–509	$Rb(NO_3)$	588–590	E_1	471	–	1971	1	
				2A.B	475				
				E_2	467	–			

1. GIMEL'SHTEIN, V.G., *Tr. Irkutsk. Politekh. Inst.*, **66**, 80 (1971)
2. DIOGENOV, G.G., and MORGEN, L.T., *Nekotorye Vopr. Khimii Rasplavlen. Solei i Produktov Destruktsii Sapropelitov*, <u>1974</u>, 27

Addendum to: TABLE 5. Reciprocal ternary systems

Comp. A	Comp. B	Comp. C	Comp. D	Type	T/K	% A	% B	% C	% D	Year	Ref.	Notes
						Characteristic point						
$(C_2H_3O_2)Na$	$(C_2H_3O_2)Rb$	$(NO_2)Rb$	$(NO_2)Na$	E_1	383	10	–	51	39	1974	1	
				E_2	377	22.5	51.0	26.5	–			
				P	389	21	45	34	–			
				(AC)	471	32						
				(BD)$_1$	471		18.5					
				(BD)$_2$	391		77					

1. DIOGENOV, G.G., and MORGEN, L.T., *Nekotorye Vopr. Khimii Rasplavlen. Solei i Produktov Destruktsii Sapropelitov*, <u>1974</u>, 27

Addendum to: TABLE 7. Systems with organic cation - a) Binaries with common anion

Anion	Cation in A	Cation in B	Type	T/K	% B	Year	Ref.	Notes
			Characteristic point					
ClO_4	$(C_2H_5)_3(C_{14}H_{29})N$	$(C_5H_{11})_4N$	E	363	–	1978	1	
ClO_4	$(C_2H_5)_2(C_8H_{17})_2N$	$(C_5H_{11})_4N$	E	332	–	1978	1	*a*
ClO_4	$(C_2H_5)(C_6H_{13})_3N$	$(C_5H_{11})_4N$	E	333	–	1978	1	
ClO_4	$(C_3H_7)_3(C_{11}H_{23})N$	$(C_5H_{11})_4N$	E	333.7	–	1978	1	
ClO_4	$(C_4H_9)_3(C_8H_{17})N$	$(C_4H_9)(C_5H_{11})_2(C_6H_{13})N$	E	318	–	1978	1	
ClO_4	$(C_4H_9)_3(C_8H_{17})N$	$(C_5H_{11})_4N$	E	329	–	1978	1	
ClO_4	$(C_4H_9)_2(C_6H_{13})_2N$	$(C_5H_{11})_4N$	m	353	–	1978	1	*b*
ClO_4	$(C_4H_9)(C_5H_{11})_2(C_6H_{13})N$	$(C_5H_{11})_4N$	m	353	–	1978	1	*b*

a: sometimes this system crystallizes in a more complex pattern that appears to involve a molecular compound melting incongruently at 324 K
b: continuous solid solutions

1. GORDON, J.E., and SUBBA RAO, G.N., *J. Amer. Chem. Soc.*, <u>100</u>, 7445 (1978)

Chapter 2.2

TRANSPORT PROPERTIES OF BINARY MIXTURES OF MOLTEN SALTS WITH ORGANIC IONS

A. Kisza

INTRODUCTION

The formation of a molten binary mixture of salts with organic ions involves incorporation of the solute ionic species into the quasi-lattice structure of the solvent. With increasing solute concentration the quasi-lattice of the solvent changes into that of the solute, at the other end of the mole fraction scale.

Depending on the nature of the interactions between the ionic species of the solute and of the solvent, the transport properties of the mixture show more or less pronounced deviations from simple additivity.

A rigorous theory of transport properties for such mixtures, based on intermolecular interactions, has not yet been developed.

VISCOSITY OF BINARY MOLTEN MIXTURES OF SALTS WITH ORGANIC IONS

DEFINITIONS

In the same way as for a pure molten salt, the viscosity of molten binary mixtures of salts with organic ions is a measure of the frictional resistance that a fluid in motion offers to an applied shearing force.

The viscosity, η, of these mixtures may be treated – to a first approximation – as an additive quantity

$$\eta_{12}^{id} = x_1\eta_1 + (1 - x_1)\,\eta_2 \tag{1}$$

where suffixes $_1$, $_2$ and $_{12}$ refer to the solvent, the solute and the mixture, respectively, x_1 being the mole fraction of the solvent.

In most cases the experimentally determined viscosity value, η_{12}, differs from the ideal one given by Eq. (1), and the difference

$$\Delta\eta_{12} = \eta_{12} - \eta_{12}^{id} \tag{2}$$

represents the deviation from additivity. In molten inorganic salts (binary mixtures with one common ion) it was found by Murgulescu & Zuca (1966, 1969) and by Zuca & Borcan (1970) that the deviation from additivity can be expressed through the semiempirical symmetrical equation

$$\Delta\eta_{12} = a\,(r_1 - r_2)\,x_1x_2 \tag{3}$$

where r_1 and r_2 are the cationic radii, x_1 and x_2 the mole fractions and a is a constant independent of the nature of the mixture and determined by the temperature, i.e.,

$$a = A \exp (B/T) \tag{4}$$

Murgulescu & Zuca (1965 a, b) showed that, starting from the Eyring theory of viscosity (1941), it is possible to give the kinetic-molecular significance of $\Delta\eta_{12}$ as follows:

$$\Delta\eta_{12} = 2^{5/2}\,\pi^{\frac{1}{2}}\,R^{\frac{1}{2}}\,(\rho_1/M_1^{\frac{1}{2}} - \rho_2/M_2^{\frac{1}{2}})\,T^{\frac{1}{2}}\,(r_1 - r_2)\,x_1x_2 \exp (E_{\eta_{12}}/RT) \tag{5}$$

Here ρ_1 and ρ_2 are the densities of the pure salts, M_1 and M_2 their molecular masses, $E_{\eta_{12}}$ is the activation energy of viscous flow, and the other symbols have their usual meaning.

As stated by the above authors, Eq. (5) predicts correctly the dependence of $\Delta\eta_{12}$ upon the ionic radii, the mole fractions and the temperature.

In the interpretation of the transport properties of binary molten salt mixtures, a correlation between the concentration dependence of the viscosity and the phase diagram is frequently used. In a qualitative way, it can be said that a complex phase diagram leads to larger deviations from additivity.

The temperature dependence of the viscosity of molten salt mixtures is described by the same type of equation as that of pure molten salts. The Arrhenius type equation

$$\eta_{12} = \eta_{12}^{\infty} \exp{(E_{\eta_{12}}/RT)} \tag{6}$$

is often employed. However, the activation energy, $E_{\eta_{12}}$, of viscous flow is in the case of a mixture far less meaningful, since it should represent the potential barrier in momentum transport of (at least) three different ionic species.

The experimental value of $E_{\eta_{12}}$ may be compared with the additive quantity

$$E_{\eta_{12}}^{id} = x_1 E_{\eta_1} + (1 - x_1) E_{\eta_2} \tag{7}$$

Any deviation from additivity

$$\Delta E_{\eta_{12}} = E_{\eta_{12}} - E_{\eta_{12}}^{id} \tag{8}$$

indicates that structural changes occur in the mixing process.

EXPERIMENTAL RESULTS

Experimental viscosity data for binary mixtures of molten salts with organic ions are not numerous. The only systematic measurements are those on molten alkylammonium picrates by Walden & Birr (1932), who studied both the concentration and the temperature dependence of the viscosity. The relevant numerical data are here presented in the subsequent tables.

The possibility of gradually changing the constitution of the components in such binary mixtures allows one to see the influence of the structure of the alkylammonium cation on the momentum transfer.

Let us first consider molten heptylammonium picrate as the solvent: this was mixed with diethylammonium picrate (Table A) and with tetrapropylammonium picrate (Table B). In both cases the difference between the heptylammonium cation and the solute cations is substantial. In the former mixture the negative deviation of the viscosity from simple additivity is accompanied by an asymmetric plot of the energy of activation of viscous flow (Fig. 1). In this case both cations have acidic properties, and specific interactions with the picrate ion may occur in the melt.

A similar picture is obtained in the molten mixture of heptylammonium picrate with tetrapropylammonium picrate (Fig. 2). As no specific interactions are possible between the tetrapropylammonium cation and the picrate anion, the addition of the latter salt destroys only the associated structure of the molten solvent. The deviation of the viscosity from additivity is unsymmetrical and is not described by the equation of Murgulescu & Zuca (1969).

The addition of tripropylammonium picrate into molten diethylammonium picrate increases the viscosity above the additive value and this is accompanied by a steep rise of the activation energy of viscous flow (Fig. 3). A similar picture is obtained with molten binary mixtures of diethylammonium picrate and tetrapropylammonium picrate (Fig. 4), even though the viscosities of the pure molten components are almost identical. Here the influence of the large tetrapropylammonium cation on the viscosity of the mixture is even smaller than that of the tripropylammonium cation in the case of associated molten diethylammonium picrate.

The behaviour observed in molten binary mixtures of tetrapropyl- and tetra*iso*amylammonium picrates is shown in Fig. 5. In spite of the difference in size between the two alkylammonium cations, the viscosity of the mixtures exhibits negative deviations from additivity which are small and can be described by a type (3) symmetrical equation. i.e.:

$$\Delta\eta_{12} = -3.964 \ 10^3 \ x_1 x_2$$

Even here, however, the activation energy of viscous flow is far from additivity.

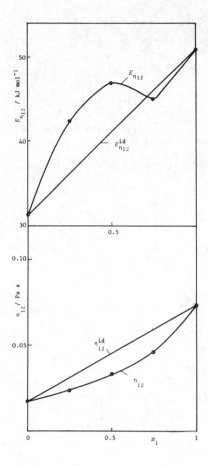

Fig. 1. Viscosity and activation energy of viscous flow in the molten binary system
$C_7H_{15}NH_3^+Pic^-$ (x_1) + $(C_2H_5)_2NH_2^+Pic^-$ (x_2) at 398 K.

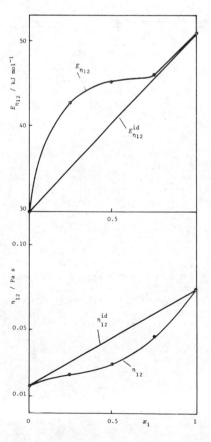

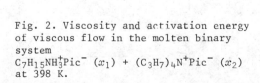

Fig. 2. Viscosity and activation energy of viscous flow in the molten binary system
$C_7H_{15}NH_3^+Pic^-$ (x_1) + $(C_3H_7)_4N^+Pic^-$ (x_2) at 398 K.

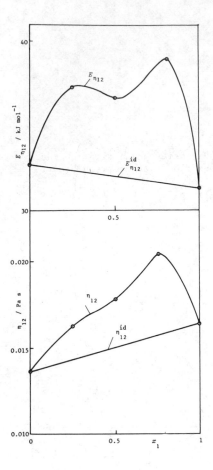

Fig. 3. Viscosity and activation energy of viscous flow in the molten binary system $(C_2H_5)NH_3^+Pic^-$ (x_1) + $(C_3H_7)_3NH^+Pic^-$ (x_2) at 398 K.

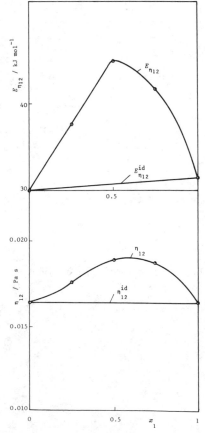

Fig. 4. Viscosity and activation energy of viscous flow in the molten binary system $(C_2H_5)_2NH_2^+Pic^-$ (x_1) + $(C_3H_7)_4N^+Pic^-$ (x) at 398 K.

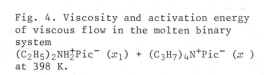

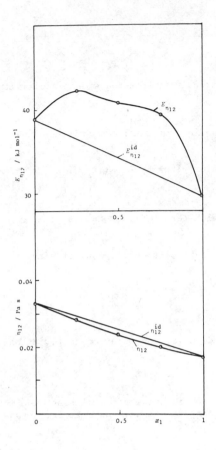

Fig. 5. Viscosity and activation energy of viscous flow in the molten binary system
$(C_3H_7)_4N^+Pic^-$ (x_1) + $(i.C_5H_{11})_4N^+Pic^-$ (x_2) at 398 K.

CONDUCTANCE OF BINARY MOLTEN MIXTURES OF SALTS WITH ORGANIC IONS

DEFINITIONS

The molar conductivity of a binary molten salt mixture, Λ_{12}, is related to the specific conductivity, κ_{12}, and the density, ρ_{12}, through

$$\Lambda_{12} = \kappa_{12} \ (M_{12}/\rho_{12}) \tag{9}$$

where

$$M_{12} = x_1 M_1 + x_2 M_2 \tag{10}$$

is the mean molecular mass of the mixture.

At a constant temperature T, the concentration dependence of the molar conductivity is of great interest. In most cases the experimental value of the molar conductivity deviates from the additive one, Λ_{12}^{id},

$$\Lambda_{12}^{id} = x_1 \Lambda_1 + x_2 \Lambda_2 \tag{11}$$

where Λ_1 and Λ_2 are the molar conductivities of the pure components

For a binary molten salt mixture with a common ion, Markov & Sumina (1956, 1957) proposed for the molar conductivity the expression

$$\Lambda_{12}^{MS} = x_1^2 \Lambda_1 + x_2^2 \Lambda_2 + 2\,x_1 x_2\,\Lambda_1 \tag{12}$$

in which $\Lambda_1 < \Lambda_2$.

Generally, the deviation of the experimental molar conductivity from the simple additivity value

$$\Delta\Lambda_{12} = \Lambda_{12} - \Lambda_{12}^{id} \tag{13}$$

may be ascribed to all the new interactions which arise in a molten binary mixture as compared with the pure components.

In particular, the deviation from the Markov-Sumina value given by Eq. (12), i.e.,

$$\Delta\Lambda_{12}^{MS} = \Lambda_{12} - \Lambda_{12}^{MS} \tag{14}$$

should be due to the strong interionic interactions which result either in association or in complex-ion formation.

The same type of concentration dependence was proposed for the specific conductivity by Mochinaga, Cho & Kuroda (1968):

$$\kappa_{12}^{K} = x_1^2 \kappa_1 + x_2^2 \kappa_2 + 2\,x_1 x_2 \kappa_1 \tag{15}$$

where $\kappa_1 < \kappa_2$.

At a given constant composition, the temperature dependence of the molar conductivity can be described by the Arrhenius type equation

$$\Lambda_{12} = \Lambda_{12}^{\infty}\,\exp(-\,E_{\Lambda_{12}}/RT) \tag{16}$$

In many cases the fit of the experimental data to Eq. (16) is quite good, although the meaning of the "activation energy of the conductance process" is not well understood. Anyway, it is not an additive quantity, both negative and positive deviations

$$\Delta E_{\Lambda_{12}} = E_{\Lambda_{12}} - E_{\Lambda_{12}}^{id} \tag{17}$$

being observed with respect to the additive value

$$E_{\Lambda_{12}}^{id} = x_1 E_{\Lambda_1} + x_2 E_{\Lambda_2} \tag{18}$$

An equation similar to Eq. (16) may be introduced for the specific conductivity

$$\kappa_{12} = \kappa_{12}^{\infty}\,\exp(-\,E_{\kappa_{12}}/RT) \tag{19}$$

Here $E_{\kappa_{12}}$ is the activation energy of the specific conductivity and κ_{12}^{∞} is the specific conductivity extrapolated to infinite temperature. The relationship between $E_{\kappa_{12}}$ and $E_{\Lambda_{12}}$ was given by Martin (1954) as follows:

$$E_{\Lambda} = E_{\kappa} + \alpha\,R\,T^2 \tag{20}$$

which involves the knowledge of the thermal expansion coefficient

$$\alpha = -\,1/\rho\,(d\rho/dT) \tag{21}$$

for a mixture of molten salts.

EXPERIMENTAL RESULTS

For the conductivity of molten binary mixtures of salts with organic ions the number of experimental data available in the literature is larger than for viscosity. In many cases, however, the molar volume was not determined and only specific conductivities were therefore reported. The most complete investigations are those by Walden & Birr (1932) on molten binary mixtures of alkylammonium picrates. Their numerical data are presented in the subsequent Tables and Figures (in the latter, the dashed curves show the trends predicted by Eq. 12).

The addition of long-chain heptylammonium picrate to diethtylammonium picrate causes a decrease in the molar conductivity (Table A and Fig. 6), which is accompanied by an obvious increase of the activation energy of the conductance process. Negative deviations from simple additivity are observed, but it is to be noted that even more negative deviations are predicted by the Markov-Sumina equation. Such increments of the experimental molar conductivities above the Markov-Sumina values confirm the existence of an associated structure within the melt.

The binary molten mixture of heptylammonium picrate with tetrapropylammonium picrate (Table B and Fig. 7) also deviates negatively from simple additivity, but the prediction of the Markov-Sumina equation is here more close to the experimentally measured isotherm.

In the molten binary mixture of diethylammonium and tripropylammonium picrates (Table C) the components are similar in chemical character although the access to the proton in the tripropylammonium is more difficult as a consequence of steric hindrance. This results in the almost ideal experimental molar conductivity plot shown in Fig. 8, from which the remarkable positive deviations from the trend predicted by the Markov-Sumina equation are also apparent.

As one may expect, the addition of tetrapropylammonium picrate to the associated structure

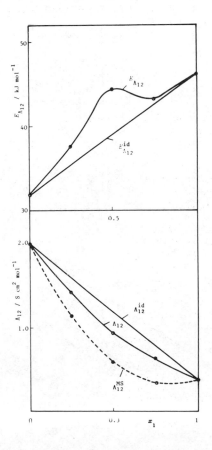

Fig. 6. Molar conductivity and activation energy of conductance in the molten binary system
$C_7H_{15}NH_3^+Pic^-$ (x_1) + $(C_2H_5)_2NH_2^+Pic^-$ (x_2) at 398 K.

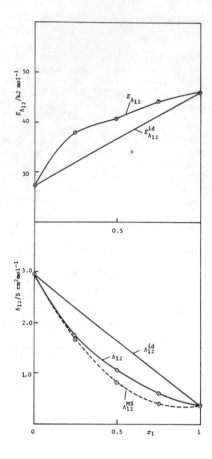

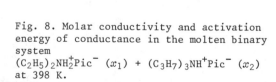

Fig. 7. Molar conductivity and activation energy of conductance in the molten binary system
$C_7H_{15}NH_3^+Pic^-$ (x_1) + $(C_3H_7)_4N^+Pic^-$ (x_2) at 398 K.

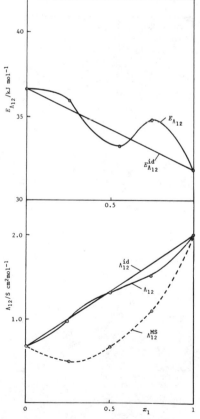

Fig. 8. Molar conductivity and activation energy of conductance in the molten binary system
$(C_2H_5)_2NH_2^+Pic^-$ (x_1) + $(C_3H_7)_3NH^+Pic^-$ (x_2) at 398 K.

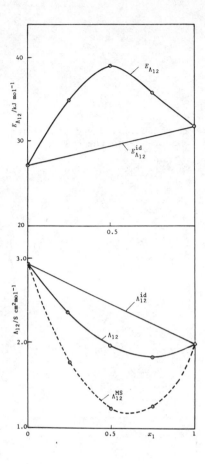

Fig. 9. Molar conductivity and activation energy of conductance in the molten binary system
$(C_2H_5)_2NH_2^+Pic^-$ (x_1) + $(C_3H_7)_4N^+Pic^-$ (x_2) at 398 K.

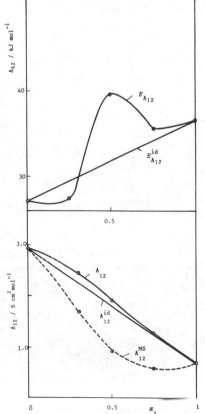

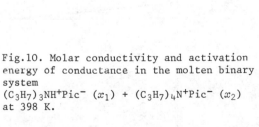

Fig.10. Molar conductivity and activation energy of conductance in the molten binary system
$(C_3H_7)_3NH^+Pic^-$ (x_1) + $(C_3H_7)_4N^+Pic^-$ (x_2) at 398 K.

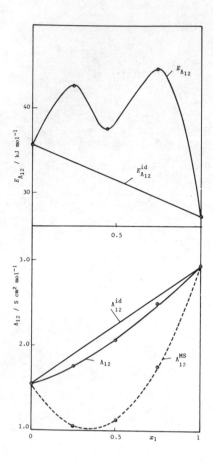

Fig.11. Molar conductivity and activation energy of conductance in the molten binary system
$(C_3H_7)_4N^+Pic^-$ (x_1) + $(i.C_5H_{11})_4N^+Pic^-$ (x_2) at 398 K.

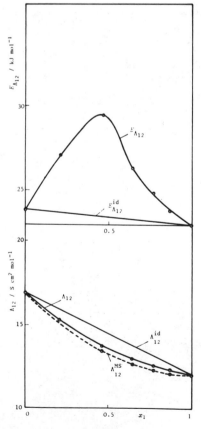

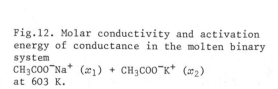

Fig.12. Molar conductivity and activation energy of conductance in the molten binary system
$CH_3COO^-Na^+$ (x_1) + $CH_3COO^-K^+$ (x_2) at 603 K.

of molten diethylammonium picrate (Table D and Fig. 9) gives rise to positive deviations of the experimental values from the Markov-Sumina curve.

The similar, although more ideal, molten binary mixture of tri- and tetrapropylammonium picrates is presented in Table E and Fig. 10. The slightly associated structure of tripropylammonium picrate is destroyed by the addition of the non-specifically interacting tetrapropylammonium salt.

The data presented for the molten binary mixture of tetrapropyl- and tetra*iso*amylammonium picrates (Table F and Fig. 11) are a good example of steric hindrance in the conductance process. Also in this mixture the positive deviations from the Markov-Sumina predictions are in concordance with the assumption that complex ions are formed within the melt.

The molten binary mixture of sodium and potassium acetates (Table G and Fig. 12) was studied carefully by Hazlewood, Rhodes & Ubbelohde (1966, density) and by Leonesi, Cingolani & Berchiesi (1973, specific conductivity). Negative deviations of the molar conductivity from simple additivity are accompanied by an increase of the activation energy of conductance. The prediction by the Markov-Sumina equation is here much closer to the experimental isotherm although even in this case small positive deviations have been found.

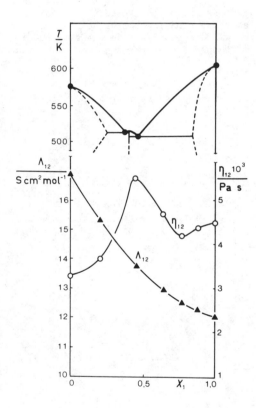

Fig. 13. Correlation between the transport properties (in the molten state at 603 K) and the phase diagram of the binary system
$CH_3COO^-Na^+$ (x_1) + $CH_3COO^-K^+$ (x_2)
Viscosities by Kisza, private communication (1979); molar conductivities by Leonesi, Cingolani & Berchiesi (1973); phase diagram as reported by Franzosini et al.

This system also offers an interesting example of correlation between the transport properties in the molten state and the phase diagram (Fig. 13). The formation of a congruently melting compound seems to have no influence on the trend of conductivity (although the activation energy, $E_{\Lambda_{12}}$, shows a pronounced maximum, see Fig. 12), while it remarkably affects viscosity.

TABLES

In the following tables the most complete data available on the transport properties of molten binary mixtures of salts with organic ions have been presented.

For each considered binary mixture, the first table (say, A, or B, etc.) summarizes values of the transport properties as a function of the mole fraction, x_1, of component 1 at a given selected temperature, whereas in the subsequent tables (i.e., A1, A2, ..., B1, B2, ..., etc.) the temperature dependence of the same properties for different x_1 values is reported.

For the sake of simplicity, the picrate anion, $C_6H_2(NO_2)_3O^-$, has been indicated in the table headings (as well as in the captions of the figures) by the abbreviation Pic⁻.

In the tables the experimental data of the original papers are reported. In addition, for numerical purposes and discussion, the least-squares fits of the experimental values to the proper equations are given. The following linear equation was used to describe the temperature dependence of the density of binary molten salts mixtures

$$\rho_{12} = a + b\, T \tag{22}$$

whereas quadratic equations were employed for the temperature dependence of the specific conductivity and of the viscosity, i.e.,

$$\kappa_{12} = a' + b'T + c'T^2 \tag{23}$$

$$\eta_{12} = a'' + b''T + c''T^2 \tag{24}$$

The validity of such equations does not extend beyond the temperature ranges given in the tables.

The Arrhenius type equations (6) and (16) were also obtained by least-squares analysis, together with the correlation coefficient ($r^2 = 1$, for ideal fit).

TABLE A. Transport properties of the molten mixtures of heptylammonium picrate and
diethylammonium picrate

$$C_7H_{15}NH_3^+Pic^- \ (x_1) \quad + \quad (C_2H_5)_2NH_2^+Pic^- \ (x_2)$$

T = 398 K

x_1	ρ_{12} g cm^{-3}	$\kappa_{12} \ 10^3$ S cm^{-1}	Λ_{12} S cm^2 mol^{-1}	η_{12} Pa s	$\Lambda_{12}\eta_{12}$	$E_{\Lambda_{12}}$ J mol^{-1}	$E_{\eta_{12}}$ J mol^{-1}
0.00	1.3082	8.696	2.009	0.01703	0.0342	31806	31259
0.25	1.2775	5.787	1.416	0.02368	0.0335	37501	42224
0.50	1.2563	3.590	0.9238	0.03319	0.0307	44353	46851
0.75	1.2353	2.302	0.6220	0.04609	0.0287	43237	44990
1.00	1.2160	1.335	0.3780	0.07386	0.0279	46058	50962

TABLE A1. $C_7H_{15}NH_3^+Pic^- \ (x_1) \quad + \quad (C_2H_5)_2NH_2^+Pic^- \ (x_2)$

x_1 = 0.25   Ref. 1

$\rho_{12} = 1.5652 - 7.224 \ 10^{-4}T$ r^2 = 0.9998   Note (a)

$\kappa_{12} = 0.113064 - 6.862 \ 10^{-4}T + 1.0472 \ 10^{-6}T^2$ Note (a)

$\eta_{12} = 6.93115 - 3.3816 \ 10^{-2}T + 4.129 \ 10^{-5}T^2$ Note (a)

T K	ρ_{12} g cm^{-3}	$\kappa_{12} \ 10^3$ S cm^{-1}	Λ_{12} S cm^2 mol^{-1}	η_{12} Pa s	$\Lambda_{12}\eta_{12}$
348	1.3141	1.065	0.2534	0.1675	0.0424
373	1.2955	2.840	0.6856	0.05314	0.0364
398	1.2775	5.787	1.4161	0.02368	0.0335
423	1.2599	10.11	2.527	0.01256	0.0317

$\Lambda_{12} = 113965.6 \ \exp\{-\dfrac{37501}{R \ T}\}$ r^2 = 0.9957   Note (a)

$\eta_{12} = 7.15 \ 10^{-8} \ \exp\{\dfrac{42224}{R \ T}\}$ r^2 = 0.9940   Note (a)

Ref. 1: Walden & Birr (1932)   data in the table given according to Ref. 1
Note (a): equations obtained by least-squares analysis of the data given in Ref. 1

TABLE A2. $C_7H_{15}NH_3^+Pic^- \ (x_1) \quad + \quad (C_2H_5)_2NH_2^+Pic^- \ (x_2)$

x_1 = 0.50   Ref. 1

$\rho_{12} = 1.5318 - 6.92 \ 10^{-4}T$ r^2 = 0.9999   Note (a)

$\kappa_{12} = 0.102182 - 6.0313 \ 10^{-4}T + 6.9352 \ 10^{-7}T^2$ Note (a)

$\eta_{12} = 13.10892 - 6.4296 \ 10^{-2}T + 7.8868 \ 10^{-5}T^2$ Note (a)

T K	ρ_{12} g cm^{-3}	$\kappa_{12} \ 10^3$ S cm^{-1}	Λ_{12} S cm^2 mol^{-1}	η_{12} Pa s	$\Lambda_{12}\eta_{12}$

348	1.2912	0.4728	0.1183	0.2921	0.0346
373	1.2736	1.609	0.4084	0.07803	0.0319
398	1.2563	3.590	0.9238	0.03319	0.0307
423	1.2393	6.960	1.8155	0.01629	0.0296

$\Lambda_{12} = 587269 \exp\{-\dfrac{44353}{R\,T}\}$ $r^2 = 0.9926$ Note (a)

$\eta_{12} = 2.46\ 10^{-8} \exp\{\dfrac{46851}{R\,T}\}$ $r^2 = 0.9920$ Note (a)

Ref. 1: Walden & Birr (1932)   data in the table given according to Ref.1

Note (a): equations obtained by least-squares analysis of the data given in Ref. 1

TABLE A3. $C_7H_{15}NH_3^+Pic^-$ (x_1) + $(C_2H_5)_2NH_2^+Pic^-$ (x_2)

x_1 = 0.75   Ref. 1

$\rho_{12} = 1.5148 - 7.02\ 10^{-4}T$ $r^2 = 0.9999$ Note (a)

$\kappa_{12} = 0.0747564 - 4.3632\ 10^{-4}T + 6.3888\ 10^{-7}T^2$ Note (a)

$\eta_{12} = 8.05491 - 3.8179\ 10^{-2}T + 4.5368\ 10^{-5}T^2$ Note (a)

$\dfrac{T}{K}$	$\dfrac{\rho_{12}}{g\ cm^{-3}}$	$\dfrac{\kappa_{12}\ 10^3}{S\ cm^{-1}}$	$\dfrac{\Lambda_{12}}{S\ cm^2\ mol^{-1}}$	$\dfrac{\eta_{12}}{Pa\ s}$	$\Lambda_{12}\eta_{12}$
373	1.2531	0.8956	0.2385	0.1261	0.0301
398	1.2353	2.302	0.6220	0.04609	0.0287
423	1.2180	4.507	1.235	0.02279	0.0281

$\Lambda_{12} = 277925 \exp\{-\dfrac{43237}{R\,T}\}$ $r^2 = 0.9964$ Note (a)

$\eta_{12} = 6.12\ 10^{-8} \exp\{\dfrac{44990}{R\,T}\}$ $r^2 = 0.9957$ Note (a)

Ref. 1: Walden & Birr (1932)   data in the table given according to Ref. 1

Note (a): equations obtained by least-squares analysis of the data given in Ref. 1

TABLE B. Transport properties of the molten mixtures of heptylammonium picrate and
         tetrapropylammonium picrate

         $C_7H_{15}NH_3^+Pic^-$ (x_1) + $(C_3H_7)_4N^+Pic^-$ (x_2)

T = 398 K

x_1	$\dfrac{\rho_{12}}{g\ cm^{-3}}$	$\dfrac{\kappa_{12}\ 10^3}{S\ cm^{-1}}$	$\dfrac{\Lambda_{12}}{S\ cm^2\ mol^{-1}}$	$\dfrac{\eta_{12}}{Pa\ s}$	$\Lambda_{12}\eta_{12}$	$\dfrac{E_{\Lambda_{12}}}{J\ mol^{-1}}$	$\dfrac{E_{\eta_{12}}}{J\ mol^{-1}}$
0.00	1.1323	8.046	2.944	0.01646	0.0485	27150	29730
0.25	1.1673	5.174	1.759	0.02283	0.0402	38354	42691
0.50	1.1745	3.296	1.064	0.02980	0.0317	40679	45080
0.75	1.1947	1.998	0.6052	0.04530	0.0274	44017	45947
1.00	1.2160	1.335	0.3780	0.07386	0.0279	46058	50962

TABLE B1. $C_7H_{15}NH_3^+Pic^-$ (x_1) + $(C_3H_7)_4N^+Pic^-$ (x_2)

x_1 = 0.25   Ref. 1

ρ_{12} = 1.4254 - 6.48 $10^{-4}T$ r^2 = 0.9997   Note (a)

κ_{12} = 0.156552 - 9.0825 $10^{-4}T$ + 1.3264 $10^{-6}T^2$ Note (a)

η_{12} = 3.43742 - 1.6224 $10^{-2}T$ + 1.9208 $10^{-6}T^2$ Note (a)

T	ρ_{12}	κ_{12} 10^3	Λ_{12}	η_{12}	$\Lambda_{12}\eta_{12}$
K	g cm^{-3}	S cm^{-1}	S cm^2 mol^{-1}	Pa s	
373	1.1839	2.314	0.7758	0.05820	0.0452
398	1.1673	5.174	1.7592	0.02283	0.0402
423	1.1515	9.692	3.3409	0.01147	0.0383

Λ_{12} = 184826 exp$\{- \dfrac{38354}{R\,T}\}$ r^2 = 0.9988   Note (a)

η_{12} = 5.98 10^{-8} exp$\{\dfrac{42691}{R\,T}\}$ r^2 = 0.9973   Note (a)

Ref. 1: Walden & Birr (1932)   data in the table given according to Ref. 1

Note (a): equations obtained by least-squares analysis of the data given in Ref. 1

TABLE B2. $C_7H_{15}NH_3^+Pic^-$ (x_1) + $(C_3H_7)_4N^+Pic^-$ (x_2)

x_1 = 0.50   Ref. 1

ρ_{12} = 1.4237 - 6.26 $10^{-4}T$ r^2 = 0.9999   Note (a)

κ_{12} = 0.118121 - 6.7727 $10^{-4}T$ + 9.768 $10^{-7}T^2$ Note (a)

η_{12} = 4.88930 - 2.3115 $10^{-2}T$ + 2.74 $10^{-5}T^2$ Note (a)

T	ρ_{12}	κ_{12} 10^3	Λ_{12}	η_{12}	$\Lambda_{12}\eta_{12}$
K	g cm^{-3}	S cm^{-1}	S cm^2 mol^{-1}	Pa s	
373	1.1903	1.400	0.4462	0.07954	0.0355
398	1.1745	3.296	1.0647	0.02980	0.0317
423	1.1590	6.413	2.0992	0.01431	0.0300

Λ_{12}= 225221 exp$\{- \dfrac{40679}{R\,T}\}$ r^2 = 0.9987   Note (a)

η_{12}= 3.78 10^{-8} exp$\{\dfrac{45080}{R\,T}\}$ r^2 = 0.9977   Note (a)

Ref. 1: Walden & Birr (1932)   data in the table given according to Ref. 1

Note (a): equations obtained by least-squares analysis of the data given in Ref. 1

TABLE B3. $C_7H_{15}NH_3^+Pic^-$ (x_1) + $(C_3H_7)_4N^+Pic^-$ (x_2)

x_1 = 0.75   Ref. 1

ρ_{12} = 1.4560 - 6.58 $10^{-4}T$ r^2 = 0.9998   Note (a)

κ_{12} = 0.098830 - 5.5395 $10^{-4}T$ + 7.8056 $10^{-7}T^2$ Note (a)

η_{12} = 7.329123 - 3.4606 $10^{-2}T$ + 4.0968 $10^{-5}T^2$ Note (a)

T	ρ_{12}	$\kappa_{12}\,10^3$	Λ_{12}	η_{12}	$\Lambda_{12}\eta_{12}$
K	g cm^{-3}	S cm^{-1}	S cm^2 mol^{-1}	Pa s	
373	1.2115	0.8017	0.2395	0.1208	0.0289
398	1.1947	1.998	0.6052	0.04530	0.0274
423	1.1787	4.170	1.2802	0.02101	0.0269

$\Lambda_{12} = 353535 \exp\{-\dfrac{44017}{R\,T}\}$ $\qquad\qquad\qquad r^2 = 0.9993 \qquad$ Note (a)

$\eta_{12} = 4.37\ 10^{-8} \exp\{\dfrac{45947}{R\,T}\}$ $\qquad\qquad\qquad r^2 = 0.9988 \qquad$ Note (a)

Ref. 1: Walden & Birr (1932)   data in the table given according to Ref. 1

Note (a): equations obtained by least-squares analysis of the data given in Ref. 1

TABLE C. Transport properties of the molten mixtures of diethylammonium picrate and
triethylammonium picrate triethylammonium picrate

tripropylammonium picrate

$(C_2H_5)_2NH_2^+Pic^-$ (x_1) + $(C_3H_7)_3NH^+Pic^-$ (x_2)

T = 398 K

x_1	ρ_{12}	$\kappa_{12}\,10^3$	Λ_{12}	η_{12}	$\Lambda_{12}\eta_{12}$	$E_{\Lambda_{12}}$	$E_{\eta_{12}}$
	g cm^{-3}	S cm^{-1}	S cm^2 mol^{-1}	Pa s		J mol^{-1}	J mol^{-1}
0.00	1.1569	2.105	0.6775	0.01366	0.00925	36591	32660
0.25	1.1950	3.294	0.9781	0.01625	0.0159	35900	37239
0.50	1.2220	4.755	1.312	0.01780	0.0233	33248	36614
0.75	1.2531	5.980	1.526	0.02048	0.0313	34898	39017
1.00	1.3082	8.270	2.009	0.01638	0.0312	31806	31259

TABLE C1. $(C_2H_5)_2NH_2^+Pic^-$ (x_1) + $(C_3H_7)_3NH^+Pic^-$ (x_2)

x_1 = 0.25 $\qquad\qquad\qquad\qquad\qquad\qquad\qquad\qquad\qquad$ Ref. 1

$\rho_{12} = 1.5014 - 7.68\ 10^{-4}T$ $\qquad\qquad\qquad r^2 = 0.9986 \qquad$ Note (a)

$\kappa_{12} = 0.101731 - 5.8387\ 10^{-4}T + 8.456\ 10^{-7}T^2$ $\qquad\qquad$ Note (a)

$\eta_{12} = 1.731178 - 8.0768\ 10^{-3}T + 9.4672\ 10^{-6}T^2$ $\qquad\qquad$ Note (a)

T	ρ_{12}	$\kappa_{12}\,10^3$	Λ_{12}	η_{12}	$\Lambda_{12}\eta_{12}$
K	g cm^{-3}	S cm^{-1}	S cm^2 mol^{-1}	Pa s	
373	1.2154	1.592	0.4648	0.03569	0.0166
398	1.1950	3.294	0.9781	0.01625	0.0159
423	1.1770	6.053	1.8248	0.008644	0.0158

$\Lambda_{12} = 49780.7 \exp\{-\dfrac{35900}{R\,T}\}$ $\qquad\qquad\qquad r^2 = 0.9997 \qquad$ Note (a)

$\eta_{12} = 2.153\ 10^{-7} \exp\{\dfrac{37239}{R\,T}\}$ $\qquad\qquad\qquad r^2 = 0.9992 \qquad$ Note (a)

Ref. 1: Walden & Birr (1932)   data in the table given according to Ref. 1

Note (a): equations obtained by least-squares analysis of the data given in Ref. 1

TABLE C2. $(C_2H_5)_2NH_2^+Pic^-$ (x_1) + $(C_3H_7)_3NH^+Pic^-$ (x_2)

x_1 = 0.50   Ref. 1

ρ_{12} = 1.4864 - 6.64 $10^{-4}T$ r^2 = 0.9998   Note (a)

κ_{12} = 0.118173 - 6.8870 $10^{-4}T$ + 1.0144 $10^{-6}T^2$ Note (a)

η_{12} = 1.743046 - 8.0979 $10^{-3}T$ + 9.4552 $10^{-6}T^2$ Note (a)

$\dfrac{T}{K}$	$\dfrac{\rho_{12}}{g\ cm^{-3}}$	$\dfrac{\kappa_{12}\ 10^3}{S\ cm^{-1}}$	$\dfrac{\Lambda_{12}}{S\ cm^2\ mol^{-1}}$	$\dfrac{\eta_{12}}{Pa\ s}$	$\Lambda_{12}\eta_{12}$
373	1.2389	2.420	0.6583	0.0380	0.0250
398	1.2220	4.755	1.312	0.0178	0.0233
423	1.2057	8.353	2.338	0.00942	0.0220

Λ_{12} = 30001 exp$\{-\dfrac{33248}{R\ T}\}$ r^2 = 0.9997   Note (a)

η_{12} = 2.818 $10^{-7}\{$exp $\dfrac{36614}{R\ T}\}$ r^2 = 0.9997   Note (a)

Ref. 1: Walden & Birr (1932)   data in the table given according to Ref. 1

Note (a): equations obtained by least-squares analysis of the data given in Ref. 1

TABLE C3. $(C_2H_5)_2NH_2^+Pic^-$ (x_1) + $(C_3H_7)_3NH^+Pic^-$ (x_2)

x_1 = 0.75   Ref. 1

ρ_{12} = 1.5447 - 7.32 $10^{-4}T$ r^2 = 0.9997   Note (a)

κ_{12} = 0.215519 - 1.2145 $10^{-3}T$ + 1.7288 $10^{-6}T^2$ Note (a)

η_{12} = 1.927592 - 8.9029 $10^{-3}T$ + 1.0329 $10^{-5}T^2$ Note (a)

$\dfrac{T}{K}$	$\dfrac{\rho_{12}}{g\ cm^{-3}}$	$\dfrac{\kappa_{12}\ 10^3}{S\ cm^{-1}}$	$\dfrac{\Lambda_{12}}{S\ cm^2\ mol^{-1}}$	$\dfrac{\eta_{12}}{Pa\ s}$	$\Lambda_{12}\eta_{12}$
373	1.2719	3.021	0.7595	0.04395	0.0334
398	1.2531	5.980	1.5260	0.02048	0.0313
423	1.2353	11.10	2.8733	0.009922	0.0285

Λ_{12} = 58407$\{$exp $-\dfrac{34898}{R\ T}\}$ r^2 = 0.9999   Note (a)

η_{12} = 1.523 10^{-7} exp$\{\dfrac{39017}{R\ T}\}$ r^2 = 0.9995   Note (a)

Ref. 1: Walden & Birr (1932)   data in the table given according to Ref. 1

Note (a): equations obtained by least-squares analysis of the data given in Ref. 1

TABLE D. Transport properties of the molten mixtures of diethylammonium picrate and
tetrapropylammonium picrate

$$(C_2H_5)_2NH_2^+Pic^- \ (x_1) \ + \ (C_3H_7)_4N^+Pic^- \ (x_2)$$

T = 398 K

x_1	$\dfrac{\rho_{12}}{\text{g cm}^{-3}}$	$\dfrac{\kappa_{12} \ 10^3}{\text{S cm}^{-1}}$	$\dfrac{\Lambda_{12}}{\text{S cm}^2 \text{ mol}^{-1}}$	$\dfrac{\eta_{12}}{\text{Pa s}}$	$\Lambda_{12}\eta_{12}$	$\dfrac{E_{\Lambda_{12}}}{\text{J mol}^{-1}}$	$\dfrac{E_{\eta_{12}}}{\text{J mol}^{-1}}$
0.00	1.1328	8.048	2.944	0.01646	0.0485	27150	29730
0.25	1.1701	7.197	2.377	0.01759	0.0418	34959	37672
0.50	1.2016	6.631	1.977	0.01890	0.0374	39027	45112
0.75	1.2416	7.078	1.883	0.01874	0.0353	35744	41918
1.00	1.3082	8.270	2.009	0.01638	0.0312	31806	31259

TABLE D1. $(C_2H_5)_2NH_2^+Pic^- \ (x_1) \ + \ (C_3H_7)_4N^+Pic^- \ (x_2)$

x_1 = 0.25   Ref. 1

$\rho_{12} = 1.4451 - 6.9 \ 10^{-4}T$ $r^2 = 0.9996$ Note (a)

$\kappa_{12} = 0.17068 - 1.0078 \ 10^{-3}T + 1.5 \ 10^{-6}T^2$ Note (a)

$\eta_{12} = 1.907382 - 8.9030 \ 10^{-3}T + 1.0439 \ 10^{-5}T^2$ Note (a)

$\dfrac{T}{\text{K}}$	$\dfrac{\rho_{12}}{\text{g cm}^{-3}}$	$\dfrac{\kappa_{12} \ 10^3}{\text{S cm}^{-1}}$	$\dfrac{\Lambda_{12}}{\text{S cm}^2 \text{ mol}^{-1}}$	$\dfrac{\eta_{12}}{\text{Pa s}}$	$\Lambda_{12}\eta_{12}$
373	1.1879	3.479	1.132	0.03895	0.0441
398	1.1701	7.197	2.377	0.01759	0.0418
423	1.1534	12.79	4.285	0.009279	0.0398

$\Lambda_{12} = 90008 \ \exp\{- \dfrac{34959}{R \ T}\}$ $r^2 = 0.9991$ Note (a)

$\eta_{12} = 2.044 \ 10^{-7} \ \exp\{\dfrac{37672}{R \ T}\}$ $r^2 = 0.9983$ Note (a)

Ref. 1: Walden & Birr (1932)   data in the table given according to Ref. 1

Note (a): equations obtained by least-squares analysis of the data given in Ref. 1

TABLE D2. $(C_2H_5)_2NH_2^+Pic^- \ (x_1) \ + \ (C_3H_7)_4N^+Pic^- \ (x_2)$

x_1 = 0.50   Ref. 1

$\rho_{12} = 1.4825 - 7.052 \ 10^{-4}T$ $r^2 = 0.9997$ Note (a)

$\kappa_{12} = 0.142153 - 8.8097 \ 10^{-4}T + 1.3545 \ 10^{-6}T^2$ Note (a)

$\eta_{12} = 6.943021 - 3.4023 \ 10^{-2}T + 4.1706 \ 10^{-5}T^2$ Note (a)

$\dfrac{T}{\text{K}}$	$\dfrac{\rho_{12}}{\text{g cm}^{-3}}$	$\dfrac{\kappa_{12} \ 10^3}{\text{S cm}^{-1}}$	$\dfrac{\Lambda_{12}}{\text{S cm}^2 \text{ mol}^{-1}}$	$\dfrac{\eta_{12}}{\text{Pa s}}$	$\Lambda_{12}\eta_{12}$
348	1.2375	1.137	0.3292	0.1572	0.0518
373	1.2192	3.102	0.9117	0.04387	0.0400

| 398 | 1.2016 | 6.631 | 1.977 | 0.01890 | 0.0374 |
| 423 | 1.1846 | 11.73 | 3.548 | 0.009835 | 0.0349 |

$$\Lambda_{12} = 250140 \ \exp\{- \frac{39027}{R\,T}\} \qquad\qquad r^2 = 0.9962 \qquad \text{Note (a)}$$

$$\eta_{12} = 2.41 \ 10^{-8} \ \exp\{\frac{45112}{R\,T}\} \qquad\qquad r^2 = 0.9908 \qquad \text{Note (a)}$$

Ref. 1: Walden & Birr (1932)   data in the table given according to Ref. 1

Note (a): equations obtained by least-squares analysis of the data given in Ref. 1

TABLE D3. $(C_2H_5)_2NH_2^+Pic^-$ (x_1) + $(C_3H_7)_4N^+Pic^-$ (x_2)

x_1 = 0.75   Ref. 1

$\rho_{12} = 1.5257 - 7.128 \ 10^{-4}T$ $r^2 = 0.9996$ Note (a)

$\kappa_{12} = 0.123278 - 7.5601 \ 10^{-4}T + 1.1664 \ 10^{-6}T^2$ Note (a)

$\eta_{12} = 5.619499 - 2.7449 \ 10^{-2}T + 3.356 \ 10^{-5}T^2$ Note (a)

T	ρ_{12}	$\kappa_{12} \ 10^3$	Λ_{12}	η_{12}	$\Lambda_{12}\eta_{12}$
K	g cm^{-3}	S cm^{-1}	S cm^2 mol^{-1}	Pa s	
348	1.2780	1.418	0.3665	0.1339	0.0491
373	1.2596	3.634	0.9529	0.04149	0.0395
398	1.2416	7.078	1.883	0.01874	0.0353
423	1.2246	12.21	3.293	0.01023	0.0337

$$\Lambda_{12} = 89709 \ \exp\{- \frac{35744}{R\,T}\} \qquad\qquad r^2 = 0.9956 \qquad \text{Note (a)}$$

$$\eta_{12} = 6.28 \ 10^{-8} \ \exp\{\frac{41918}{R\,T}\} \qquad\qquad r^2 = 0.9916 \qquad \text{Note (a)}$$

Ref. 1: Walden & Birr (1932)   data in the table given according to Ref. 1

Note (a): equations obtained by least-squares analysis of the data given in Ref. 1

TABLE E. Transport properties of the molten mixtures of tripropylammonium picrate and
tetrapropylammonium picrate

$(C_3H_7)_3NH^+Pic^-$ (x_1) + $(C_3H_7)_4N^+Pic^-$ (x_2)

T = 398 K

x_1	ρ_{12}	$\kappa_{12} \ 10^3$	Λ_{12}	η_{12}	$\Lambda_{12}\eta_{12}$	$E_{\Lambda_{12}}$	$E_{\eta_{12}}$
	g cm^{-3}	S cm^{-1}	S cm^2 mol^{-1}	Pa s		J mol^{-1}	J mol^{-1}
0.00	1.1328	8.048	2.944	0.01646	0.0485	27150	29730
0.25	1.1526	6.972	2.443	0.01576	0.0386	27433	38083
0.50	1.1541	5.618	1.915	0.01573	0.0301	39600	37240
0.75	1.1587	3.884	1.283	0.01483	0.0190	35542	36786
1.00	1.1569	2.105	0.6775	0.01366	0.00925	36591	32660

TABLE E1. $(C_3H_7)_3NH^+Pic^-$ (x_1) + $(C_3H_7)_4N^+Pic^-$ (x_2)

x_1 = 0.25   Ref. 1

ρ_{12} = 1.4100 - 6.46 $10^{-4}T$ r^2 = 0.9997   Note (a)

κ_{12} = 0.208194 - 1.1615 $10^{-3}T$ + 1.6480 $10^{-6}T^2$ Note (a)

η_{12} = 1.87699 - 8.8016 $10^{-3}T$ + 1.0364 $10^{-5}T^2$ Note (a)

T	ρ_{12}	$\kappa_{12}\ 10^3$	Λ_{12}	η_{12}	$\Lambda_{12}\eta_{12}$
K	g cm^{-3}	S cm^{-1}	S cm^2 mol^{-1}	Pa s	
373	1.1692	4.244	1.466	0.03602	0.0528
398	1.1526	6.972	2.443	0.01576	0.0385
423	1.1369	11.76	4.179	0.008456	0.0353

Λ_{12} = 10037.3 exp$\{-\dfrac{27433}{R\ T}\}$ r^2 = 0.9974   Note (a)

η_{12} = 1.643 10^{-7} exp$\{\dfrac{38083}{R\ T}\}$ r^2 = 0.9979   Note (a)

Ref. 1: Walden & Birr (1932)   data in the table given according to Ref. 1

Note (a): equations obtained by least-squares analysis of the data given in Ref. 1

TABLE E2. $(C_3H_7)_3NH^+Pic^-$ (x_1) + $(C_3H_7)_4N^+Pic^-$ (x_2)

x_1 = 0.50   Ref. 1

ρ_{12} = 1.4435 - 7.264 $10^{-4}T$ r^2 = 0.9998   Note (a)

κ_{12} = 0.103693 - 6.3597 $10^{-4}T$ + 9.788 $10^{-7}T^2$ Note (a)

η_{12} = 1.68599 - 7.8687 $10^{-3}T$ + 9.2264 $10^{-6}T^2$ Note (a)

T	ρ_{12}	$\kappa_{12}\ 10^3$	Λ_{12}	η_{12}	$\Lambda_{12}\eta_{12}$
K	g cm^{-3}	S cm^{-1}	S cm^2 mol^{-1}	Pa s	
348	1.1909	0.909	0.3003		
373	1.1725	2.655	0.8908	0.03461	0.0308
398	1.1541	5.618	1.915	0.01573	0.0301
423	1.1365	9.811	3.396	0.008383	0.0285

Λ_{12} = 284670 exp$\{-\dfrac{39600}{R\ T}\}$ r^2 = 0.9927   Note (a)

η_{12} = 2.087 10^{-7} exp$\{\dfrac{37240}{R\ T}\}$ r^2 = 0.9991   Note (a)

Ref. 1: Walden & Birr (1932)   data in the table given according to Ref. 1

Note (a): equations obtained by least-squares analysis of the data given in Ref. 1

TABLE E3. $(C_3H_7)_3NH^+Pic^-$ (x_1) + $(C_3H_7)_4N^+Pic^-$ (x_2)

x_1 = 0.75   Ref. 1

ρ_{12} = 1.4264 - 6.72 $10^{-4}T$ r^2 = 0.9998   Note (a)

κ_{12} = 0.098809 - 5.7951 $10^{-4}T$ + 8.568 $10^{-7}T^2$ Note (a)

η_{12} = 1.52701 - 7.1143 $10^{-3}T$ + 8.328 $10^{-6}T^2$ Note (a)

T	ρ_{12}	$\kappa_{12}\,10^3$	Λ_{12}	η_{12}	$\Lambda_{12}\eta_{12}$
K	g cm^{-3}	S cm^{-1}	S cm^2 mol^{-1}	Pa s	
373	1.1759	1.857	0.6047	0.03215	0.0194
398	1.1587	3.884	1.283	0.01483	0.0190
423	1.1423	6.982	2.340	0.007921	0.0185

$\Lambda_{12} = 57999 \exp\{-\dfrac{35542}{R\,T}\}$ $\qquad\qquad r^2 = 0.9991 \qquad$ Note (a)

$\eta_{12} = 2.247\ 10^{-7} \exp\{\dfrac{36786}{R\,T}\}$ $\qquad\qquad r^2 = 0.9994 \qquad$ Note (a)

Ref. 1: Walden & Birr (1932) $\qquad\qquad$ data in the table given according to Ref. 1

Note (a): equations obtained by least-squares analysis of the data given in Ref. 1

TABLE F. Transport properties of the molten mixtures of tetrapropylammonium picrate and
tetra*iso*amylammonium picrate

$\qquad$ $(C_3H_7)_4N^+Pic^-$ (x_1) + $(i.C_5H_{11})_4N^+Pic^-$ (x_2)

T = 398 K

x_1	ρ_{12}	$\kappa_{12}\,10^3$	Λ_{12}	η_{12}	$\Lambda_{12}\eta_{12}$	$E_{\Lambda_{12}}$	$E_{\eta_{12}}$
	g cm^{-3}	S cm^{-1}	S cm^2 mol^{-1}	Pa s		J mol^{-1}	J mol^{-1}
0.00	1.0275	3.029	1.552	0.03237	0.0502	35802	38747
0.25	1.0561	3.740	1.758	0.02759	0.0485	42989	42257
0.50	1.0819	4.752	2.062	0.02347	0.0484	37674	40797
0.75	1.1112	6.152	2.449	0.01959	0.0480	44947	39334
1.00	1.1328	8.048	2.944	0.01646	0.0485	27150	29730

TABLE F1. $(C_3H_7)_4N^+Pic^-$ (x_1) + $(i.C_5H_{11})_4N^+Pic^-$ (x_2)

$\qquad$ x_1 = 0.25 $\qquad\qquad\qquad\qquad\qquad\qquad\qquad\qquad$ Ref. 1

$\qquad$ $\rho_{12} = 1.3081 - 6.32\ 10^{-4}T$ $\qquad\qquad\qquad$ $r^2 = 0.9997 \qquad$ Note (a)

$\qquad$ $\kappa_{12} = 0.109351 - 6.4279\ 10^{-4}T + 9.4884\ 10^{-7}T^2$ $\qquad\qquad$ Note (a)

$\qquad$ $\eta_{12} = 3.687120 - 1.7314\ 10^{-2}T + 2.04\ 10^{-5}T^2$ $\qquad\qquad$ Note (a)

T	ρ_{12}	$\kappa_{12}\,10^3$	Λ_{12}	η_{12}	$\Lambda_{12}\eta_{12}$
K	g cm^{-3}	S cm^{-1}	S cm^2 mol^{-1}	Pa s	
348	1.0883	0.5401	0.2464		
373	1.0725	1.677	0.7765	0.06723	0.0522
398	1.0561	3.740	1.758	0.02759	0.0485
423	1.0411	7.249	3.457	0.01345	0.0465

$\Lambda_{12} = 745395 \exp\{-\dfrac{42989}{R\,T}\}$ $\qquad\qquad r^2 = 0.9958 \qquad$ Note (a)

$\eta_{12} = 8.04 \ 10^{-8} \exp\{\frac{42257}{R\,T}\}$ $\qquad\qquad\qquad\qquad$ $r^2 = 0.9993$ $\qquad$ Note (a)

Ref. 1: Walden & Birr (1932) $\qquad\qquad\qquad$ data in the table given according to Ref. 1

Note (a): equations obtained by least-squares analysis of the data given in Ref. 1

TABLE F2. $(C_3H_7)_4N^+Pic^-$ (x_1) $\ +\ $ $(i.C_5H_{11})_4N^+Pic^-$ (x_2)

$\qquad x_1 \ = 0.50$ $\qquad\qquad\qquad\qquad\qquad\qquad\qquad\qquad\qquad$ Ref. 1

$\qquad \rho_{12} = 1.3352 - 6.36 \ 10^{-4}T$ $\qquad\qquad\qquad$ $r^2 = 0.9998$ $\qquad$ Note (a)

$\qquad \kappa_{12} = 0.132833 - 7.7587 \ 10^{-4}T + 1.1408 \ 10^{-6}T^2$ $\qquad\qquad\qquad$ Note (a)

$\qquad \eta_{12} = 3.098516 - 1.4562 \ 10^{-2}T + 1.7176 \ 10^{-5}T^2$ $\qquad\qquad\qquad$ Note (a)

T	ρ_{12}	$\kappa_{12} \ 10^3$	Λ_{12}	η_{12}	$\Lambda_{12}\eta_{12}$
K	g cm^{-3}	S cm^{-1}	S cm^2 mol^{-1}	Pa s	
373	1.0981	2.150	0.9213	0.05646	0.0520
398	1.0819	4.752	2.062	0.02347	0.0484
423	1.0663	8.760	3.866	0.01195	0.0462

$\Lambda_{12} = 176307 \ \exp\{- \frac{37674}{R\,T}\}$ $\qquad\qquad\qquad\qquad$ $r^2 = 0.9987$ $\qquad$ Note (a)

$\eta_{12} = 1.075 \ 10^{-7} \exp\{\frac{40797}{R\,T}\}$ $\qquad\qquad\qquad\qquad$ $r^2 = 0.9984$ $\qquad$ Note (a)

Ref. 1: Walden & Birr (1932) $\qquad\qquad\qquad$ data in the table given according to Ref. 1

Note (a): equations obtained by least-squares analysis of the data given in Ref. 1

TABLE F3. $(C_3H_7)_4N^+Pic^-$ (x_1) $\ +\ $ $(i.C_5H_{11})_4N^+Pic^-$ (x_2)

$\qquad x_1 \ = 0.75$ $\qquad\qquad\qquad\qquad\qquad\qquad\qquad\qquad\qquad$ Ref. 1

$\qquad \rho_{12} = 1.3605 - 6.26 \ 10^{-4}T$ $\qquad\qquad\qquad$ $r^2 = 0.9999$ $\qquad$ Note (a)

$\qquad \kappa_{12} = 0.094741 - 6.3141 \ 10^{-4}T + 1.0272 \ 10^{-6}T^2$ $\qquad\qquad\qquad$ Note (a)

$\qquad \eta_{12} = 2.419316 - 1.1349 \ 10^{-2}T + 1.3368 \ 10^{-5}T^2$ $\qquad\qquad\qquad$ Note (a)

T	ρ_{12}	$\kappa_{12} \ 10^3$	Λ_{12}	η_{12}	$\Lambda_{12}\eta_{12}$
K	g cm^{-3}	S cm^{-1}	S cm^2 mol^{-1}	Pa s	
373	1.1271	2.138	0.8394	0.04567	0.0383
398	1.1112	6.152	2.449	0.01959	0.0480
423	1.0958	11.45	4.624	0.01022	0.0473

$\Lambda_{12} = 1740446 \ \exp\{- \frac{44947}{R\,T}\}$ $\qquad\qquad\qquad\qquad$ $r^2 = 0.9879$ $\qquad$ Note (a)

$\eta_{12} = 1.394 \ 10^{-7} \ \exp\{\frac{39334}{R\,T}\}$ $\qquad\qquad\qquad\qquad$ $r^2 = 0.9984$ $\qquad$ Note (a)

Ref. 1: Walden & Birr (1932) $\qquad\qquad\qquad$ data in the table given according to Ref. 1

Note (a): equations obtained by least-squares analysis of the data given in Ref. 1

TABLE G. Transport properties of the molten mixtures of sodium ethanoate and potassium ethanoate

$$CH_3COO^-Na^+ \ (x_1) \ + \ CH_3COO^-K^+ \ (x_2)$$

$T = 603$ K

x_1	ρ_{12} g cm^{-3}	κ_{12} S cm^{-1}	Λ_{12} S cm^2 mol^{-1}	η_{12} Pa s	$\Lambda_{12}\eta_{12}$	$E_{\Lambda_{12}}$ J mol^{-1}	$E_{\eta_{12}}$ J mol^{-1}
0.000	1.3702	0.2367	16.95	0.003295	0.0558	23889	15463
0.205	1.3494	0.2181	15.33	0.003659	0.0560	27061	22647
0.463	1.3254	0.2012	13.76	0.005534$^+$	0.0761	29446	23444
0.648	1.3037	0.1931	12.99	0.004680$^+$	0.0643	26267	24089
0.772	1.2920	0.1895	12.56	0.004201	0.0528	24818	25316
0.874	1.2828	0.1877	12.30	0.004461	0.0548	23761	26856
1.000	1.2729	0.1865	12.02	0.004561	0.0548	22850	29515

($^+$) Calculated from the pertinent equations given in Tables G2 and G3 (beyond the experimental temperature range)

TABLE G1. $CH_3COO^-Na^+ \ (x_1) \ + \ CH_3COO^-K^+ \ (x_2)$

$x_1 = 0.205$

$\rho_{12} = 1.7135 - 6.035 \ 10^{-4}T$ $r^2 = 0.9998$ Note (a)

$\kappa_{12} = -0.77082 + 1.64 \ 10^{-3}T$ Note (b)

$\eta_{12} = 0.229529 - 7.4799 \ 10^{-4}T + 6.1926 \ 10^{-7}T^2$ Note (c)

T K	ρ_{12} g cm^{-3}	κ_{12} S cm^{-1}	Λ_{12} S cm^2 mol^{-1}	η_{12} Pa s	$\Lambda_{12}\eta_{12}$	
543	1.3859	0.1197	8.190	0.05875	0.0481	ρ: Ref. 1
553	1.3797	0.1361	9.354	0.05265	0.0492	κ: Ref. 2
563	1.3735	0.1525	10.53	0.004696	0.0494	η: Ref. 3
573	1.3674	0.1689	11.71	0.004251	0.0497	
583	1.3613	0.1853	12.91	0.003930	0.0507	
593	1.3553	0.2017	14.11	0.003732	0.0526	
603	1.3494	0.2181	15.33	0.003659	0.0560	
613	1.3435	0.2345	16.55			
623	1.3376	0.2509	17.79			

$\Lambda_{12} = 3375.4 \ \exp\{-\dfrac{27061}{R\ T}\}$ $r^2 = 0.9962$ Note (b)

$\eta_{12} = 3.796 \ 10^{-5} \ \exp\{\dfrac{22647}{R\ T}\}$ $r^2 = 0.9628$ Note (c)

Ref. 1: Hazlewood, Rhodes & Ubbelohde (1966)

Ref. 2: Leonesi, Cingolani & Berchiesi (1973)

Ref. 3: Kisza, private communication (1979)

Note (a): equation obtained by least-squares analysis of the data given in Ref. 1

Note (b): equations obtained by least-squares analysis of the data given in Ref. 2

Note (c): equations obtained by least-squares analysis of the data given in Ref. 3

TABLE G2. $CH_3COO^-Na^+$ (x_1) + $CH_3COO^-K^+$ (x_2)

x_1 = 0.463

$\rho_{12} = 1.6553 - 5.4338 \ 10^{-4}T$ $r^2 = 0.9924$ Note (a)

$\kappa_{12} = -0.69509 + 1.4863 \ 10^{-3}T$ Note (b)

$\eta_{12} = 0.4207025 - 1.416149 \ 10^{-3}T + 1.211648 \ 10^{-6}T^2$ Note (c)

$\dfrac{T}{K}$	$\dfrac{\rho_{12}}{g\ cm^{-3}}$	$\dfrac{\kappa_{12}}{S\ cm^{-1}}$	$\dfrac{\Lambda_{12}}{S\ cm^2\ mol^{-1}}$	$\dfrac{\eta_{12}}{Pa\ s}$	$\Lambda_{12}\eta_{12}$
513	1.3785	0.06742	4.435	0.01308	0.0580
523	1.3724	0.08228	5.437	0.01147	0.0624
533	1.3663	0.09715	6.448	0.01011	0.0651
543	1.3603	0.1120	7.466	0.008986	0.0671
553	1.3544	0.1269	8.496	0.008104	0.0688
563	1.3485	0.1417	9.529	0.007465	0.0711
573	1.3425	0.1566	10.57	0.007068	0.0747
583	1.3368	0.1715	11.63	0.006913	0.0804
593	1.3311	0.1863	12.69		
603	1.3254	0.2012	13.76		
613	1.3197	0.2309	15.86		
623	1.3141	0.2458	16.96		

ρ: Ref. 1
κ: Ref. 2
η: Ref. 3

$\Lambda_{12} = 4891 \ \exp\{-\dfrac{29446}{R\ T}\}$ $r^2 = 0.9858$ Note (b)

$\eta_{12} = 5.15418 \ 10^{-5} \ \exp\{\dfrac{23444}{R\ T}\}$ $r^2 = 0.9747$ Note (c)

Ref. 1: Hazlewood, Rhodes & Ubbelohde (1966)

Ref. 2: Leonesi, Cingolani & Berchiesi (1973)

Ref. 3: Kisza, private communication (1979)

Note (a): equation obtained by least-squares analysis of the data given in Ref. 1

Note (b): equations obtained by least-squares analysis of the data given in Ref. 2

Note (c): equations obtained by least-squares analysis of the data given in Ref. 3

TABLE G3. $CH_3COO^-Na^+$ (x_1) + $CH_3COO^-K^+$ (x_2)

x_1 = 0.648

$\rho_{12} = 1.6535 - 5.8 \ 10^{-4}T$ $r^2 = 0.9998$ Note (a)

$\kappa_{12} = -0.66108 + 1.4165 \ 10^{-3}T$ Note (b)

$\eta_{12} = 0.3013490 - 9.8092 \ 10^{-4}T + 8.1203 \ 10^{-7}T^2$ Note (c)

T	ρ_{12}	κ_{12}	Λ_{12}	η_{12}	$\Lambda_{12}\eta_{12}$	ρ: Ref. 1
$\overline{K}$	$g\ cm^{-3}$	$S\ cm^{-1}$	$S\ cm^2\ mol^{-1}$	Pa s		κ: Ref. 2 η: Ref. 3
543	1.3388	0.1081	7.081	0.008134	0.0576	
553	1.3328	0.1222	8.041	0.007225	0.0581	
563	1.3269	0.1364	9.015	0.006478	0.0584	
573	1.3210	0.1506	9.998	0.005893	0.0589	
583	1.3152	0.1647	10.98	0.005471	0.0600	
593	1.3094	0.1789	11.98	0.005212	0.0624	
603	1.3037	0.1931	12.99			
613	1.2980	0.2072	14.00			
623	1.2924	0.2214	15.02			

$\Lambda_{12} = 2441.57\ \exp\{-\dfrac{26267}{R\ T}\}$ $r^2 = 0.9966$ Note (b)

$\eta_{12} = 3.83327\ 10^{-5}\ \exp\{\dfrac{24089}{R\ T}\}$ $r^2 = 0.9808$ Note (c)

Ref. 1: Hazlewood, Rhodes & Ubbelohde (1966)

Ref. 2: Leonesi, Cingolani & Berchiesi (1973)

Ref. 3: Kisza, private communication (1979)

Note (a): equation obtained by least-squares analysis of the data given in Ref. 1

Note (b): equations obtained by least-squares analysis of the data given in Ref. 2

Note (c): equations obtained by least-squares analysis of the data given in Ref. 3

TABLE G4. $CH_3COO^-Na^+$ (x_1) + $CH_3COO^-K^+$ (x_2)

$x_1 = 0.772$

$\rho_{12} = 1.6534 - 5.9928\ 10^{-4}T$ $r^2 = 0.9999$ Note (a)

$\kappa_{12} = -0.64989 + 1.3921\ 10^{-3}T$ Note (b)

$\eta_{12} = 0.1999918 - 6.2399\ 10^{-4}T + 4.9635\ 10^{-7}T^2$ Note (c)

T	ρ_{12}	κ_{12}	Λ_{12}	η_{12}	$\Lambda_{12}\eta_{12}$	ρ: Ref. 1
$\overline{K}$	$g\ cm^{-3}$	$S\ cm^{-1}$	$S\ cm^2\ mol^{-1}$	Pa s		κ: Ref. 2 η: Ref. 3
563	1.3162	0.1339	8.717	0.006011	0.0524	
573	1.3100	0.1478	9.668	0.005409	0.0523	
583	1.3040	0.1617	10.62	0.004907	0.0521	
593	1.2979	0.1756	11.59	0.004505	0.0522	
603	1.2920	0.1895	12.56	0.004201	0.0528	
613	1.2861	0.2034	13.55			
623	1.2802	0.2173	14.54			

$\Lambda_{12} = 1767.17\ \exp\{-\dfrac{24818}{R\ T}\}$ $r^2 = 0.9986$ Note (b)

$\eta_{12} = 2.66946\ 10^{-5}\ \exp\{\dfrac{25316}{R\ T}\}$ $r^2 = 0.9940$ Note (c)

Ref. 1: Hazlewood, Rhodes & Ubbelohde (1966)

Ref. 2: Leonesi, Cingolani & Berchiesi (1973)

Ref. 3: Kisza, private communication (1979)

Note (a): equation obtained by least-squares analysis of the data given in Ref. 1
Note (b): equations obtained by least-squares analysis of the data given in Ref. 2
Note (c): equations obtained by least-squares analysis of the data given in Ref. 3

TABLE G5. $CH_3COO^-Na^+$ (x_1) + $CH_3COO^-K^+$ (x_2)

x_1 = 0.874

$\rho_{12} = 1.6513 - 6.11 \ 10^{-4}T$	$r^2 = 0.9999$ Note (a)
$\kappa_{12} = -0.64325 + 1.378 \ 10^{-4}T$	Note (b)
$\eta_{12} = 0.1859467 - 5.6684 \ 10^{-4}T + 4.4092 \ 10^{-7}T^2$	Note (c)

T — K	ρ_{12} — g cm^{-3}	κ_{12} — S cm^{-1}	Λ_{12} — S cm^2 mol^{-1}	η_{12} — Pa s	$\Lambda_{12}\eta_{12}$	
583	1.2952	0.1601	10.39	0.005339	0.0554	ρ: Ref. 1
593	1.2890	0.1739	11.34	0.004856	0.0550	κ: Ref. 2
603	1.2828	0.1877	12.30	0.004461	0.0548	η: Ref. 3
613	1.2769	0.2015	13.26	0.004154	0.0551	
623	1.2707	0.2152	14.24			

$\Lambda_{12} = 1402.59 \exp\{-\frac{23761}{R\ T}\}$ $r^2 = 0.9994$ Note (b)

$\eta_{12} = 2.09959 \ 10^{-5} \exp\{\frac{26856}{R\ T}\}$ $r^2 = 0.9967$ Note (c)

Ref. 1: Hazlewood, Rhodes & Ubbelohde (1966)

Ref. 2: Leonesi, Cingolani & Berchiesi (1973)

Ref. 3: Kisza, private communication (1979)

Note (a): equation obtained by least-squares analysis of the data given in Ref. 1
Note (b): equations obtained by least-squares analysis of the data given in Ref. 2
Note (c): equations obtained by least-squares analysis of the data given in Ref. 3

TABLE H. Transport properties of the molten mixtures of sodium ethanoate and rubidium
        ethanoate

$CH_3COO^-Na^+$ (x_1) + $CH_3COO^-Rb^+$ (x_2)

T = 602 K

x_1	ρ_{12} — g cm^{-3}	κ_{12} — S cm^{-1}	Λ_{12} — S cm^2 mol^{-1}	η_{12} — Pa s	$\Lambda_{12}\eta_{12}$	$E_{\Lambda_{12}}$ — J mol^{-1}	$E_{\eta_{12}}$ — J mol^{-1}
0.00	1.9300	0.2229	16.69			22729	
0.25	1.7737	0.2072	15.06			23495	
0.50	1.6124	0.1973	13.86			23997	
0.75	1.4352	0.1889	12.85			25224	
1.00	1.2658	0.1852	12.01			23477	

TABLE H1. $CH_3COO^-Na^+$ (x_1) + $CH_3COO^-Rb^+$ (x_2)

x_1 = 0.25

ρ_{12} = 2.2583 - 8.0500 $10^{-4}T$ Ref. 1

κ_{12} = - 0.7397 + 1.573 $10^{-3}T$ Ref. 1

T	ρ_{12}	κ_{12}	Λ_{12}	η_{12}	$\Lambda_{12}\eta_{12}$
K	g cm^{-3}	S cm^{-1}	S cm^2 mol^{-1}	Pa s	
602	1.7737	0.2072	15.06		
608	1.7688	0.2167	15.79		
614	1.7640	0.2261	16.52		
620	1.7592	0.2356	17.26		

Λ_{12} = 1647.141 exp$\{-\dfrac{23495}{R\,T}\}$ r^2 = 0.9999

Ref. 1: Leonesi, Cingolani & Berchiesi (1976)

TABLE H2. $CH_3COO^-Na^+$ (x_1) + $CH_3COO^-Rb^+$ (x_2)

x_1 = 0.50

ρ_{12} = 2.1141 - 8.3333 $10^{-4}T$ Ref. 1

κ_{12} = - 0.7177 + 1.520 $10^{-3}T$ Ref. 1

T	ρ_{12}	κ_{12}	Λ_{12}	η_{12}	$\Lambda_{12}\eta_{12}$
K	g cm^{-3}	S cm^{-1}	S cm^2 mol^{-1}	Pa s	
602	1.6124	0.1973	13.86		
608	1.6074	0.2065	14.55		
614	1.6024	0.2156	15.24		
620	1.5974	0.2247	15.93		

Λ_{12} = 1675.899 exp$\{-\dfrac{23997}{R\,T}\}$ r^2 = 0.9998

Ref. 1: Leonesi, Cingolani & Berchiesi (1976)

TABLE H3. $CH_3COO^-Na^+$ (x_1) + $CH_3COO^-Rb^+$ (x_2)

x_1 = 0.75

ρ_{12} = 1.8666 - 7.1666 $10^{-4}T$ Ref. 1

κ_{12} = - 0.6876 + 1.456 $10^{-3}T$ Ref. 1

T	ρ_{12}	κ_{12}	Λ_{12}	η_{12}	$\Lambda_{12}\eta_{12}$
K	g cm^{-3}	S cm^{-1}	S cm^2 mol^{-1}	Pa s	
602	1.4352	0.1889	12.85		
608	1.4309	0.1976	13.17		

614   1.4266   0.2064   14.13
620   1.4223   0.2151   14.77

$$\Lambda_{12} = 1966.481 \exp\{-\frac{25224}{R\,T}\}$$ $r^2 = 0.9694$

Ref. 1: Leonesi, Cingolani & Berchiesi (1976)

TABLE I. Transport properties of the molten mixtures of sodium ethanoate and cesium ethanoate

$$CH_3COO^-Na^+ \ (x_1) \ + \ CH_3COO^-Cs^+ \ (x_2)$$

T = 602 K

x_1	ρ_{12} $g\ cm^{-3}$	κ_{12} $S\ cm^{-1}$	Λ_{12} $S\ cm^2\ mol^{-1}$	η_{12} $Pa\ s$	$\Lambda_{12}\eta_{12}$	$E_{\Lambda_{12}}$ $J\ mol^{-1}$	$E_{\eta_{12}}$ $J\ mol^{-1}$
0.00	2.6297	0.1852	22.18			18592	
0.20	2.1277	0.2181	17.42			18826	
0.50	1.8454	0.1926	14.29			20357	
0.75	1.5710	0.2211	13.01			21744	
1.00	1.2658	0.2419	12.01			23477	

TABLE I1. $CH_3COO^-Na^+ \ (x_1) \ + \ CH_3COO^-Cs^+ \ (x_2)$

x_1 = 0.20
$\rho_{12} = 2.5290 - 6.6667 \ 10^{-4}T$ Ref. 1
$\kappa_{12} = - 0.5802 + 1.326 \ 10^{-3}T$ Ref. 1

T K	ρ_{12} $g\ cm^{-3}$	κ_{12} $S\ cm^{-1}$	Λ_{12} $S\ cm^2\ mol^{-1}$	η_{12} $Pa\ s$	$\Lambda_{12}\eta_{12}$
602	2.1277	0.2181	17.42		
608	2.1237	0.2260	18.09		
614	2.1197	0.2340	18.76		
620	2.1157	0.2419	19.43		

$$\Lambda_{12} = 749.361 \exp\{-\frac{18826}{R\,T}\}$$ $r^2 = 0.9999$

Ref. 1: Leonesi, Cingolani & Berchiesi (1976)

TABLE I2. $CH_3COO^-Na^+ \ (x_1) \ + \ CH_3COO^-Cs^+ \ (x_2)$

x_1 = 0.50
$\rho_{12} = 2.1735 - 5.4500 \ 10^{-4}T$ Ref. 1
$\kappa_{12} = - 0.5708 + 1.268 \ 10^{-3}T$ Ref. 1

T	ρ_{12}	κ_{12}	Λ_{12}	η_{12}	$\Lambda_{12}\eta_{12}$
K	g cm^{-3}	S cm^{-1}	S cm^2 mol^{-1}	Pa s	
602	1.8454	0.1926	14.29		
608	1.8421	0.2001	14.88		
614	1.8388	0.2078	15.48		
620	1.8356	0.2154	16.08		

$$\Lambda_{12} = 834.636 \ \exp\{-\frac{20357}{R\,T}\}$$ $r^2 = 0.9999$

Ref. 1: Leonesi, Cingolani & Berchiesi (1976)

TABLE I3. $CH_3COO^-Na^+$ (x_1) + $CH_3COO^-Cs^+$ (x_2)

$$x_1 = 0.75$$
$$\rho_{12} = 1.8719 - 5.0000 \ 10^{-4}T$$ Ref. 1
$$\kappa_{12} = -0.6091 + 1.322 \ 10^{-3}T$$ Ref. 1

T	ρ_{12}	κ_{12}	Λ_{12}	η_{12}	$\Lambda_{12}\eta_{12}$
K	g cm^{-3}	S cm^{-1}	S cm^2 mol^{-1}	Pa s	
602	1.5710	0.1867	13.01		
608	1.5679	0.1947	13.60		
614	1.5649	0.2026	14.18		
620	1.5620	0.2105	14.76		

$$\Lambda_{12} = 1003.098 \ \exp\{-\frac{21744}{R\,T}\}$$ $r^2 = 0.9988$

Ref. 1: Leonesi, Cingolani & Berchiesi (1976)

REFERENCES

FRANZOSINI, P., FERLONI, P., and SPINOLO, G., *"Molten Salts with Organic Anions - An Atlas of Phase Diagrams"*, CNR - Institute of Physical Chemistry, University of Pavia (Italy), 1973

GLASSTONE, S., LAIDLER, K.J., and EYRING, H., *"The Theory of Rate Processes"*, McGraw-Hill, New York, 1941, p. 485

HAZLEWOOD, F.J., RHODES, E., and UBBELOHDE, A.R., *Trans. Faraday Soc.*, **62**, 3101 (1966)

LEONESI, D., CINGOLANI, A., and BERCHIESI, G., *J. Chem. Eng. Data*, **18**, 391 (1973)

LEONESI, D., CINGOLANI, A., and BERCHIESI, G., *Z. Naturforsch.*, **31a**, 1609 (1976)

MARKOV, B.F., and SUMINA, L.A., *Dokl. Akad. Nauk SSSR*, **110**, 411 (1956)

MARKOV, B.F., and SUMINA, L.A., *Zh. Fiz. Khim.*, **31**, 1767 (1957)

MARTIN, R.L., *J. Chem. Soc.*, **1954**, 3246

MOCHINAGA, J., CHO, K., and KURODA, T., *Denki Kagaku*, **36**, 746 (1960)

MURGULESCU, I.G., and ZUCA, S., *Rev. Roum. Chim.*, <u>10</u>, 123 (1965)

MURGULESCU, I.G., and ZUCA, S., *Rev. Roum. Chim.*, <u>10</u>, 129 (1965)

MURGULESCU, I.G., and ZUCA, S., *Electrochim. Acta*, <u>11</u>, 1383 (1966)

MURGULESCU, I.G., and ZUCA, S., *Electrochim. Acta*, <u>14</u>, 519 (1969)

WALDEN, P., and BIRR, E.J., *Z. physik. Chem.*, <u>A160</u>, 161 (1932)

ZUCA, S., and BORCAN, R., *Electrochim. Acta*, <u>15</u>, 1817 (1970)

Part 3
SOLUTIONS

TRANSPORT PROPERTIES OF ORGANIC SALTS IN AQUEOUS SOLUTIONS

P. Stenius, S. Backlund and P. Ekwall

MOLAR CONDUCTIVITIES OF ORGANIC IONS AT INFINITE DILUTION

DEFINITIONS

The molar conductivity of an electrolyte in solution is defined by

$$\Lambda_m = \kappa/c \tag{1}$$

where κ is the specific conductivity of the solution and c is the concentration of the electrolyte. The equivalent conductivity is obtained by dividing the molar conductivity by the number z of positive or negative charges formed when the electrolyte dissociates:

$$\Lambda_{eq} = \kappa/zc \tag{2}$$

Since the current through an electrolyte solution is the result of motion of oppositely charged ions into opposite directions, the molar conductivity of the salt is the sum of the molar conductivities of the individual ions:

$$\Lambda_m = \Lambda_+ + \Lambda_- \tag{3}$$

In giving Λ_m, the formula unit whose concentration is c must be specified. For strong electrolytes in aqueous solution, Λ_m depends on the concentration even in extremely dilute solution. Up to ≈ 0.01 M solutions this dependence follows Kohlrausch's law

$$\Lambda_m = \Lambda_m^\infty + A\, c^{\frac{1}{2}} \tag{4}$$

where A is a constant. Λ_m^∞ is called the molar conductivity at infinite dilution, and may be expressed as the sum of ionic molar conductivities at infinite dilution:

$$\Lambda_m^\infty = \Lambda_+^\infty + \Lambda_-^\infty \tag{5}$$

CONCENTRATION DEPENDENCE OF MOLAR CONDUCTIVITIES

The literature on conductivities in aqueous solutions, whether concerning organic or inorganic ions, is very voluminous. Conductivity measurements are considered to be one of the most reliable methods to detect the critical micelle concentration of surfactant ions. This is very lucidly discussed by Mukerjee & Mysels and their tables include such determinations up to 1966. The low values of critical micelle concentrations for ions with more than 8 carbon atoms in the chain implies that deviation from Kohlrausch's law for such molecules will be due not only to limitations of the model of ion-ion interactions predicted by the Debye-Hückel-Onsager theory of electrolyte conductivity but also to the formation of pre-micellar aggregates and micelles (Mukerjee). Here, we limit ourselves to giving tables of molar conductivities for organic ions at infinite dilution (Tables 1 and 2).

VISCOSITIES OF AQUEOUS SOLUTIONS OF ORGANIC IONS

Viscosity is defined as the force required to produce unit rate of shear between two layers separated by unit distance. For solutions it is very usual to use the relative viscosity

$$\eta_{rel} = \eta/\eta_o \tag{6}$$

(where η_o is the viscosity of the pure solvent at the same temperature as η) or the specific viscosity

$$\eta_{sp} = \eta/\eta_o - 1 \tag{7}$$

Viscosities of dilute electrolyte solutions increase with concentration due to the electrical forces between ions in adjacent layers. This effect has been theoretically investigated by Falkenhagen & Coworkers {Falkenhagen & Dole; Falkenhagen & Vernon (1932 a,b); Falkenhagen & Kelbg}. They showed that the relative viscosity in the limit of very dilute solutions is given by

$$\eta_{rel} = 1 + A_1\sqrt{c} \tag{8}$$

However, even at very moderate concentrations, a linear term has to be added to this equation (Jones & Dole)

$$\eta_{rel} = 1 + A_1\sqrt{c} + A_2c \tag{9}$$

While A_1 is above all a function of solvent properties, of the charge of the ions, and of temperature, A_2 is specific to the electrolyte and very sensitive to temperature. In solutions of salts that associate to form micelles a small increase in A_2 is often observed already below the critical micelle concentration (cmc). Marked deviations from Eq. (9) are observed above the cmc, as is to be expected due to the rapid increase in the mean size of the ionic units in this region. As has been shown by Ekwall & Coworkers {Ekwall, Mandell & Solyom; Ekwall & Holmberg (1965 a,b); Solyom & Ekwall}, it is possible to treat solutions above the cmc on the basis of the extended Einstein theory for the viscosity according to which (Einstein; Vand)

$$\log \eta_{rel} = A_3c/(1 - Q'c) \tag{10}$$

where Q' is a constant related to the interaction between the spherical micelles and, for spherical aggregates, $A_3 = 2.5\ V_m/\ln 10$, where V_m is the molar volume of the solute, including hydration water that is so firmly bound to the aggregates that it does not participate in the viscous shearing process. For sodium octanoate Ekwall et al.(1965a) found that $A_3 = 0.344\ 1\ mol^{-1}$, giving $V_m = 0.317\ 1\ mol^{-1}$ for the hydrated micelles while the partial molar volume of the salt is $0.1414\ 1\ mol^{-1}$. This indicates moderate hydration of the micelles, whereas for cetyltrimethylammonium bromide (CTAB) $A_3 = 1.58\ 1\ mol^{-1}$, $V_m = 1.46\ 1\ mol^{-1}$ and the partial molar volume is $0.36\ 1\ mol^{-1}$ at corresponding concentrations which indicates a very strong binding of water.

At high micellar concentrations the viscosity increases rapidly and for many salts is described by an equation of the type

$$\log (\eta_{sp}/c) = k_1 + k_2c \tag{11}$$

where k_1 and k_2 are constants.

The temperature dependence of the viscosity is very different in different concentration ranges of solutions of micelle-forming salts. Ekwall et al.(1965b) have shown that the activation energy of the viscous flow, E, undergoes only small variations not only below the cmc but also in a large concentration range in which micellar association occurs. E, however, increases rapidly at high micellar concentrations. E has been calculated from the Arrhenius equation

$$\eta = B\ e^{E/RT} \tag{12}$$

In Table 3 are given viscosities according to Eq. (9) for several organic salts, some of which undoubtedly form micelles and for which, hence, care should be taken in utilizing Eq. (9) at concentrations above the cmc. Viscosity data at higher concentrations are given in Tables 4, 5 and 6. Tables 7 and 8 give values of η_{rel} for organic salts in aqueous solution for which there is not enough data to merit fitting to an equation of types (9–11).

TABLES

TABLE 1. Molar conductivities of ions at infinite dilution – Salts with organic anion *(Note a)*

Anion	T —— K	$\dfrac{\Lambda_i^\infty}{\Omega^{-1}cm^2mol^{-1}}$	Year	Ref.	Notes
Methanoate	291.2	47.0	1876	1	
(Formate)		47.3	1921	2	
CHO_2		47.5	1923	3	
		48.0	1960	4	
	298.2	46.0			
		54.59	1940	5	
		$\approx$52	1942	6	
	301.2	47.0	1970	7	
	308.2	67.69	1965	8	
	323.2	86	1960	4	

a: for references, see the list at the end of Table 2

TABLE 1 (Continued)

Methylsulfonate CH_3SO_3	273.2 293.2 298.2	24.1 42.5 48.8	1951	9	
Ethanoate (Acetate) $C_2H_3O_2$	273.2 291.2 298.2 301.2 308.2 323.2	20.3 34.6 32.48 35.0 40.8 40.90 41.4 42.07 35 50.71 67	1909 1927 1960 1909 1932 1941 1942 1970 1965 1909	10 11 4 10 12 13 14 7 8 10	
Chloroethanoate (Chloroacetate) $C_2H_2ClO_2$	288.2 298.2 308.2	33.70 39.8 42.20 52.34	1955, 1959 1950, 1952 1955, 1959	15, 16 17, 18 15, 16	a a a
Dichloroethanoate (Dichloroacetate) $C_2HCl_2O_2$	298.2	38.3	1924	19	
Trichloroethanoate (Trichloroacetate) $C_2Cl_3O_2$		36.6			
Bromoethanoate (Bromoacetate) $C_2H_2BrO_2$	288.2 298.2 308.2	30.84 39.22 48.42	1955, 1959	15, 16	a a a
Iodoethanoate (Iodoacetate) $C_2H_2IO_2$	288.2 298.2 308.2	31.85 40.60 50.10			a a a
Fluoroethanoate (Fluoroacetate) $C_2H_2FO_2$	288.2 298.2 308.2	35.28 44.39 53.70	1955, 1959	15, 16	b b b
Cyanoethanoate (Cyanoacetate) $C_2H_2(CN)O_2$	278.2 288.2 298.2 308.2 318.2	25.77 34.17 41.5 41.8 43.2 53.21 64.08	1956, 1959 1927 1940 1956, 1959	20, 16 11 5 20, 16	b b b b b
Ethylsulfonate $C_2H_5SO_3$	273.2 293.2 298.2	19.3 34.7 39.6 39.57	1951 1967	9 21	
Propanoate (Propionate) $C_3H_5O_2$	291.2 298.2 308.2	31.0 35.80 44.42	1876 1938 1965	1 22 8	
Propylsulfonate $C_3H_7SO_3$	298.2	37.1	1951	9	

a: $\Lambda_{H^+}^{\infty}$ from Ref. 16

b: $\Lambda_{Na^+}^{\infty} = 50.1 \ \Omega^{-1}cm^2mol^{-1}$

TABLE 1 (Continued)

Butanoate (Butyrate) $C_4H_7O_2$	298.2 308.2	32.6 30.8 40.29	1938 1943 1965	22 23 8	
Trichlorobutanoate (Trichlorobutyrate) $C_4H_4Cl_3O_2$	298.2	30.9	1927	11	
Butylsulfonate $C_4H_9SO_3$		32.47	1967	21	
Pentanoate (Valerate) $C_5H_9O_2$	291.2 298.2	25.7 ≈29 28.8 28.6	1876 1942 1943 1978	1 6 23 24	 a
3-Methylbutanoate (Isovalerate) $C_5H_9O_2$		37.2			a
2,2-Dimethylpropanoate (Trimethylacetate) $C_5H_9O_2$		33.4			a
Hexanoate (Caproate) $C_6H_{11}O_2$	 308.2	≈28 27.4 27.37 34.69	1942 1943 1960	6 23 25	
Hexylsulfonate $C_6H_{13}SO_3$	298.2	28.72	1967	21	
Octanoate (Caprylate) $C_8H_{15}O_2$	 308.2	23.08 29.09	1962	26	
Octylsulfonate $C_8H_{17}SO_3$	298.2	26.15	1967	21	
Octylsulfate $C_8H_{17}SO_4$		29	1942	27	
Decanoate (Caprate) $C_{10}H_{19}O_2$	 308.2	22.01 27.56	1965	8	
Decylsulfonate $C_{10}H_{21}SO_3$	298.2 313.2 333.2 353.2	23.76 30.7 43.0 56.5	1967 1939	21 28	
Decylsulfate $C_{10}H_{21}SO_4$	283.2 298.2 318.2 338.2	15.4 26 23.4 35.5 49.0	1962 1942 1962	29 27 29	
Dodecanoate (Laurate) $C_{12}H_{23}O_2$	298.2 308.2	20.66 26.32	1965	8	
Dodecylsulfonate $C_{12}H_{25}SO_3$	298.2 313.2 333.2 353.2	21.63 28.5 39.5 53	1967 1939	21 28	

a: $\Lambda^{\infty}_{Na^+} = 50.1\ \Omega^{-1}cm^2mol^{-1}$

TABLE 1 (Continued)

Dodecylsulfate	283.2	14.3	1962	29
$C_{12}H_{25}SO_4$	298.2	24.0	1942	27
		21.59	1958	30
		21.3	1962	29
	318.2	32.5		
	338.2	45		
Tetradecanoate	298.2	20.11	1965	8
(Myristate)	318.2	25.23		
$C_{14}H_{27}O_2$				
Tetradecylsulfonate	313.2	26.5	1939	28
$C_{14}H_{29}SO_3$	333.2	37.5		
	353.2	50.0		
Tetradecylsulfate	298.2	21.0	1962	29
$C_{14}H_{29}SO_4$	318.2	31.9		
	338.2	44.0		
2-Tetradecylsulfate	298.2	22.1		
$C_{14}H_{29}SO_4$	318.2	34.8		
	338.2	50.0		
4-Tetradecylsulfate	298.2	29.9		
$C_{14}H_{29}SO_4$	318.2	45.8		
	338.2	65.0		

TABLE 2. Molar conductivities of ions at infinite dilution - Salts with organic cation

Cation	T / K	$\dfrac{\Lambda_i^\infty}{\Omega^{-1}cm^2mol^{-1}}$	Year	Ref.	Notes
Methylammonium	273.2	30.6	1933	31	
CH_3NH_3	298.2	58.7			
		58.72	1942	32	
		57.82	1951	33	
	363.2	150.0	1933	31	
Ethylammonium	273.2	23.1			
$C_2H_5NH_3$	298.2	47.2			
	363.2	125.7			
Dimethylammonium	233.2	27.15			
$(CH_3)_2NH_2$	298.2	52.45			
		51.87	1942	32	
		51.45	1951	33	
	363.2	135.3	1933	31	
Propylammonium	273.2	19.9			
$C_3H_7NH_3$	298.2	40.8			
	363.2	112.9			
Trimethylammonium	273.2	24.35			
$(CH_3)_3NH$	298.2	48.0			
		47.2	1941	13	
		47.25	1942	32	
		46.62	1951	33	
	363.2	123.3	1933	31	

TABLE 2 (Continued)

Butylammonium $C_4H_9NH_3$	291.2	31.6	1932	34
*Iso*butylammonium $i.C_4H_9NH_3$	298.2	31.4 38.0	1924	19
Diethylammonium $(C_2H_5)_2NH_2$	273.2 291.2 298.2 363.2	20.15 32.17 41.95 110.3	1933	31
Tetramethylammonium $(CH_3)_4N$	273.2	23.2 23.0 24.1	1941 1959	13 16
	291.2	39.72 40.0	1933 1959	31 16
	298.2	45.8 44.82 44.92	1933 1951 1959	31 33 16
	363.2	121.0	1933	31
Ethyltrimethylammonium $(C_2H_5)(CH_3)_3N$	298.2	40.50	1951	33
Dipropylammonium $(C_3H_7)_2NH_2$	273.2 291.2 298.2 363.2	14.15 25.3 30.1 90.4	1933 1932 1933	31 34 31
Triethylammonium $(C_2H_5)_3NH$	273.2 298.2 363.2	17.05 34.3 98.0		
Trimethylpropylammonium $(CH_3)_3(C_3H_7)N$	291.2	32.4	1932	34
Butyltrimethylammonium $(C_4H_9)(CH_3)_3N$	298.2	33.25		
Octylammonium $C_8H_{17}NH_3$	293.2 313.2 333.2	27.6 51.7 71.6	1942	35
Dimethyldipropylammonium $(CH_3)_2(C_3H_7)_2N$	291.8	27.1	1932	34
Tetraethylammonium $(C_2H_5)_4N$	273.2	16.2 15.8	1933 1941	31 13
	291.2	27.89 28.0	1928 1933	36 31
	298.2	33.3 32.66	1951	37
	363.2 373.2	92.0 100.0	1933 1941	31 13
Hexyltrimethylammonium $(C_6H_{13})(CH_3)_3N$	298.2	29.22	1951	33
Phenyltrimethylammonium $(C_6H_5)(CH_3)_3N$		34.31		
Tripropylammonium $(C_3H_7)_3NH$	273.2 298.2 363.2	12.15 26.1 82.1	1933	31

TABLE 2 (Continued)

Triethylpropylammonium $(C_2H_5)_3(C_3H_7)N$	291.2	26.0	1932	34
Decylammonium $C_{10}H_{21}NH_3$	293.2 313.2 333.2	26.7 49.9 64.2	1942	35
Diethyldipropylammonium $(C_2H_5)_2(C_3H_7)_2N$	291.2	24.3	1932	34
Methyltripropylammonium $(CH_3)(C_3H_7)_3N$		23.6		
Octyltrimethylammonium $(C_8H_{17})(CH_3)_3N$	298.2	26.20	1951	33
Ethyltripropylammonium $(C_2H_5)(C_3H_7)_3N$	291.2	22.5	1932	34
Dodecylammonium $C_{12}H_{25}NH_3$	298.2	23.83	1950	38
Tetrapropylammonium $(C_3H_7)_4N$	273.2 291.2 298.2	11.25 10.4 20.7 23.0 23.45	1933 1941 1932 1941 1951	31 13 34 13 37
Decyltrimethylammonium $(C_{10}H_{21})(CH_3)_3N$	298.2	24.02	1951	33
Octylpyridinium $(C_8H_{17})C_5H_5N$	333.2	41.5 63.0	1937	39
Dodecyltrimethylammonium $(C_{12}H_{25})(CH_3)_3N$	298.2	22.28	1951	33
Decylpyridinium $(C_{10}H_{21})C_5H_5N$	313.2 323.2 333.2	29.5 37.0 44.0 51.0	1937	39
Tetrabutylammonium $(C_4H_9)_4N$	273.2 298.2	9.2 8.8 19.13	1933 1941 1951	31 13 37
Tetradecyltrimethylam= monium $(C_{14}H_{29})(CH_3)_3N$		21.12	1951	40
Dodecylpyridinium $(C_{12}H_{25})C_5H_5N$	313.2 323.2 333.2	27.5 37.0 44.0 51.0	1937	39
Hexadecyltrimethylam= monium $(C_{16}H_{33})(CH_3)_3N$	298.2	20.82	1951	40
Tetradecylpyridinium $(C_{14}H_{29})C_5H_5N$	313.2 323.2 333.2	36.0 43.0 50.0	1937	39
Tetrapentylammonium $(C_5H_{11})_4N$	273.2 298.2	8.0 17.13	1933 1951	31 37

TABLE 2 (Continued)

Octadecyltrimethylam= monium $(C_{18}H_{37})(CH_3)_3N$	298.2	19.89	1951	40
Hexadecylpyridinium $(C_{16}H_{33})C_5H_5N$	313.2 323.2 333.2	35.0 42.0 50.0	1937	39
Octadecylpyridinium $(C_{18}H_{37})C_5H_5N$	323.2 333.2	42.0 50.0		
Octadecyltriethylam= monium $(C_{18}H_{37})(C_2H_5)_3N$	298.2	18.0	1938	41
Octadecyltripropylam= monium $(C_{18}H_{37})(C_3H_7)_3N$		17.3		
Hexadecyltributylammonium $(C_{16}H_{33})(C_4H_9)_3N$		16.9		

1. Kohlrausch
2. Kraus
3. Auerbach & Zeglin
4. Landolt-Börnstein
5. Saxton & Darken
6. Glasstone
7. Moelwyn-Hughes
8. Campbell & Lakshminarayanan
9. Dawson, Golben, Leader & Zimmermann
10. Johnston
11. Ferguson & Vogel
12. MacInnes & Shedlovsky
13. Lange
14. Li & Brüll
15. Ives & Pryor
16. Robinson & Stokes
17. Harned & Owen
18. Conway
19. Walden
20. Feates & Ives
21. Clunie, Goodman & Symons
22. Belcher
23. Creighton & Koehler
24. Friman & Stenius
25. Campbell & Friesen
26. Campbell, Kartzmark & Lakshminarayanan
27. Haffner, Piccione & Rosenblum
28. Wright, Abbott, Sivertz & Tartar
29. Flockhart
30. Mukerjee, Mysels & Dulin
31. Ekwall
32. Jones, Spuhler & Felsing
33. McDowell & Kraus
34. Walden & Birr
35. Ralston & Hoerr
36. Lattey
37. Daggett, Bair & Kraus
38. Kuhn & Kraus
39. Lottermoser & Frotscher
40. Bair & Kraus
41. McInnes

TABLE 3. Viscosities of salts with organic anion or cation given as A_1 and A_2 in the equation $\eta/\eta_o = 1 + A_1 c^{\frac{1}{2}}_{salt} + A_2 c_{salt}$. Concentration ranges below the cmc.

Substance	T / K	conc. range mol dm^{-3} (a) or mol kg^{-1} (b)	A_1 M$^{-\frac{1}{2}}$ or m$^{-\frac{1}{2}}$	A_2 M^{-1} or m^{-1}	Year	Ref.	Notes
Li ethanoate $C_2H_3O_2Li$	298.2	0.0111÷0.09393	0.0066	0.397	1934	1	a
Li octanoate $C_8H_{15}O_2Li$	293.2	0÷0.4	0.005	0.94÷0.99	1968	3	a
	323.2	0÷0.33	0.005	0.71÷0.82			a
Na ethanoate $C_2H_3O_2Na$	298.2	0.125÷1.0	0.005	0.331	1963	2	a
Na heptanoate $C_7H_{13}O_2Na$	293.2	0.066÷0.67	0.005	0.73÷0.96	1968	3	a
Na octanoate $C_8H_{15}O_2Na$	293.2	0.04÷0.350	0.005	0.94	1965	4, 5	a
	298.2	0.04÷0.356	0.005	0.90			
	303.2	0.04÷0.363	0.005	0.87			
	308.2	0.04÷0.371	0.005	0.84			
	313.2	0.04÷0.380	0.005	0.82			
Na nonanoate $C_9H_{17}O_2Na$	293.2	0.05÷0.222	0.005	0.83÷0.98	1968	3	a
		0.05÷0.222	0.005	0.87÷0.97			a, c
		0.05÷0.209	0.005	0.86÷0.97			a, d
		0.05÷0.151	0.005	0.92÷0.95			a, e
Na decanoate $C_{10}H_{19}O_2Na$	293.2	0÷0.10	0.005	0.91÷0.95	1968	3	a
			0.005	0.87÷0.97			a, d
			0.005	0.84÷0.99			a, e
Na hendecanoate $C_{11}H_{21}O_2Na$	298.2	0÷0.048	0.005	0.95÷1.05	1968	3	a
			0.005	0.90÷1.06			a, c
			0.005	0.91÷1.00			a, d
Na dodecanoate $C_{12}H_{23}O_2Na$	323.2	0÷0.025	0.005	0.90÷1.05	1968	3	a
Na ethylsulfate $C_2H_5SO_4Na$	288.2	0.005÷0.058	0.01	0.26	1973	6	b
	298.2		0.01	0.25			
	308.2		0.01	0.24			
	328.2		0.01	0.23			
Na butylsulfate $C_4H_9SO_4Na$	288.2	0.005÷0.068	0.01	0.43	1973	6	b
	298.2		0.01	0.41			b
	308.2		0.01	0.39			b
	328.2		0.01	0.37			b
Na hexylsulfate $C_6H_{13}SO_4Na$	288.2	0.005÷0.068	0.01	0.61	1973	6	b
	298.2		0.01	0.56			b
	308.2		0.01	0.52			b
	328.2		0.01	0.49			b
Na octylsulfate $C_8H_{17}SO_4Na$	278.2	0.005÷0.068	0.01	0.85	1973	6	b
	288.2		0.01	0.77			
	293.2	0÷0.131	0.005	0.61÷0.75	1968	3	a

a: concentration
b: molality
c: in 0.05 M NaCl
d: in 0.1 M NaCl
e: in 0.2 M NaCl

TABLE 3 (Continued)

	298.2	0.005÷0.068	0.01	0.71	1973	6	*b*
	303.2	0÷0.131	0.005	0.61÷0.75	1968	3	*a*
	308.2	0.005÷0.068	0.01	0.65	1973	6	*b*
	313.2	0÷0.131	0.005	0.61÷0.75	1968	3	*a*
	328.2	0.005÷0.068	0.01	0.56	1973	6	*b*
Na decylsulfate $C_{10}H_{21}SO_4Na$	278.2	0.005÷0.03	0.01	0.90	1973	6	*b*
	288.2		0.01	0.83			*b*
	298.2		0.01	0.77			*b*
	308.2		0.01	0.71			*b*
Na dodecylsulfate $C_{12}H_{25}SO_4Na$	278.2		0.01	0.95	1973	6	*b, e*
	288.2		0.01	0.89			*b, e*
	298.2	0.002÷0.010	0.005	1.06÷1.09	1968	3	*a*
	308.2		0.005	1.06÷1.09			*a*
	318.2		0.005	1.06÷1.09			*a*
	298.2		0.005	1.06÷1.09			*a, c*
Na octylsulfonate $C_8H_{17}SO_3Na$	313.2	0÷0.15	0.005	0.65÷0.67	1968	3	*a*
	318.2		0.005	0.65÷0.67			*a*
	323.2		0.005	0.65÷0.67			*a*
Na cholate $C_{24}H_{39}O_5Na$	293.2	0.025÷0.1	0.005	2.04÷2.4	1971	7	*a*
	303.2		0.005	2.6			*a*
	313.2		0.005	2.1			*a*
	293.2		0.005	2.05			*a, c*
Na desoxycholate $C_{24}H_{39}O_4Na$	293.2	0÷0.05	0.005	2.2	1971	7	*a*
	303.2		0.005	2.1			*a*
	313.2		0.005	1.9			*a*
	293.2		0.005	2.1			*a, c*
	293.2		0.005	2.1			*a, d*
Na dehydrocholate $C_{24}H_{37}O_5Na$	293.2	0÷0.18	0.005	1.67	1971	7	*a*
	303.2		0.005	1.6			*a*
	313.2		0.005	1.5			*a*
	293.2		0.005	1.7			*a, c*
K ethanoate $C_2H_3O_2K$	298.2	0.00932÷0.06296	0.0038	0.238	1934	1	*a*
		0.125÷1.0	0.005	0.238	1963	2	*a*
K octanoate $C_8H_{15}O_2K$	293.2	0.04÷0.38	0.005	0.73÷0.82	1968	3	*a*
	303.2		0.005	0.73÷0.82			*a*
	313.2		0.005	0.73÷0.82			*a*
Mg ethanoate $(C_2H_3O_2)_2Mg$	298.2	0.125÷1.0	0.005	0.875	1963	2	*a*
Ba ethanoate $(C_2H_3O_2)_2Ba$	298.2	0.125÷1.0	0.005	0.710	1963	2	*a*
Octylammonium chloride $C_8H_{17}NH_3Cl$	293.2	0.09÷0.30	0.005	0.65÷0.74	1968	3	*a*
	307.2		0.005	0.65÷0.74			
	313.2		0.005	0.65÷0.74			
Octylmethylammonium chloride $(C_8H_{17})(CH_3)NH_2Cl$	293.2	0÷0.30	0.005	0.71÷0.74	1968	3	*a*
	303.2		0.005	0.71÷0.74			
	313.2		0.005	0.71÷0.74			

a: concentration
b: molality
c: in 0.05 M NaCl
d: in 0.10 M NaCl
e: concentration interval not given by the authors

TABLE 3 (Continued)

Octyldimethylammonium chloride	293.2	0÷0.28	0.005	0.69÷0.76	1968	3	*a*
$(C_8H_{17})(CH_3)_2NHCl$	303.2		0.005	0.69÷0.76			
	313.2		0.005	0.69÷0.76			
Ethyltrimethylammonium bromide	288.2	0.011÷0.073	0.01	0.16	1973	6	*b*
$(C_2H_5)(CH_3)_3NBr$	298.2		0.01	0.16			*b*
	313.2		0.01	0.16			*b*
Butyltrimethylammonium bromide	288.2	0.006÷0.044	0.01	0.36	1973	6	*b*
$(C_4H_9)(CH_3)_3NBr$	298.2		0.01	0.31			*b*
	313.2		0.01	0.28			*b*
Hexyltrimethylammonium bromide	288.2	0.008÷0.044	0.01	0.53	1973	6	*b*
$(C_6H_{13})(CH_3)_3NBr$	298.2		0.01	0.45			*b*
	313.2		0.01	0.43			*b*
Octyltrimethylammonium bromide	288.2	0.01÷0.044	0.01	0.70	1973	6	*b*
$(C_8H_{17})(CH_3)_3NBr$	298.2		0.01	0.61			*b*
	313.2		0.01	0.57			*b*
Decyltrimethylammonium bromide	288.2	0.023÷0.044	0.01	0.88	1973	6	*b*
$(C_{10}H_{21})(CH_3)_3NBr$	298.2		0.01	0.76			*b*
	313.2		0.01	0.70			*b*
Dodecyltrimethylammonium bromide	288.2	0.006÷0.017	0.01	1.07	1973	6	*b*
$(C_{12}H_{25})(CH_3)_3NBr$	298.2		0.01	0.91			
	313.2		0.01	0.76			

a: concentration
b: molality

1. Laurence & Wolfenden
2. Padova
3. Solyom & Ekwall
4. Ekwall & Holmberg (1965 a)

5. Ekwall & Holmberg (1965 b)
6. Tanaka, Kaneshina, Nishimoto & Takabatake
7. Fontell

TABLE 4. Viscosities of salts with organic anion or cation that form micelles given as A_3 and Q' in the equation $\log \eta_{rel}/c_{salt} = A_3 + Q' \log \eta_{rel}$. Concentration ranges above the cmc.

Substance	T — K	conc. range mol dm^{-3}	A_3 M^{-1}	Q' M^{-1}	Year	Ref.	Notes
Li octanoate $C_8H_{15}O_2Li$	323.2	0.35÷0.56	0.235	0.66	1968	1	
Na heptanoate $C_7H_{13}O_2Na$	293.2	0.88÷1.9	0.260	0.236	1968	1	
	303.2	0.9÷1.9	0.252	0.242			
	313.2	0.9÷1.9	0.244	0.246			
Na octanoate $C_8H_{15}O_2Na$	293.2	0.45÷1.2	0.324	0.274	1965	2, 3	
		1.2÷1.8	0.344	0.241			
	298.2	0.5÷1.1	0.314	0.279			
		1.2÷1.8	0.343	0.242			
	303.2	0.5÷1.1	0.306	0.290			
		1.2÷1.8	0.341	0.243			

TABLE 4 (Continued)

	308.2	0.5÷1.1	0.298	0.310		
		1.2÷1.8	0.342	0.242		
	313.2	0.57÷1.1	0.294	0.318		
		1.2÷1.8	0.343	0.240		
Na nonanoate $C_9H_{17}O_2Na$	293.2	0.25÷0.69	0.367	0.354	1968	1
		0.70÷1.25	0.396	0.253		
	303.2	0.25÷0.69	0.33	0.56		
		0.70÷1.25	0.40	0.27		
	313.2	0.25÷0.65	0.31	0.55		
		0.72÷1.25	0.40	0.27		
	323.2	0.28÷0.70	0.31	0.51		
		0.8÷1.4	0.42	0.23		
	293.2	0.25÷0.5	0.368	0.36		*a*
		0.25÷0.5	0.372	0.27		*b*
		0.25÷0.5	0.372	0.27		*c*
Na decanoate $C_{10}H_{19}O_2Na$	293.2	0.11÷0.3	0.374	0.88	1968	1
		0.66÷1.26	0.504	0.195		
	303.2	0.26÷0.5	0.42	0.50		
		0.7÷1.3	0.54	0.19		
	313.2	0.26÷0.46	0.41	0.57		
		0.7÷1.3	0.52	0.20		
	323.2	0.26÷0.46	0.40	0.64		
		0.7÷1.3	0.52	0.21		
	293.2	0.15÷0.42	0.41	0.44		*b*
		0.2÷0.42	0.41	0.29		*c*
Na hendecanoate $C_{11}H_{21}O_2Na$	298.2	0.04÷0.12	0.41	2.52	1968	1
		0.48÷0.78	0.662	0.10		
		0.05÷0.09	0.41	1.51		*a*
		0.024÷0.14	0.41	1.39		*b*
Na dodecanoate $C_{12}H_{23}O_2Na$	323.2	0.027÷0.056	0.385	7.05	1968	1
		0.27÷0.99	0.77	0.09		
Na octylsulfate $C_8H_{17}SO_4Na$	293.2	0.24÷0.37	0.310	0.767	1968	1
		0.48÷1.70	0.405	0.260		
	303.2	0.60÷1.53	0.412	0.26		
	313.2	0.66÷1.25	0.412	0.28		
Na dodecylsulfate $C_{12}H_{25}SO_4Na$	298.2	0.015÷0.026	0.56	16.7	1968	1
		0.40÷0.91	0.956	0		
	308.2	0.40÷0.96	0.97	0		
	318.2	0.35÷0.95	0.98	0		
	298.2	0.02÷0.045	0.56	1.33		*a*
Na octylsulfonate $C_8H_{17}SO_3Na$	313.2	0.93÷1.5	0.416	0.238	1968	1
	318.2	0.93÷1.5	0.414	0.237		
	323.2	0.2÷0.37	0.245	1.04		
		0.93÷1.5	0.412	0.238		
Na cholate $C_{24}H_{39}O_5Na$	293.2	0.1÷0.48	0.86	0.88	1971	4
	303.2		0.85	0.74		
	313.2		0.84	0.70		
	293.2		0.87	0.99		*a*
Na desoxycholate $C_{24}H_{39}O_4Na$	293.2	0.06÷0.3	1.03	1.73	1971	4
	303.2		0.94	1.75		
	313.2		0.87	1.49		
	293.2		0.88	2.32		*a*
			0.83	2.64		*b*

a: in 0.05 M NaCl
b: in 0.1 M NaCl
c: in 0.2 M NaCl

TABLE 4 (Continued)

TABLE 4 (Continued)

Na dehydrocholate $C_{24}H_{37}O_5Na$	293.2	0.18÷0.66	0.62	0.54	1971	4	
	303.2		0.62	0.58			
	313.2		0.54	0.62			
	293.2		0.64	0.65			*a*
K octanoate $C_8H_{15}O_2K$	293.2	0.45÷1.53	0.268	0.321	1968	1	*a*
		1.55÷2.00	0.316	0.266			
	303.2	0.70÷1.50	0.266	0.331			
		1.55÷1.92	0.31	0.27			
	313.2	0.7÷1.48	0.266	0.334			
		1.48÷1.90	0.31	0.27			
Octylammonium chloride $C_8H_{17}NH_3Cl$	293.2	0.30÷0.90	0.238	0.340	1968	1	
		1.14÷2.06	0.298	0.178			
	307.2	0.4÷0.9	0.24	0.34			
		1.2÷1.8	0.304	0.17			
	313.2	0.4÷0.9	0.24	0.34			
		1.2÷1.8	0.304	0.17			
Octylmethylammonium chloride $(C_8H_{17})(CH_3)NH_2Cl$	293.2	0.30÷0.82	0.263	0.398	1968	1	
		0.82÷1.37	0.302	0.22			
	303.2	0.8÷1.4	0.300	0.22			
	313.2	0.8÷1.4	0.296	0.22			
Octyldimethylammonium chloride $(C_8H_{17})(CH_3)_2NHCl$	293.2	0.31÷0.82	0.277	0.314	1968	1	
		1.00÷1.54	0.292	0.268			
	303.2	0.90÷1.51	0.28	0.27			
	313.2	0.90÷1.50	0.28	0.28			
Hexadecyltrimethylammonium bromide $(C_{16}H_{33})(CH_3)_3NBr$	298.2	0.0045÷0.01	1.58	6.7	1971	5	
		0.135÷0.27	1.05	0			

a: in 0.05 M NaCl

1. Solyom & Ekwall   4. Fontell

2. Ekwall & Holmberg (1965 a)   5. Ekwall, Mandell & Solyom

3. Ekwall & Holmberg (1965 b)

TABLE 5. Viscosities of salts with organic anion or cation that form micelles, given as k_1 and k_2 in the equation log $(\eta_{sp}/c_{salt}) = k_1 + k_2 c_{salt}$. Concentration ranges above the cmc.

Substance	T ___ K	conc. range _____ mol dm^{-3}	k_1	k_2 ___ M^{-1}	Year	Ref.	Notes
Na heptanoate $C_7H_{13}O_2Na$	293.2	1.8÷2.5	−0.93	0.79	1968	1	
	303.2	1.85÷2.6	−0.95	0.79			
	313.2	1.85÷2.6	−0.97	0.79			
Na octanoate $C_8H_{15}O_2Na$	293.2	1.8÷2.55	−0.99	0.99	1965	2, 3	
	298.2	1.65÷2.45	−0.89	0.93			
	303.2	1.60÷2.35	−0.83	0.92			
	308.2	1.60÷2.32	−0.84	0.92			
	313.2	1.65÷2.25	−0.80	0.85			
Na nonanoate $C_9H_{19}O_2Na$	293.2	1.35÷1.83	−0.43	0.77	1968	1	
	303.2	1.2÷1.83	−0.38	0.75			

TABLE 5 (Continued)

	313.2	1.2÷1.83	−0.33	0.71		
	323.2	1.3÷1.83	−0.33	0.71		
Na decanoate $C_{10}H_{19}O_2Na$	303.2	1.47÷1.90	−0.69	1.1	1968	1
	313.2	1.48÷1.88	−0.76	1.2		
	323.2	1.50÷1.86	−0.90	1.4		
Na dodecanoate $C_{12}H_{23}O_2Na$	323.7	1.16÷1.35	−0.89	1.45	1968	1
Na octylsulfate $C_8H_{17}SO_4Na$	293.2	1.62÷2.13	−1.05	1.16	1968	1
	303.2	1.61÷2.12	−0.84	1.05		
	313.2	1.60÷2.11	−0.56	0.93		
Na dodecylsulfate $C_{12}H_{25}SO_4Na$	298.2	1.08÷1.27	−4.05	4.57	1968	1
	308.2	1.07÷1.26	−3.30	3.96		
	318.2	1.06÷1.26	−3.23	4.00		
Na octylsulfonate $C_8H_{17}SO_3Na$	313.2	1.4÷2.0	−0.53	0.83	1968	1
	318.2	1.4÷2.0	−0.51	0.83		
	323.2	1.35÷2.0	−0.49	0.82		
Na cholate $C_{24}H_{39}O_5Na$	293.2	0.48÷1.3		2.6	1971	4
Na desoxycholate $C_{24}H_{39}O_4Na$	293.2	0.3÷0.65		4.24	1971	4
K octanoate $C_8H_{15}O_2K$	293.2	1.70÷2.45	−1.23	1.11	1968	1
	303.2	1.60÷2.35	−1.0	0.91		
	313.2	1.60÷2.25	−0.90	0.88		
Octylammonium chloride $C_8H_{17}NH_3Cl$	293.2	2.0÷2.9	−0.70	0.64	1968	1
	307.2	2.0÷2.9	−0.53	0.54		
	313.2	2.0÷2.9	−0.47	0.50		
Octylmethylammonium chloride $(C_8H_{17})(CH_3)NH_2Cl$	293.2	1.35÷1.90	−0.39	0.52	1968	1
	303.2	1.35÷1.9	−0.38	0.51		
	313.2	1.25÷1.9	−0.37	0.50		
Octyldimethylammonium chloride $(C_8H_{17})(CH_3)_2NHCl$	293.2	1.30÷2.0	−0.705	0.785	1968	1
	303.2	1.25÷2.0	−0.640	0.737		
	313.2	1.25÷2.0	−0.565	0.680		
Hexadecyltrimethylammonium bromide $(C_{16}H_{33})(CH_3)_3NBr$	298.2	0.30÷0.37	−2.65	5.73	1971	5

1. Solyom & Ekwall

2. Ekwall & Holmberg (1965 a)

3. Ekwall & Holmberg (1965 b)

4. Fontell

5. Ekwall, Mandell & Solyom

TABLE 6. The activation energy of viscous flow, E, for some micelle-forming organic salts

Substance	Temp. range K	Conc. range mol dm^{-3}	E kJ mol^{-1}	Year	Ref.	Notes
Na heptanoate $C_7H_{13}O_2Na$	293.2÷303.2	0÷1.85	16.6÷20.5	1968	1	a
		1.8÷2.92	28.1÷34.7			b
	303.2÷313.2	0÷1.80	16.6÷20.5			a
		1.8÷2.96	28.1÷34.7			b
Na octanoate $C_8H_{15}O_2Na$	293.2÷298.2	0÷1.79	16.6÷17.4	1965	2, 3	a
		1.82÷2.55	17.6÷32.1			b
	298.2÷303.2	0÷1.78	16.6÷17.4			a
		1.81÷2.54	17.6÷32.1			b
	303.2÷308.2	0÷1.77	16.6÷17.4			a
		1.81÷2.53	17.6÷32.1			b
	308.2÷313.2	0÷1.77	16.6÷17.4			a
		1.80÷2.52	17.6÷32.1			b
Na nonanoate $C_9H_{17}O_2Na$	303.2÷313.2	0÷1.49	16.1÷15.5	1968	1	a
		1.59÷2.0	15.9÷18.8			b
	313.2÷323.2	0÷1.46	15.3÷15.5			a
		1.53÷2.0	18.8÷20.9			b
Na decanoate $C_{10}H_{19}O_2Na$	303.2÷313.2	0÷1.47	16.6÷13.0	1968	1	a
		1.57÷1.85	13.0÷16.3			b
	313.2÷323.2	0÷1.41	15.3÷13.8			a
		1.56÷1.87	13.8÷17.2			b
Na octylsulfate $C_8H_{17}SO_4Na$	293.2÷303.2	0÷1.74	16.6÷11.4	1968	1	a
		1.74÷2.13	11.4÷18.2			b
	303.2÷313.2	0÷1.70	16.1÷12.1			a
		1.70÷2.13	12.1÷20.5			b
Na dodecylsulfate $C_{12}H_{25}SO_4Na$	298.2÷308.2	0÷0.97	16.3÷12.6	1968	1	a
		0.97÷1.23	12.6÷28.5			b
	308.2÷318.2	0÷0.97	15.9÷12.3			a
		0.97÷1.23	12.6÷23.3			b
Na octylsulfonate $C_8H_{17}SO_3Na$	313.2÷323.2	0÷1.38	15.9÷12.6	1968	1	a
		1.38÷1.76	12.6÷14.6			b
Na cholate $C_{24}H_{39}O_5Na$	293.2÷313.2	0÷0.48	16.3÷23.0	1971	4	a
		0.48÷1.32	23.0÷66.9			b
Na desoxycholate $C_{24}H_{39}O_4Na$	293.2÷313.2	0÷0.3	16.3÷25.1	1971	4	a
		0.3÷0.76	25.1÷58.6			b
K octanoate $C_8H_{15}O_2K$	293.2÷303.2	0÷1.61	16.6÷18.0÷15.9	1968	1	a
		1.73÷2.45	16.1÷17.2			b
	303.2÷313.2	0÷1.43	15.3÷16.7÷25.1			a
		1.55÷2.45	16.1÷26.4			b
Octylammonium chloride $C_8H_{17}NH_3Cl$	293.2÷307.2	0÷1.54	16.6÷17.6	1968	1	a
		1.97÷2.9	17.6÷33.1			b
	307.2÷313.2	0÷1.48	15.9÷16.3			a
		1.77÷2.9	17.6÷31.8			b
Octylmethylammonium chloride $(C_8H_{17})(CH_3)NH_2Cl$	293.2÷303.2	0÷1.4	16.3			a
	303.2÷313.2	1.4÷1.9	16.7÷18.8			b
Octyldimethylammonium chloride $(C_8H_{17})(CH_3)_2NHCl$	293.2÷303.2	0÷1.37	16.6÷17.6	1968	1	a
		1.4÷2.0	17.6÷23.0			b

a: concentration range with small variations in activation energy
b: concentration range with rapid variations in activation energy

TABLE 6 (Continued)

	303.2÷313.2	0÷1.31	16.1÷17.6			*a*
		1.4÷2.0	17.6÷23.8			*b*
Hexadecyltrimethylammonium bromide	298.2÷308.2	0÷0.27	15.9÷17.6	1971	5	*a*
		0.25÷0.38	17.6÷66.9			*b*
$(C_{16}H_{33})(CH_3)_3NBr$	308.2÷318.2	0÷0.28	15.1÷16.7			*a*
		0.25÷0.38	16.7÷38.9			*b*

a: concentration range with small variations in activation energy
b: concentration range with rapid variations in activation energy

1. Solyom & Ekwall	4. Fontell
2. Ekwall & Holmberg (1965 a)	5. Ekwall, Mandell & Solyom
3. Ekwall & Holmberg (1965 b)	

TABLE 7. Viscosities of salts with organic anion (other than ethanoates) given as η/η_o where η_o is the viscosity of water

Substance	T —— K	Concentration —————— mol dm^{-3} *(a)* or weight % *(b)*	η/η_o	Year	Ref.	Notes
Na hexanoate	293.2	10	1.77	1967	1	*b*
$C_6H_{11}O_2Na$		30	6.10			*b*
		40	21.6			*b*
	298.2	0.019937	1.014	1960	2	*a*
		0.099588	1.067			*a*
		0.98536	1.875			*a*
		2.6215	11.654			*a*
		3.4487	39.27			*a*
	313.2	10	1.57	1967	1	*b*
		30	4.60			*b*
		40	11.8			*b*
	333.2	10	1.34			*b*
		30	4.00			*b*
		40	8.50			*b*
Na octanoate	293.2	10	2.14	1967	1	*b*
$C_8H_{15}O_2Na$		20	4.50			*b*
		30	13.10			*b*
		40	51.80			*b*
	298.2	10	1.74			*b*
		20	3.40			*b*
		30	9.20			*b*
		40	35.80			*b*
	333.2	10	1.57			*b*
		20	3.30			*b*
		30	9.20			*b*
		40	24.30			*b*
Na dodecanoate	303.2	5	1.29	1967	1	*b*
$C_{12}H_{23}O_2Na$		10	2.02			*b*
		15	3.16			*b*
	313.2	5	1.29			*b*

a: concentration
b: weight %

TABLE 7 (Continued)

		10	2.00			*b*
		15	3.00			*b*
	323.2	5	1.29			*b*
		10	2.00			*b*
		15	2.90			*b*
		20	4.53			*b*
	333.2	5	1.29			*b*
		10	2.00			*b*
		15	2.95			*b*
		20	4.50			*b*
	333.2	0.2	1.224	1969	3	*a*
			1.225			*a, c*
			1.212			*a, d*
			1.207			*a, e*
	343.2	0.2	1.214			*a*
			1.204			*a, c*
			1.206			*a, d*
			1.180			*a, e*
	333.2	0.2	1.212			*a, f*
			1.200			*a, g*
			1.193			*a, h*
	343.2	0.2	1.194			*a, f*
			1.181			*a, g*
			1.190			*a, h*
Na tetradecanoate $C_{14}H_{27}O_2Na$	313.2	1.25	1.10	1967	1	*b*
		2.5	1.32			*b*
		3.0	1.55			*b*
	323.2	1.25	1.06			*b*
		2.5	1.28			*b*
		3.0	1.46			*b*
	333.2	1.25	1.05			*b*
		2.5	1.20			*b*
		3.0	1.38			*b*
Na hexadecanoate $C_{16}H_{31}O_2Na$	333.2	0.2	1.386	1969	3	*a*
			1.367			*a, c*
			1.355			*a, d*
			1.421			*a, e*
	343.2	0.2	1.359			*a*
			1.345			*a, c*
			1.327			*a, d*
			1.359			*a, e*
	353.2	0.2	1.351			*a*
			1.333			*a, c*
			1.314			*a, d*
			1.332			*a, e*
	333.2	0.2	1.348			*a, i*
			1.329			*a, j*
			1.353			*a, k*
	343.2	0.2	1.319			*a, i*
			1.298			*a, j*
			1.316			*a, k*
	353.2	0.2	1.305			*a, i*
			1.299			*a, j*
			1.303			*a, k*
Na octadecanoate $C_{18}H_{35}O_2Na$	343.2	0.2	1.720	1969	3	*a*
			1.519			*a, c*
			2.313			*a, d*

a: concentration	*e*: in 0.1 M Na_2SO_4	*i*: in 0.02 M Na_2CO_3
b: weight %	*f*: in 0.02 M $NaNO_3$	*j*: in 0.06 M Na_2CO_3
c: in 0.02 M Na_2SO_4	*g*: in 0.06 M $NaNO_3$	*k*: in 0.1 M Na_2CO_3
d: in 0.06 M Na_2SO_4	*h*: in 0.1 M $NaNO_3$	

TABLE 7 (Continued)

	353.2	0.2	1.695			a
			1.504			a, c
			2.035			a, d
	343.2	0.2	1.528			a, e
			2.165			a, f
	353.2	0.2	1.516			a, e
			1.874			a, f
Na 9.10-octadecenoate (cis) (Na oleate) $C_{18}H_{33}O_2Na$	293.2	5.0	1.51	1976	4	b
		10.0	41.8			b
		14.0	677.0			b
	303.2	5.0	1.44			b
		10.0	15.2			b
		14.0	164.0			b
	313.2	5.0	1.43			b
		10.0	7.80			b
		14.0	49.3			b
		16.5	530			b
	323.2	5.0	1.42			b
		10.0	3.80			b
		14.0	21.7			b
		16.5	120			b
		18.0	437			b
	333.2	5.0	1.34			b
		10.0	2.92			b
		14.0	11.6			b
		16.5	48.0			b
		18.0	178			b
	343.2	5.0	1.34			b
		10.0	2.00			b
		14.0	9.30			b
		16.5	21.2			b
		18.0	65.5			b
K methanoate CHO_2K	323.8	1.516	1.146	1953	5	a
		3.125	1.339			a
		4.199	1.519			a
		8.224	2.685			a
		12.26	6.245			a
		15.64	15.091			a
Cs methanoate CHO_2Cs	323.8	0.7496	1.0730	1953	5	a
		1.784	1.1989			a
		2.808	1.3421			a
		5.819	2.0500			a
		8.203	3.3199			a

a: concentration
b: weight %
c: in 0.02 M Na_2SO_4

d: in 0.06 M Na_2SO_4
e: in 0.02 M $NaNO_3$
f: in 0.06 M $NaNO_3$

1. Chinnikova & Markina

2. Campbell & Friesen

3. Angelescu & Radu

4. Markina, Chinnikova & Rehbinder

5. Rice & Kraus

TABLE 8. Viscosities of ethanoates given as η/η_o where η_o is the viscosity of water

T	Concentration	η/η_o				Year	Ref.
K	mol dm^{-3}						
		Na ethanoate $C_2H_3O_2Na$	K ethanoate $C_2H_3O_2K$	Mg ethanoate $(C_2H_3O_2)_2Mg$	Ba ethanoate $(C_2H_3O_2)_2Ba$		
298.2	0.125	1.0432	1.0290	1.1131	1.0930	1963	1
	0.25	1.0859	1.0580	1.2323	1.1349		
	0.375	1.1299	1.0860	1.3754	1.2876		
	0.500	1.1756	1.1152	1.5305	1.4053		
	0.75	1.2732	1.1740	1.9196	1.6777		
	1.0	1.3753	1.2330	2.4184	2.0163		

1. Padova

REFERENCES

ANGELESCU, E., and RADU, M., *Rev. Roum. Chim.*, 14, 441 (1969)

AUERBACH, F.R., and ZEGLIN, H., *Z. physik. Chem.*, 103, 178 (1923)

BAIR, E.J., and KRAUS, C.A., *J. Amer. Chem. Soc.*, 73, 1129 (1951)

BELCHER, D., *J. Amer. Chem. Soc.*, 60, 2744 (1938)

CAMPBELL, A.N., and FRIESEN, J.I., *Can. J. Chem.*, 38, 1939 (1960)

CAMPBELL, A.N., KARTZMARK, E.M., and LAKSHMINARAYANAN, G.R., *Can. J. Chem.*, 40, 839 (1962)

CAMPBELL, A.N., and LAKSHMINARAYANAN, G.R., *Can. J. Chem.*, 43, 1729 (1965)

CHINNIKOVA, A.V., and MARKINA, Z.N., *Colloid J. USSR*, 29, 542 (1967)

CLUNIE, J.S., GOODMAN, J.F., and SYMONS, P.C., *Trans. Faraday Soc.*, 63, 754 (1967)

CONWAY, B.E., *"Electrochemical Data"*, Elsevier Pub. Co., London, 1952

CREIGHTON, H.J., and KOEHLER, W.A., *"Electrochemistry"*, J. Wiley, New York, 1943

DAGGETT, H.M., Jr., BAIR, E.J., and KRAUS, C.A., *J. Amer. Chem. Soc.*, 73, 799 (1951)

DAWSON, L.R., GOLBEN, M., LEADER, G.R., and ZIMMERMANN, H.K., Jr., *J. Phys. Colloid Chem.*, 55, 1499 (1951)

EINSTEIN, A., *Ann. Phys.*, 19, 289 (1906)

EKWALL, P., *Z. physik. Chem.*, 165, 331 (1933)

EKWALL, P., and HOLMBERG, P., *Acta Chem. Scand.*, 19, 455 (1965)

EKWALL, P., and HOLMBERG, P., *Acta Chem. Scand.*, 19, 573 (1965)

EKWALL, P., MANDELL, L., and SOLYOM, P., *J. Colloid Interface Sci.*, 35, 519 (1971)

FALKENHAGEN, H., and DOLE, M., *Physik. Z.*, 30, 611 (1929)

FALKENHAGEN, H., and KELBG, G., *Z. Elektrochem.*, 56, 834 (1952)

FALKENHAGEN, H., and VERNON, E.L., *Physik. Z.*, 33, 140 (1932)

FALKENHAGEN, H., and VERNON, E.L., *Phil. Mag.*, 14, 537 (1932)

FEATES, F.S., and IVES, D.J.C., *J. Chem. Soc.*, 1936, 2798

FERGUSON, A., and VOGEL, I., *Phil. Mag.*, 4, 233 (1927)

FLOCKHART, B.D., *J. Colloid Sci.*, 17, 305 (1962)

FONTELL, K., *Kolloid-Z. Polym.*, 246, 614 (1971)

FRIMAN, R., and STENIUS, P., *Acta Chem. Scand.*, A32, 289 (1978)

GLASSTONE, S., *"An Introduction to Electrochemistry"*, Van Nostrand Co. Inc., New York, 1942

HAFFNER, F.D., PICCIONE, G.A., and ROSENBLUM, C., *J. Phys. Chem.*, 46, 662 (1942)

HARNED, H.S., and OWEN, B.B., *"The Physical Chemistry of Electrolytic Solutions"*, Reinhold Pub. Co., New York, 1950

IVES, D.J.G., and PRYOR, J.H., *J. Chem. Soc.*, 1955, 2104

JOHNSTON, A., *J. Amer. Chem. Soc.*, 31, 1010 (1909)

JONES, G., and DOLE, M., *J. Amer. Chem. Soc.*, 51, 2950 (1929)

JONES, J.H., SPUHLER, F.J., and FELSING, W.A., *J. Amer. Chem. Soc.*, 64, 965 (1942)

KOHLRAUSCH, F., *Gött. Nachrichten*, 1876, 213; *Z. Elektrochem.*, 13, 333 (1907)

KRAUS, C.A., *"The Properties of Electrically Conducting Systems"*, Chemical Catalog Co., New York, 1921

KUHN, D.W., and KRAUS, C.A., *J. Amer. Chem. Soc.*, 72, 3676 (1950)

LANDOLT-BÖRNSTEIN, *"Physikalisch-Chemische Tabellen"*, Springer Verlag, II 7, 266 (1960)

LANGE, J., *Z. physik. Chem.*, A188, 284 (1941)

LATTEY, R.T., *Phil. Mag.*, 6, 258 (1928)

LAURENCE, V.D., and WOLFENDEN, J.H., *J. Chem. Soc.*, 1934, 1144

LI, N.C.C., and BRÜLL, W., *J. Amer. Chem. Soc.*, 64, 1635 (1942)

LOTTERMOSER, A., and FROTSCHER, H., *Kolloid-Beih.*, 45, 303 (1937)

MacINNES, D.A., *J. Franklin Inst.*, 225, 661 (1938)

MacINNES, D.A., and SHEDLOVSKY, T., *J. Amer. Chem. Soc.*, 54, 1429 (1932)

MARKINA, Z.N., CHINNIKOVA, A.V., and REHBINDER, P., *Dokl. Akad. Nauk SSSR* (English ed.), 174, 319 (1976)

McDOWELL, M.J., and KRAUS, C.A., *J. Amer. Chem. Soc.*, 73, 2170 (1951)

MOELWYN-HUGHES, E.A., *"Physikalische Chemie"*, Georg Thiele Verlag, Stuttgart, 1970

MUKERJEE, P., *Adv. Colloid Interface Sci.*, 1, 241 (1967)

MUKERJEE, P., and MYSELS, K.J., *"Critical Micelle Concentrations of Aqueous Surfactant Systems"*, NSRDS-NBS 36, Washington, 1971

MUKERJEE, P., MYSELS, K.J., and DULIN, C.I., *J. Phys. Chem.*, 62, 1390 (1958)

PADOVA, J., *J. Chem. Phys.*, 38, 2635 (1963)

RALSTON, A.W., and HOERR, C.W., *J. Amer. Chem. Soc.*, 64, 772 (1942)

RICE, M.J., Jr., and KRAUS, C.A., *Proc. Nat. Acad. Sci. U.S.*, 39, 802 (1953)

ROBINSON, R.A., and STOKES, R.H., *"Electrolyte Solutions"*, Butterworths, London, 1959

SAXTON, B., and DARKEN, L.S., *J. Amer. Chem. Soc.*, 62, 846 (1940)

SOLYOM, P., and EKWALL, P., *Chim. Phys. Appl. Prat. Ag. Surface, C.R. Congr. Int. Deterg.*, 5th, 2, 1041 (1968)

TANAKA, M., KANESHINA, S., NISHIMOTO, W., and TAKABATAKE, H., *Bull. Chem. Soc. Japan*, 46, 364 (1973)

VAND, V., *J. Phys. Chem.*, 52, 277 (1948)

WALDEN, P., *"Das Leitvermögen der Lösungen"*, Akad. Verlagsges., Leipzig, 1924

WALDEN, P., and BIRR, E.J., *Z. physik. Chem.*, A160, 327 (1932)

WRIGHT, K.A., ABBOTT, A.D., SIVERTZ, V., and TARTAR, H.W., *J. Amer. Chem. Soc.*, 61, 549 (1939)

THERMODYNAMIC QUANTITIES OF MICELLE FORMATION

P. Stenius, S. Backlund and P. Ekwall

INTRODUCTION

A characteristic property of amphiphilic salts is their tendency to form micellar aggregates in aqueous solution , i.e., the amphiphilic ions associate reversibly to large aggregates, the total charge of the ions in the aggregates being partly neutralized by the binding of counterions at the micellar surface.

A complete description of the thermodynamics of this aggregation process would take into account a series of association equilibria

$$p_iA + n_iB \rightleftarrows A_{p_i}B_{n_i} \tag{1}$$

where A denotes the counterion, B the amphiphilic ion and the stoichiometric numbers p_i, n_i range from 1 to around 100 with the most probable values of n_i (according to massive experimental evidence) between 25 and 80 (depending on the length and chemical structure of the hydrocarbon moiety of the ion) and those of p_i between 40 and 80 % of n_i (the actual percentage varies with the nature of the polar head group and the counterion but may also vary considerably depending on the actual experimental method used to determine p_i/n_i).

However, it is in general not possible to obtain experimental data with a precision that would allow determination of each of the equilibria described by Eq. (1), nor would such information be very interesting from a practical point of view. Instead, the thermodynamics of micelle formation is usually described in terms of average quantities that are defined on the basis of a choice of a suitable model of micelle formation that allows a clear definition of such average quantities. The relationship of these definitions to experimentally determined quantities is still a matter of much discussion and it is necessary to clearly state the way a given thermodynamic quantity has been determined in order to be able to compare it with other quantities. In particular, attention should be given to

(i) the association model that forms the basis for the calculation of thermodynamic quantities;

(ii) the actual process taking place in the experiment in which a given thermodynamic quantity was determined;

(iii) the choice of standard states;

(iv) the concentration units used.

Apart from being of very uneven quality, the data published so far on the thermodynamics of micelle formation also represent a broad spectrum of different choices with respect to points (i)-(iv). Hence a table of thermodynamic quantities for micelle formation cannot be given without systematic reference to the various equations and experimental conditions pertaining to the cited values. In the following, we give a short survey of the equations most frequently used in the literature. For a full discussion, we refer to the voluminous literature on the subject, a review of which is given in, e.g., Wennerström & Lindman (1979), Kresheck (1975), Ekwall, Danielsson & Stenius (1972), Ekwall & Stenius (1975).

THE CRITICAL MICELLE CONCENTRATION (CMC)

From a physico-chemical point of view the formation of micelles can be considered as the formation of large molecular complexes. In aqueous solution the amphiphilic ions arrange themselves in such a way that their hydrocarbon parts form an inner core that is surrounded by hydrophilic groups and counterion. Thus, the very structure of the aggregates indicates that the driving force of micelle formation is the replacement of energetically unfavourable water/ hydrocarbon interactions with more favourable hydrocarbon/hydrocarbon contacts. On the other hand, the factor limiting the growth of the micellar aggregates obviously is the loss in free energy when the hydrophilic groups are brought into close contact with one another. This is due to hydration forces and, in the case of ionic amphiphiles, electrostatic repulsion between the charged end-groups. It is also obvious that there must be little tendency for the polar endgroups or water to penetrate very deep into the hydrocarbon part of the micelle, although the actual extension of such contacts in micelles is still a matter of discussion {Wennerström &

Lindman (1979), Mukerjee & Mysels (1975), Muller (1977), Rosenholm, Stenius & Danielsson (1976)}.
Nevertheless, it is evident that the tendency to simultaneously minimize the hydrocarbon/water
contacts and the repulsive energy between the polar end-groups must for purely geometrical reasons favour a spherical or oblate shape of the micelles, since there is considerable evidence
that there is no extensive penetration of water into the micellar core. The hydrocarbon chains
in the core are much more mobile than in crystals; their state may be described as "semi-liquid".
Thus, there will be a range of small aggregation numbers, n_i, that will be energetically unfavourable since these two requirements cannot be properly met. Recent studies of micelle formation kinetics {Aniansson & Wall (1974, 1975), Aniansson, Wall et al. (1976)} confirm that this
is the case; the formation kinetics of typical ionic micelles can be described by assuming that
(a) the aggregation numbers, n_i, have a Gaussian distribution around a mean aggregation number,
$<n>$, with a standard deviation, σ, that is fairly large ($\simeq <n>/2$) for short-chain amphiphiles
and small ($\sigma/<n>$ $<<1$) for long-chain compounds, and (b) $<n>$ is so large that there will be a
range of intermediate n_i values for which the concentrations of $A_{p_i} B_{n_i}$ are very low.

A consequence of this type of association behaviour is that appreciable amounts of micelles are
formed only above a more or less well-defined concentration, the critical micelle concentration, cmc. This is readily understood on the basis of a very simple argument. A reasonable first
approximation is to assume that micelle formation can be described by a simple aggregation number, n:

$$nB \underset{\rightarrow}{\leftarrow} B_n \tag{2}$$

with the equilibrium constant

$$K = a_{B_n}/a_B^n \simeq [B_n]/[B]^n \tag{3}$$

It can be easily shown that even for rather small K and n, $[B_n]$ will start to increase rapidly above a total concentration of B given by

$$c_B = \text{cmc} \simeq K^{1/(n-1)} \tag{4}$$

The critical micelle concentration is by far the most frequently determined characteristic property of amphiphilic ions and an enormous volume of cmc's for different compounds occurs in the
literature. Since most physico-chemical properties of the solution change rapidly around the
cmc, experimental methods used most frequently probably are surface tension, conductivity and
solubilizing capacity.

It should be evident that since micelles are in rapid equilibrium with monomers the cmc is not
an exactly defined quantity but rather will depend upon the physico-chemical property utilized
to detect the formation of micelles as well as the way this property is being represented when
determining the cmc. Also, the use of any method that will tend to shift the equilibrium in
either direction for the determination of cmc should be viewed critically. The most notable of
such methods, and one very frequently used, is the solubilization of various compounds in the
micelles.

A very complete list of cmc values up to 1966 as well as an excellent discussion of these points
has been given by Mukerjee & Mysels (1971). As will be shown below, the vast majority of thermodynamic quantities for micelle formation are based on determinations of the cmc and the uncertainty associated with such determinations, hence, will be directly reflected in the values
reported. To give a complete list of cmc values, however, would have been far beyond the scope
of this review. When utilizing the thermodynamic values in Table 1, however, one should be
aware of the difficulty in comparing values reported by different workers not only because of
definitions based on different equations but also because of the widely varying methods of determining the cmc. In Tables 1 and 2 we have included only thermodynamic values calculated from
cmc's in papers where the authors have specifically aimed at elucidation of the thermodynamics
of micelle formation. As shown below, a measure of the Gibbs' free energy of micelle formation
can be obtained from almost any cmc value that has been determined with reasonable accuracy; to
include all such cmc values in these tables would, however, have been completely impractical.

GIBBS' ENERGY OF MICELLIZATION

The Gibbs' energy of micellization can formally be unambiguously defined as the change in Gibbs' energy of the system when monomers and counterions are taken from a chosen standard state in solution to a solution in which they form the equilibrium distribution of micellar species, also in a chosen standard state. The difficulties in how such a definition is to be associated with experimentally determined quantities, however, are obvious.

PHASE SEPARATION MODEL

The most easily visualized treatment is the so-called (pseudo) phase separation model. In this model, we assume that micelle formation is akin to the precipitation of a separate phase, which then, in order to avoid the additional complication of assuming the formation of a charged phase, has to include the same number of counterions as amphiphilic monomers in the phase, i.e., the phase consists of micelles with a mean aggregation number, $<n>$, and an equivalent amount of counterions surrounding the micelles. The formation of micelles is described as the "precipitation" of micellar phase at a saturation concentration of monomers that may be calculated from a "solubility product"

$$K_s^{-1} = f_A^n \, x_A^n \, f_B^n \, x_B^n \tag{5}$$

where f_i and x_i are activity coefficients and mole fractions, respectively. Choosing the standard state of the solute ions as the hypothetical pure solute in which the environment of each molecule is the same as at infinite dilution ($f_i = 1$, $x_i = 1$), the standard Gibbs'energy of micelle formation per monomer is given by

$$\Delta G_m^{\ominus} = RT \, \ln\{f_{\pm} x_c(AB)\}^2 \tag{6}$$

where $x_c(AB)$ is the mole fraction of monomers when micelles "precipitate" and $f_{\pm} = \sqrt{f_A f_B}$ is the mean activity coefficient. In the presence of an electrolyte with the ion A in common with the amphiphilic monomer, the solubility of micelles is shifted towards lower concentrations according to Eq. (5) and $\Delta G_m^{\ominus}$ is given by

$$\Delta G_m^{\ominus} = RT \, \ln\{f_A x_c(A) \, f_B x_c(B)\} \tag{7}$$

where $x_c(A)$ and $x_c(B)$ are the concentrations of A and B in equilibrium with the micelles.

$x_c(AB)$, of course, in this model is identified with the cmc. If it is assumed that $f_{\pm} \approx 1$, Eq. (6) becomes

$$\Delta G_m^{\ominus} = 2 \, RT \, \ln\{x_c(AB)\} \tag{8}$$

MULTIPLE EQUILIBRIUM MODEL

In the multiple equilibrium model of micelle formation it is assumed that micelle formation is described by a number of equilibria of the type given by Eq. (1) with the equilibrium constants

$$K_{pn} = \frac{a_{pn}}{a_A^p \, a_B^n} \tag{9}$$

(a_i = activity of aggregate i) or by a stepwise aggregation

$$A + B + A_{p-1} B_{n-1} \rightleftarrows A_p B_n \qquad \begin{array}{l} p = 1, 2, \ldots \\ n = 2, 3, \ldots \end{array} \tag{10}$$

with the equilibrium constants

$$K'_{pn} = \frac{a_{pn}}{a_{p-1,n-1} \, a_A \, a_B} \tag{11}$$

the K_{pn} and K'_{pn} being connected by the relationship

$$K_{pn} = \Pi \ K'_{ij} \qquad\qquad \begin{array}{l} i = 0, \ \dots, \ p \\ j = 1, \ \dots, \ n \end{array} \tag{12}$$

As has already been stated, it is not experimentally feasible to obtain detailed information on all the K_{pn}. Hence, it is usually assumed that all the equilibria can be described approximately by assuming that a single micellar species, of aggregation number n (and binding p counterions) is in equilibrium with the monomers:

$$pA + nB \ \overrightarrow{\leftarrow} \ A_p B_n \tag{13}$$

(we omit charge signs for simplicity; it is understood that A is the counterion and B is the amphiphilic ion, and $n \geq p$). We now choose the hypotethical standard state $f = 1$, $x = 1$ for each of the species in the equilibrium. Then

$$\Delta G_{\mathrm{m}}^{\ominus} = RT \ \left[\frac{p}{n} \ \ln(f_A x_A) + \ln(f_B x_B) - \frac{1}{n} \ \ln(f_{\mathrm{m}} \frac{x_{\mathrm{m}}}{n}) \right] \tag{14}$$

where x_{m} is the mole fraction of the micellar species. For concentrations close to the cmc and in the absence of any additives the following assumptions are usually made:

(a) all activity coefficients are neglected, i.e., $\dfrac{RT}{n} \ \ln \dfrac{f_A^{p} f_B^{n}}{f_{\mathrm{m}}} = 0$;

(b) the term $(1/n)\ln(x_{\mathrm{m}}/n)$ will give a very small contribution to $\Delta G_{\mathrm{m}}^{\ominus}$ and is neglected (the rationale for this is that n is a fairly large number);

(c) $x_A = x_B = x_{\mathrm{c}}(AB)$, i.e., the amount of counterions that is not bound to micelles gives a negligible contribution to x_A.

An approximate value of $\Delta G_{\mathrm{m}}^{\ominus}$ then is obtained by introducing these approximations into Eq. (14):

$$\Delta G_{\mathrm{m}}^{\ominus} = (1 + \frac{p}{n}) \ RT \ \ln x_{\mathrm{c}}(AB) \tag{15}$$

If it is assumed that $p = n$, i.e., the micellar surface is neutralized by counterions or, alternatively, the micellar aggregate is thought of as including all the counterions in the diffuse double layer, Eq. (15) becomes

$$\Delta G_{\mathrm{m}}^{\ominus} = 2 \ RT \ \ln x_{\mathrm{c}}(AB) \tag{16}$$

$x_{\mathrm{c}}(AB)$ is commonly identified with the cmc, which in this case is best defined operationally, i.e., as the intersection between two lines describing the change of some physico-chemical property of the solution well below and well above the region in which the concentration of micelles starts to increase rapidly (see Fig. 1).

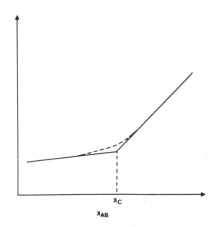

Fig. 1. The cmc is determined by extrapolation from regions below and above the region of rapid change

If the identification $x_c(AB) =$ cmc is made, Eq. (16) takes the same form as Eq. (8).

If it is assumed that the micellar surface contains no counterions, i.e., $p = 0$, Eq. (15) becomes

$$\Delta G_m^\ominus = RT \ln x_c(AB) \tag{17}$$

which, of course, is valid for the process

$$nB \rightleftarrows B_n \tag{18}$$

and, hence, will be valid for non-ionic surfactants. It can also be shown that Eq. (17) is valid for the process

$$B + B_n \rightleftarrows B_{n+1} \tag{19}$$

i.e., the incorporation of a monomer in an n-micelle in equilibrium with the monomer solution.

If an "inert" electrolyte is present in the solution it is best to start with Eq. (14). If it is assumed that $f_m = 1$, i.e., the micelles are in very dilute solution with no intermicellar interactions, Eq. (14) gives

$$\Delta G_m^\ominus = RT \left[\frac{p}{n} \ln a_A(cmc) + \ln a_B(cmc) - \frac{1}{n} \ln \frac{x_M}{n} \right] \tag{20}$$

where $a_A(cmc)$, $a_B(cmc)$ are the ionic activities at the cmc, which again is preferably operationally defined.

A reasonably complete description of $\Delta G_m^\ominus$ values for micelle formation, hence, should include a clear statement of which of the above equations that has been used and the rationale for doing so. Experimental results very clearly indicate that one should assume that only a fraction of the counterions are bound to the micelles and hence a complete determination of $\Delta G_m^\ominus$ requires the knowledge of p/n, enough information to justify the neglect of the term $(1/n)\ln(x_M/n)$, and, in the case of added electrolyte, a_A and a_B at the cmc.

OTHER CONCENTRATION UNITS

The discussion above has been carried out using mole fractions as the measure of the amounts of the materials in the solution. Very frequently, however, molalities {mol (kg solvent)$^{-1}$} or volumetric concentrations (mol dm^{-3}) are used. This will not formally change the arguments above, but implies the choice of other standard states and hence, of course, the $\Delta G_m^\ominus$ values will be different. In Table 1 we have made no attempt to recalculate all thermodynamic properties to obtain for a unified concentration scale. Instead, we have included a column for the concentration units and a reference to the equation used {Eq.s (8), (15), (16), (17) or (20)}.

ENTHALPIES AND ENTROPIES OF MICELLE FORMATION

Enthalpies of micelle formation have mainly been determined by three different methods:

(a) from the temperature dependence of the cmc;

(b) through direct calorimetric measurement;

(c) by calculation from kinetic data for micelle formation.

Again, comparison of enthalpies determined by different methods and by different authors requires very careful consideration of the actual process involved in the measurement and the equation to be used in the calculations.

ENTHALPIES FROM THE TEMPERATURE DEPENDENCE OF THE CMC

Formally, the pseudo-phase model gives an unambiguous definition of the enthalpy of micelle formation; it is the enthalpy of phase transition from monomer solution to micellar phase. Introducing Eq. (6) into the Gibbs-Helmholtz equation gives

$$\Delta H_m^{\ominus} = -RT^2 \frac{d\left[\ln\{f_{\pm}x_c(AB)\}^2\right]}{dT} \tag{21}$$

or, in the presence of another electrolyte with the ion A in common with the amphiphile

$$\Delta H_m^{\ominus} = -RT^2 \frac{d\left[\ln\{f_{\pm}x_c(AB)x_A\}\right]}{dT} \tag{22}$$

Usually, it is assumed that $f_{\pm} = 1$ which gives the following equation for the determination of molar enthalpies of micelle formation, calculated per monomer:

$$\Delta H_m^{\ominus} = -2RT^2 \frac{d\{\ln x_c(AB)\}}{dT} = -2R\frac{d\{\ln x_c(AB)\}}{d(1/T)} \tag{23}$$

This is probably the most frequently used equation for the calculation of $\Delta H_m^{\ominus}$ from the pseudo-phase model. It should be stressed, however, that the aggregation number n may vary with temperature, i.e., even if it is assumed that the cmc is sharply defined as the point of precipitation of micelles the composition of the "precipitated" phase will vary with temperature. Hence, a variation in the slope of a plot of $\ln x_c$ vs T^{-1} may be due not only to an actual change in the $\Delta H_m^{\ominus}$ but also to a variation of the cmc due to a change in n with temperature of the micelles.

Then enthalpy of micelle formation in the multiple equilibrium model can be obtained by insertion of Eq. (15) into a Gibbs-Helmholtz equation which gives

$$\Delta H_m^{\ominus} = -(1 + \frac{p}{n})\, RT^2 \frac{d\{\ln x_c(AB)\}}{dT} \tag{24}$$

p/n probably does not vary very rapidly with temperature, but the possibility that this may be the case should be considered. In addition, this definition of $\Delta H_m^{\ominus}$, of course, relies heavily upon the way that the cmc is determined and, hence, on all the factors connected with this determination discussed above.

Very often, the binding of counterions is neglected in calculations of $\Delta H_m^{\ominus}$ {i.e., Eq. (17) is used for $\Delta G_m^{\ominus}$} which gives

$$\Delta H_m^{\ominus} = -RT^2 \frac{d\{\ln x_c(AB)\}}{dT} \tag{25}$$

This, then, is the enthalpy change when a monomer is transferred from its standard state in solution into a micelle, also in its standard state, neglecting the binding of counterions. Eq. (25) is by far the most frequently used equation for the calculation of enthalpies of micelle formation.

CALORIMETRIC ENTHALPIES

Calorimetric measurements of the enthalpy of micelle formation are usually performed so that one obtains the enthalpy changes for formation or dissolution of whole micellar aggregates, i.e., $n\Delta H_m^{\ominus}$. It should be observed, however, that solutions of amphiphilic substances below the cmc behave far from ideally and hence, that a proper determination of $\Delta H_m^{\ominus}$ requires enough data to allow extrapolation to infinite dilution. The values for $\Delta H_m^{\ominus}$ thus obtained are obviously quite different from the enthalpy for the incorporation of a monomer in the micelles at the cmc. This point is clearly seen if one considers the partial molar enthalpies which have been determined for a few systems. In Fig. 2 {Danielsson et al. (1976)} we show the partial molar enthalpy of sodium octanoate at 298 K. There is a rapid increase in the enthalpy below the cmc (which for a short-chain compound as this is not very sharply defined) which continues above the cmc and shows a maximum at about 2(cmc). The enthalpy then decreases again. Obviously, to take a monomer from infinite dilution into a micelle is a strongly endothermic process, while doing so at the cmc probably involves a very small change in enthalpy. Thus, determinations of enthalpies of dilution of micellar solutions to concentrations below the cmc will be strongly dependent on the final concentration and care should be taken when comparing enthalpies that have not been obtained by extrapolation to infinite dilution.

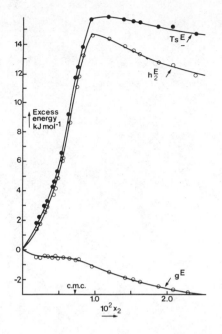

Fig. 2. Partial molal excess quantities for sodium octanoate at 298 K {Danielsson et al. (1976)}

ENTROPIES

Entropies of micelle formation are obtained once $\Delta G_m^{\ominus}$ and $\Delta H_m^{\ominus}$ are known from

$$\Delta S_m^{\ominus} = \frac{\Delta H_m^{\ominus} - \Delta G_m^{\ominus}}{T} \qquad (26)$$

Sometimes, it is stated that if the pseudo-phase model is applied $\Delta G_m^{\ominus}$ must be zero since monomer solution and micelles are in equilibrium, and, hence, $\Delta S_m^{\ominus} = \Delta H_m^{\ominus}/T$. This, of course, is due to a misunderstanding of the definition of the standard state for the monomers which is certainly not a solution in equilibrium with the micellar phase.

However, if it is assumed that the change in partial molar enthalpies from infinite dilution to the cmc is negligible, the quantity $\Delta H_m^{\ominus}/T$ can be taken as a measure of the change in entropy when a monomer in a solution in equilibrium with the micelles is taken into the micelle. For long-chain compounds it is experimentally difficult to show whether such an assumption is correct due to the very low cmc's. Studies of short-chain compounds (Stenius, Rosenholm & Hakala) as well as available data on compounds with longer chains {Birch & Hall (1972)} indicate, however, that it is questionable whether the assumption is justified (see Fig. 2). For this reason we have limited the values of the entropies given in Tables 1 and 2 to those calculated from Eq. (26).

HEAT CAPACITIES

The change in heat capacity on micelle formation can be defined simply as

$$\Delta C_{p,m}^{\ominus} = (\partial \Delta H_m^{\ominus}/\partial T)_p \qquad (27)$$

This definition has been utilized in a few cases where $\Delta H_m^{\ominus}$ is known as a function of T (see Table 3). Since the micellar aggregation number may vary with T, $\Delta C_{p,m}^{\ominus}$ values determined this way are subject to the same criticism as $\Delta H_m^{\ominus}$ values that are determined from the temperature dependence of the cmc.

Another definition has been used by Desnoyers and coworkers (Musbally, Perron & Desnoyers) who measure the heat capacity of the solutions as a function of the concentration and calculate $\Delta C_{p,m}$ from

$$\Delta C_{p,m} = \frac{\Delta(m\phi_c)}{\Delta m} - \phi_c(cmc) \qquad (28)$$

where ϕ_c is the apparent molal heat capacity of the solute and m the molality. Thus, $\Delta C_{p,m}$ is defined as the difference between the heat capacity of micellar solute {which is taken to be $\Delta(m\phi_c)/\Delta m$ at concentrations far above the cmc} and the heat capacity of the monomers (which is ϕ_c at the cmc).

TABLES

TABLE 1. Thermodynamics of micellization of salts with organic anion (*a*)

Cation	Anion	T / K	Conc. scale	Medium	$\Delta G^{\ominus}_m$ / kJ mol^{-1}	Eq.	$\Delta H^{\ominus}_m$ / kJ mol^{-1}	Eq.	$\Delta S^{\ominus}_m$ / J K^{-1}mol^{-1}	Eq.	Year	Ref.	Notes
Li	dodecylsulfate	298	x	aq	−35.70	17					1967	1	
Na	hexanoate	278 293 333 363	x	aq	−10.04 −10.29 −11.13 −11.55	17	−4.56 −5.06 −4.52 −5.31	25	19.70 17.84 19.84 17.16	26	1975	2	
		298	c	aq			10.50	cal			1976	3	*b*
Na	octanoate	293 313 333	x	aq	−12.30 −12.80 −13.31	17	−4.10 −4.60 −4.39	25	27.97 26.18 26.75	26	1975	2	
Na	nonanoate	293 313 333 363	x	aq	−14.23 −15.23 −15.94 −17.07	17	−3.31 −2.30 −2.09 −2.18	25	37.25 41.29 41.56 41.01	26	1975	2	
Na	decanoate	278 363	x	aq	−14.48 −18.98	17	0.00 0.00	25	52.05 52.06	26	1975	2	
Na	dodecanoate	303÷ 363	x	aq	−19.54÷ −23.26	17	0.00	25	64.45÷ 64.04	26	1975	2	
		293÷ 323	x	aq			6.82	25	86.61	26	1974	4	
Na	tetradecanoate	303÷ 363	x	aq	−23.64÷ −28.33	17	0.00	25	77.98	26	1975	2	
Na	9-octadecenoate (*cis*) (oleate)	278 298 333 363	x	aq	−25.65 −26.82 −28.33 −29.20	17	−14.56 −14.43 −14.48 −15.06	25	39.85 41.53 41.56 38.94	26	1975	2	

a: It should be noted that if concentrations (mol dm^{-3}) are used instead of mole fractions or molalities cmc will vary with T also because of changes in the density of the solution and, thus, application of Eq. (25) will include the heat effects associated with this change. None of the $\Delta H^{\ominus}_m$ given in Table 1 or 2 have been corrected for this effect.

b: NaC$_8$ (monomer at the cmc) → NaC$_8$ (in the micelles)

TABLE 1 (Continued.)

Ion	Surfactant	T	sym	medium	ΔG	[ref]	ΔH	[ref]	ΔS	[ref]	year	[ref]	note
Na	12-hydroxy-9-octadecenoate (ricinoleate)	293	x	aq	−16.02	17	−37.87	25	183.80	26	1975	2	
		313			−18.54		−0.84		61.85				
		333			−19.87		−0.84		62.16				
Na	octylsulfate	283	c	aq			4.60	25			1961	5	
		293					2.09						
		303					−0.42						
		313					−2.93						
		323					−5.02						
		283	c	aq			9.16	24			1975	6	*a*
		293					5.82						
		303					−3.31						
		313					−4.81						
		323					−8.24						
		298	c	aq			−4.85	cal			1978	7	
		294	x	aq	−16.53	17	5.94	cal	18.25	26	1974	8	
		294	x	aq	−20.03	17					1964	9	
		298	x	aq	−16.54	17	4.81	cal	71.56	26	1974	8	*b*
		303			−16.53		3.77		66.93				
		308			−16.53		1.46		58.38				
		298	x	aq			−2.83	24	−4.60	26	1972	10	*c*
		294	x	0.03 M NaCl	−20.30	20					1964	9	
				0.1 M NaCl	−20.77								
				0.3 M NaCl	−21.11								
				1.0 M NaCl	−22.45								
		294	x	aq	−15.06	17	6.28	cal	71.56	26	1964	11	*d*
		294	m	aq			3.35	cal			1957	12	
				0.694 M NaCl			1.26						
Na	nonylsulfate	294	x	aq	−23.19	20					1964	9	
		298	c	aq			3.05	cal			1978	7	
		294	x	0.03 M NaCl	−23.29	20					1964	9	*c*
				0.10 M NaCl	−23.97								
				0.30 M NaCl	−24.00								

a: *p/n* given graphically as a function of ionic strength
b: *p/n* = 0.7
c: variation of *p/n* with electrolyte concentration taken into account
d: $\Delta H_m^{\ominus}$ estimated from data in Ref. 12

TABLE 1 (Continued)

Ion	Compound													note
Na	decylsulfate	x	294	aq	−26.37	20					9	1964	
		x	294	aq	−19.20	17	4.94	cal	82.06	26	8	1974	
			298		−19.20		3.14		74.92				
		x	298	aq			−2.13	24	7.11	26	10	1972	a
		x	303	aq	−19.20	17	1.30	cal	67.62	26	8	1974	
			308		−19.20		0.42		63.67				
		x	273÷338	aq			1.46	25	66.94	26	4	1974	b
		x	294	0.1 M NaCl	−27.27	20					9	1964	c
			294	0.3 M NaCl	−27.86								
		c	283	aq			5.02	25			5	1961	
			293				2.51						
			303				0.00						
			313				−2.51						
			323				−5.02						
			333				−7.53						
		c	283	aq			8.74	24			6	1975	
			293				3.18						
			303				−0.75						
			313				−5.15						
			323				−9.87						
		m	298	aq			2.09	cal			12	1957	
		m	298	aq	−18.83	17	4.18	cal	71.56	26	11	1964	
Na	2-decylsulfate	c	283	aq			6.69	25			5	1961	
			293				4.18						
			303				1.67						
			313				−0.42						
			323				−2.93						
			333				−5.44						
Na	hendecylsulfate	x	294	aq	−29.82	20					9	1964	c
				0.03 M NaCl	−29.52								
				0.10 M NaCl	−29.79								
				0.30 M NaCl	−29.74								

a: $p/n = 0.7$
b: ΔH_m^Θ calculated at 298 K
c: variation of p/n with electrolyte concentration taken into account

TABLE 1 (Continued)

Compound	T (K)		Medium								Year		
Na dodecylsulfate	293	x	aq	-21.13	17	-4.56	25	40.96	26		1975	2	a
	294	x	aq	-31.90	20						1964	9	
	298	x	aq	-21.13	15	0.36	cal	71.96	26		1970	13	b
	298	x	aq	-21.76	17	-1.26	cal	68.75	26		1964	11	c
	298	x	aq	-39.25	15						1978	14	
	298	x	aq	-39.17							1955	15	d
	303	x	aq	-21.67	17	-2.76	25	53.27	26		1975	2	
	333			-23.81		-2.30		57.64					
	363			-25.61		-2.72		55.62					
	283÷343	x	aq			-0.21	25	72.38	26		1974	4	e
	293÷323	x	aq			-1.67	25				1958	16	
	294	x	0.01 M NaCl	-32.70	20						1964	9	a
			0.03 M NaCl	-33.24									
			0.10 M NaCl	-33.90									
			0.30 M NaCl	-34.68									
	298	x	0.023 M NaCl	-21.76	15	-0.64	cal	71.13	26		1970	13	f
	298	x	0.05 m NaCl	-25.03	17						1967	17	
			0.201 m NaCl	-27.24									
			0.506 m NaCl	-28.78									
	298	x	0.02 M NaCl	-39.67	15						1955	15	d
			0.03 M NaCl	-40.16									
	298	x	3 M urea	-19.58	15	-6.04	cal	45.61	26		1970	13	g
	278	c	aq			3.35	25				1953	18	
	283	c	aq			5.02	25				1961	5	
	293					1.67							
	298	c	aq			2.51	25				1953	18	
	303	c	aq			-1.26	25				1961	5	
	318	c	aq			-7.95	25				1953	18	

a: variation of p/n with electrolyte concentration taken into account
b: p/n = 0.771
c: p/n = 0.77
d: for equation used, see Ref. 15
e: $\Delta H_m^{\ominus}$ calculated at 298 K
f: p/n = 0.86
g: p/n = 0.679

TABLE 1 (Continued)

T		solvent			value				year	ref	
323	c	aq			−7.53	25			1961	5	
333					−10.88						
298÷303	c	aq			−4.2	25			1948	19	
308÷318					−7.5						
333÷353					−21						
283	c	aq			7.49	24			1975	6	a
293					3.85						
303					−3.31						
313					−7.32						
323					−12.09						
294	c	aq	−23.84	17	1.09	cal	84.76	26	1974	8	
298			−23.84		−2.13		72.82				
303			−23.85		−4.02		65.41				
308			−23.85		−7.07		54.44				
278	c	5.03% 1-propanol			10.46	25			1953	18	
298					2.09						
318					−11.72						
278		5.03% 2-propanol			9.62						
298					0.00						
318					−11.72						
296	m	aq			1.46	cal			1969	20	
297	m	aq			−1.05	cal			1951	21	
298	m	aq			0.68	cal			1976	22	
298	m	aq			0.36	cal			1969	20	
298	m	aq			≃0.42	cal			1956	23	
303	m	aq			−2.55	cal			1969	20	
293	m	aq			1.67	25			1957	24	
298					−0.04						
303					−2.09						
298	m	aq			−0.38	24	−1.26	26	1972	10	b
298	m	0.001 M NaCl			0.59	cal			1976	22	
		0.01 M NaCl			−0.80						

a: p/n given graphically as a function of T in Ref. 6

b: $p/n = 0.7$

TABLE 1 (Continued)

Ion	Compound	T				cal					
Na	tetradecylsulfate	298÷348	x	aq	−2.85		75.31	26	1974	4	
		303	c	aq	−2.93	25			1961	5	
		313			−6.28						
		323			−9.62						
		333			−12.97						
		343			−16.32						
		303	c	aq	−5.36	24			1975	6	a
		313			−10.33						
		323			−16.82						
		308÷318	c	aq	−6.28	25			1937	25	
		333÷353			−23.01						
Na	2-tetradecylsulfate	303	c	aq	−0.84	25			1961	5	
		313			−4.60						
		323			−7.95						
		333			−11.30						
		343			−16.32						
Na	4-tetradecylsulfate	303	c	aq	0.84	25			1961	5	
		313			−2.93						
		323			−6.28						
		333			−9.62						
		343			−13.39						
Na	hexadecylsulfate	298÷363	x	aq	−34.77	25	−15.06	26	1974	4	b
Na	decylsulfonate	298÷308	c	aq	−2.09	25			1957	12	
		308÷318			−5.86						
		333÷353			−12.55						
Na	dodecylsulfonate	308÷318	c	aq	−7.53	25			1948	19	
		333÷353			−16.74						

a: $\supset/n$ given graphically as a function of T

b: $\Delta S_m^{\ominus}$ calculated at 298 K

TABLE 1 (*Continued*)

Cation	Compound	T (K)		Medium	$\Delta S_m^{\ominus}$	t	$\Delta H_m^{\ominus}$	t	$\Delta H^{\ominus}$		Year	Ref	Note
Na	octyl-1-benzene sulfonate	295÷323	x	aq			0.13	25	49.32	26	1974	4	a
Na	p-octyl-benzene sulfonate	295÷323	x	aq			−8.66	25	41.42	26	1974	4	a
Na	decyl-1-benzene sulfonate	296÷353	x	aq			−20.33	25	9.20	26	1974	4	a
Na	dodecyl-benzene sulfonate	298÷318	x	aq			22.68	23	154.81	26	1974	4	a
Na	cholate	293÷323	x	aq	−19.79÷−21.80	17	0	25	67.50÷67.45	26	1975	2	
Na	desoxycholate	293÷323	x	aq	−22.67÷−24.98	17	0	25	77.34÷77.28	26	1975	2	
K	octanoate	298	x	aq	−12.13	17	13.81	cal	86.99	26	1964	11	b
		288	m	0.033 m KOH			9.62	25			1959	26	
		293					6.69						
		298					6.27						
		308					3.77						
		318					−2.09						
		323					4.18	cal					
		328					−9.62	25					
		298	m	0.042 m KOH			7.11	cal			1957	12	
		308	m	1.80 m KCl+ 0.042 m KOH			6.27	25			1958	27	
K	decanoate	298÷318	x	aq			0.25	25	53.14	26	1974	4	
K	dodecanoate	288÷333	x	aq			1.09	25	68.20	26	1974	4	
		293	c	aq			11.72	25			1961	5	
		298	c	aq			10.89	25			1948	28	
		313	c	aq			3.35	25			1961	5	
		323	c	aq			0	25			1948	28	

a: $\Delta S_m^{\ominus}$ calculated at 298 K   b: $\Delta H_m^{\ominus}$ estimated from data in Ref. 12

TABLE 1 (Continued)

	Surfactant	T (K)									Year	
		333					-5.44					
		343					-10.04					
K	tetradecanoate	298÷308	c	aq			-4.18	25			1948	19
		308÷318					-10.04					
		333÷353					-20.92					
K		298÷308	c	aq			-4.2	25			1948	19
		308÷318					-4.2					
		333÷353					-8.4					
		298÷338	x	aq			-1.13	25	71.55	26	1974	4
K	dodecylsulfate	305	x	aq	-36.39	17					1967	1
Cs	dodecylsulfate	298	x	aq	-38.18	17					1967	1
NH_4	9-octadecenoate (cis) (oleate)	293÷333	x	aq	-31.97÷ -36.40	17	0.00	25	109.02÷ 109.25	26	1975	2

1. Mukerjee, Mysels & Kapauan
2. Markina, Bovkun, Tsikurina & Sinyova
3. Stenius, Rosenholm & Hakala
4. Jolicoeur & Philip
5. Flockhart
6. Moroi, Nishikodo, Uehara & Matuura
7. Dörfler & Sackmann
8. Kresheck & Hargraves
9. Huisman
10. Shinoda
11. Benjamin
12. Goddard, Hoeve & Benson
13. Jones, Pilcher & Espada
14. Barry & Wilson
15. Phillips
16. Matijevic & Pethica
17. Emerson & Holtzer
18. Flockhart & Ubbelohde
19. Klevens
20. Pilcher, Jones, Espaĉa & Skinner
21. Goddard & Pethica
22. Paredes, Tribout, Ferreira & Leonis
23. Goddard & Benson (1956)
24. Goddard & Benson (1957)
25. Powney & Addison
26. White & Benson (1959)
27. White & Benson (1958)
28. Brady & Huff

TABLE 2. Thermodynamics of micellization of salts with organic cation

Cation	Anion	T/K	Conc. scale	Medium	$\Delta G_m^{\ominus}$ kJ mol⁻¹	Eq.	$\Delta H_m^{\ominus}$ kJ mol⁻¹	Eq.	$\Delta S_m^{\ominus}$ J K⁻¹mol⁻¹	Eq.	Year	Ref.	Notes
Decylammonium	Cl	293÷333	x	aq			-6.95	25	37.24	26	1974	1	a
Octyltrimethylammonium	Br	293÷333	x	aq			26.65	25	136.82	26	1974	1	a
Dodecylammonium	Cl	308	c	aq			-5.4	25			1947	2	
		288÷333	x	aq			1.6	25	74.09	26	1974	1	a
Decyltrimethylammonium	Cl	298	c	aq	-27.4	14					1972	3	b, c
Decyltrimethylammonium	Br	298÷333	x	aq			-3.89	25	43.10	26	1974	1	a
		298	x	aq	-28.26	14					1970	4	a
Tetradecylammonium	Cl	298÷333	x	aq			-0.38	25	79.91	26	1974	1	a
2-Ethoxydodecylammonium	Cl	278	c	aq			6.19	23			1968	5	
		288					5.19						
		293					3.05						
		298					-0.84						
		303					-6.65						
		308					-8.08						
		313					-12.13						
		323					-15.08						
2-Ethoxydodecylammonium	Br	278	c	aq			9.75	23			1968	5	
		288					5.94						
		293					3.89						
		298					-0.50						
		303					-7.03						
		308					-9.96						
		313					-15.27						
		323					-23.56						

a: $\Delta S_m^{\ominus}$ calculated at 298 K b: assuming all activity coefficients = 1 c: $p/n = 0.75$

TABLE 2 (Continued)

Surfactant		medium	T (K)		ΔG°		ΔH°		ΔS°		year	ref	note
2-Ethoxydodecylammonium I	c	aq	288				12.5				1968	5	
			293				5.10						
			298			23	-1.72	23					
			303				-7.99						
			308				-11.0						
			313				-19.3						
			323			cal	-29.0						
2-Ethoxydodecylammonium NO₃	c	aq	288				6.78				1968	5	
			293				1.80						
			298			23	-1.34	23					
			303				-6.95						
			308				-9.08						
			313				-12.6						
			323				-20.3						
Dodecyltrimethylammonium Cl	x	aq	278	14	-31.0	25	10.5	25	151	26	1972	6	a
			298		-33.7		2.30		121				
			318		-35.7		-8.41		83.7				
			338		-37.2		-12.7		71.1				
	x	aq	298	14	-33.7						1972	3	a
	x	0.05 m NaCl	298	17	-23.0						1967	7	a
		0.201 m NaCl			-24.9								
		0.506 m NaCl			-26.3								
Decylpyridinium Cl	x	aq	298	14	-28.1						1972	3	a
Docecyltrimethylammonium Br	x	aq	278	14	-32.7				80.33	26	1970	8	b
	x	aq	298	20	-20.1						1970	9	c
				17	-20.8								
	x	aq	298	15	-17.9	cal	-1.39		55.65	26	1970	10	d
	x	aq	298	14	-35.5						1970	4	e
	x	aq	313						45.19	26	1970	8	e
	x	aq	274÷343			25	-0.67		66.11	26	1974	1	f
	x	0.0175 m NaBr	288		-20.33						1972	11	
	x		298	17	-21.25	cal	-1.58	cal					

a: p/n = 0.76
d: p/n = 0.744

b: p/n estimated graphically
e: assuming all activity coefficients = 1

c: assuming no binding of counterions
f: ΔS°_m calculated at 298 K

TABLE 2 (Continued)

Compound	Ion	Scale	T/K	Medium	$-\Delta H$	n	method	ΔG	value	n	Year	Ref	Note
			308		−21.76								
			318		−22.26								
			328		−22.64								
		x	298	0.023 m NaBr	−18.83	15	cal	−1.63	57.74	26	1970	8	*a*
		x	298	0.05 m NaBr	−22.76	17					1967	7	
		x	298	0.05 m NaBr			cal	−1.85			1970	8	*b*
		x	298	0.1 M NaBr	−32.5 / −23.5 / −24.3	14 / 20 / 17					1970	9	*c*
		x	288	0.1 m NaBr	−22.54	17					1972	11	
		x	298	0.1 m NaBr	−23.7	17					1967	7	
		x	298	0.1 m NaBr	−22.8	17	cal	−2.05			1970, 1972	8, 11	*d*
		x	298	0.502 M NaBr	−32.0 / −26.0 / −26.8	14 / 20 / 17					1970	9	*b* *c*
		x	298	0.508 m NaBr	−26.2	17					1967	7	
		x	298	3.0 M urea	−15.9	15	cal	−2.64	44.4	26	1970	8, 10	*e*
Dodecyltrimethylammonium NO$_3$		x	298	0.05 M NaNO$_3$ / 0.101 M NaNO$_3$ / 0.253 M NaNO$_3$ / 0.509 M NaNO$_3$	−23.3 / −24.1 / −25.3 / −26.3	17					1967	7	
Tetradecyltrimethylammonium	Br	x	298	aq	−44.1	14					1970	4	*f*
		x	303÷343	aq			25	−10.9	44.4	26	1974	1	
		m	278 / 298 / 318 / 338 / 358	aq	−26.1 / −27.9 / −28.9 / −29.3 / −29.3	8	23	1.83 / −6.66 / −18.3 / −27.8 / −26.2	100.4 / 71.1 / 33.5 / 4.18 / 8.37		1967	12	*g*
Dodecylpyridinium	Cl	x	278	aq	−32.0	8	25	8.70	146	26	1972	6	*h*

a: $p/n = 0.85$
b: p/n estimated graphically
c: assuming no binding of counterions
d: $p/n = 0.723$
e: $p/n = 0.744$
f: assuming all activity coefficients = 1; $p/n = 0.78$
g: $-\Delta S_m^{\ominus} = \{\partial(\Delta G_m^{\ominus})/\partial T\}_p$
h: assuming all activity coefficients = 1

TABLE 2 (Continued)

		T	conv.	medium										
		298	x	aq	−34.7	8						1972	3	*a*
		298	x	aq	−34.2	20						1966	13	*b*
		298	x	aq	−20.6	17	1.88	cal	75.5	26	1974	14	*b*
		303			−20.6		−1.55		62.9					
		308			−20.6		−2.93		57.4					
		318	x	aq	−36.5	20	−8.7	25	87.8	26	1972	6	*b*
		338			−38.2		−15.2		66.9					
		298÷323	x	aq	−21.4		−21.4	25	−3.35	26	1974	1	*b*
		298	x	0.02 M KCl	−33.7	20					1966	13	*b*
				0.05 M KCl	−34.0									
				0.08 M KCl	−33.7									
Dodecylpyridinium	Br	278	x	aq	−19.62	17	5.56	25	90.4	26	1970	9	*b*
		283			−20.0		2.5		79.5					
		288			−20.4		0.08		71.1					
		293			−20.8		−2.3		62.8					
		294	x	aq	−21.1		−2.05	cal	61.7	26	1974	14	*b*
		298	x	aq		17	−4.06	25	56.9	26	1970	9	*b*
		298	x	aq			−3.47	cal	56.1	26	1974	14	*b*
		303	x	aq	−21.3	17	−5.48	25	52.3	26	1970	9	*b*
		303	x	aq			−5.56	cal	48.3	26	1974	14	*b*
		308	x	aq	−21.6	17	−6.57	25	48.5	26	1970	9	*b*
		308	x	aq			−7.32	cal	41.8	26	1974	14	*b*
		313	x	aq	−21.8	17	−7.41	25	46.0	26	1970	9	*b*
		318			−22.0		−7.95		44.3					
		323			−22.3		−8.45		42.7					
		328			−22.5		−8.83		41.6					
		333			−22.7		−9.25		40.3					
		338			−22.9		−9.79		38.6					
		343			−23.1		−10.63		36.3					
		298	x	0.02 M KBr	−36.4	20					1966	13	*b*
				0.04 M KBr	−37.4									
				0.06 M KBr	−36.9									
				0.08 M KBr	−36.2									
				0.10 M KBr	−36.7									

a: assuming all activity coefficients = 1; p/n = 0.78

b: variation of p/n with electrolyte concentration taken into account

TABLE 2 (Continued)

Compound		T (K)		Medium							Year	Ref	
Dodecylpyridinium	I	298	x	aq	−24.2	15	−9.3	cal	39.4	26	1971	15	a
		298	x	aq			−13.5	cal		26	1972	11	
		303			−24.2	15	−14.5		31.8				
		308			−24.3		−16.7		24.7				
		293÷313	x	aq			−12.43	25	39.33	26	1974	1	b
											1966	13	c
		298	x	0.0C25 M KI	−42.6	20							
				0.0C5 M KI	−42.6								
				0.010 M KI	−44.6								
		298	x	0.01 M KI	−25.2	15	−11.97	cal	44.5	26	1971	15	d
		303		0.01 M KI	−25.3		−13.93		37.66				
		308		0.01 M KI	−25.5		−15.86		31.30				
		298		0.5 M urea	−23.7		−12.6		37.1				a
				1.0 M urea	−23.1		−12.6		35.4				
				1.5 M urea	−22.6		−12.9		32.6				
				2.0 M urea	−22.1		−13.1		30.5				
		303		2.0 M urea	−22.5		−14.9		25.1				
		308		2.0 M urea	−22.8		−17.6		16.8				
2-[2-(decyloxy)ethoxy]ethyltrimethylammonium	Cl	298	x	aq	−33.6	20					1972	3	e
1-[2-(decyloxy)ethyl]pyridinium	Cl	298	x	aq	−32.6	20					1972	3	f
(2-dodecyloxyethyl)trimethylammonium	Cl	298	x	aq	−39.2	20					1972	3	g
1-dodecyl-4-methoxy-pyridinium	Cl	278	m	aq	−20.0	16	13	23	117	26	1969	16	
		283			−20.5		9.2		105				
		288			−21.0		5.9		92				
		293			−21.5		3.0		84				
		298			−21.8		0.6		75				
		303			−22.2		−1.5		67				
		308			−22.5		−3.4		63				
		313			−22.8		−5.0		54				
		318			−23.1		−6.7		50				
		323			−23.4		−8.8		46				

a: $p/n = 0.87$
d: $p/n = 1$
f: $p/n = 0.77$

b: $\Delta S^{\ominus}$ calculated at 298 K
e: $p/n = 0.79$
g: $p/n = 0.775$

c: variation of p/n with electrolyte concentration taken into account

TABLE 2 (Continued)

Compound	X	T (K)	Scale	Medium	Ref.	ΔG°	Ref.	ΔH°	ΔS°	Ref.	Year	Ref.
		328			16	−23.6	23	−10.9	39	26	1969	16
		333				−23.7		−13.8	30			
1-iodecyl-4-methoxy-pyridinium	Br	278	m	aq	16	−22.0		11	121			
		283				−22.5		5.0	96			
		288				−22.9		0.3	80			
		293				−23.3		−3.3	67			
		298				−23.6		−6.3	59			
		303				−23.9		−8.4	50			
		308				−24.1		−11	42			
		313				−24.3		−14	34			
		318				−24.5		−18	22			
		323				−24.5		−23	4.6			
		328				−24.5		−31	−18.4			
1-methyl-4-dodecyloxy-pyridinium	Br	288	m	aq	16	−27.2	23	0.67	96	26	1969	16
		293				−27.6		−3.2	84			
		298				−28.0		−7.1	71			
		303				−28.3		−11	59			
		308				−28.6		−14	50			
		313				−28.8		−17	38			
		318				−29.0		−20	29			
		323				−29.1		−22	22			
		328				−29.2		−24	15			
		333				−29.3		−26	10			
Hexadecyltrimethylammonium [a]	Br	298÷353	x	aq		−9.25	25		60.7	26	1974	1
	Br	298	m	aq		−9.29	cal				1976	17
				0.001 m NaCl		−8.83						
				0.010 m NaCl		−5.90						
				0.100 m NaCl		−3.36						
2-[2-(decyloxy)ethoxy]ethylpyridinium [b]	Cl	298	x	aq	20	−34.1					1972	3
2-[2-(dodecyloxy)ethoxy]ethyltrimethylammonium [c]	Cl	298	x	aq	20	−40.3					1972	3
2-[2-(dodecyloxy)ethoxy]ethylpyridinium [d]	Cl	298	x	aq	20	−39.5					1972	3

a: $\Delta S^{\ominus}_m$ calculated at 298 K
b: p/n = 0.79
c: p/n = 0.78
d: p/n = 0.80

TABLE 2 (Continued)

1-tetradecyl-4-methoxy-pyridinium	Br	293	m	aq	-30.2	16	-8.4	23	75	26	1969	16	
		298			-30.5		-13		59				
		303			-30.8		-15		50				
		308			-31.0		-16		50				
		313			-31.3		-17		46				
		318			-31.5		-18		42				
		323			-31.7		-21		34				
		328			-31.8		-27		15				
		333			-31.8		-38		18				
Hexadecylpyridinium	Cl	298	x	aq	-51.0	20					1972	6	[a]
		286÷353	x	aq			-5.4	25	73.6	26	1974	1	[b]
Hexadecylpyridinium	Br	298÷328	x	aq			-24.1	25	15.48	26	1974	1	[b]
2-{2-(dodecyloxy)ethoxy} ethylpyridinium	Cl	298	x	aq	-41.5	20					1972	3	[c]
2-{2-(dodecyloxy)ethoxy} ethylpyridinium	I	298	x	aq	-47.9	20					1972	3	[d]
1-{2-(hexadecyloxy) ethyl}pyridinium	Cl	298	x	aq	-56.5	20					1972	3	[a]
2-{2-(hexadecyloxy) ethoxy}ethylpyridinium	Cl	298	x	aq	-59.8	20					1972	3	[d]

a: $p/n = 0.88$
c: $p/n = 0.82$
b: $\Delta S^{\ominus}_m$ calculated at 298 K
d: $p/n = 0.92$

1. Jolicoeur & Philip
2. Weissler, Fitzgerald & Resnick
3. Mandru (1972 a)
4. Wasik & Roscher
5. Robins & Thomas
6. Mandru (1972 b)
7. Emerson & Holtzer
8. Espada, Jones & Pilcher
9. Anacker
10. Jones, Pilcher & Espada
11. Jones & Piercy
12. Adderson & Taylor
13. Ford, Ottewill & Parreira
14. Kresheck & Hargraves
15. Jones, Agg & Pilcher
16. Stead & Taylor
17. Paredes, Tribout, Ferreira & Leonis

TABLE 3. Heat capacities of micelle formation – *Salts with organic anion*

Cation	Anion	$\dfrac{T}{K}$	Conc. scale	Medium	$\dfrac{\Delta C_p}{J\ K^{-1}mol^{-1}}$	Eq.	Year	Ref.
Na	octanoate	298	*m*	aq	−360	28	1973	1
Na	octylsulfate	298	*c*	aq	−310	27	1974	2
		298	*m*	aq	−325	28	1974	3
Na	decylsulfate	298	*m*	aq	−425	28	1974	3
		298	*c*	aq	−402	27	1974	2
Na	dodecylsulfate	298	*m*	aq	−530	28	1974	3
		298	*c*	aq	−560	27	1974	2
				0.1 M NaCl	−448			
				0.5 M NaCl	−406			
				0.1 M NaBr	−389			

– Salts with organic cation

Cation	Anion	$\dfrac{T}{K}$	Conc. scale	Medium	$\dfrac{\Delta C_p}{J\ K^{-1}mol^{-1}}$	Eq.	Year	Ref.
Octylammonium	Br	298	*m*	aq	−320	28	1974	4
Nonyltrimethylammonium	Br	278	*m*	aq	−345	28	1976	5
		288			−325			
		298	*m*	aq	−295	28	1974	3
		308	*m*	aq	−280	28	1976	5
		318			−250			
		278		3 M urea	−240			
		288			−230			
		298			−205			
		308			−190			
		322.4			−170			
Decyltrimethylammonium	Br	298	*c*	aq	−365	28	1974	2
Dodecyltrimethylammonium	Br	298	*x*	aq	−423	27	1970	6
				0.0175 M NaBr	−464			
				0.05 M NaBr	−444			
				0.10 M NaBr	−417			
Tetradecyltrimethylammonium	Br	278	*m*	aq	−209	27	1967	7
		298			−544			
		318			−628			
		338			−293			
		358			586			
Dodecylpyridinium	Cl	298	*c*	aq	−490	27	1974	2
Dodecylpyridinium	Br	298	*x*	aq	−385	27	1974	2
Tetradecylpyridinium	Br	298	*m*	aq	−669	27	1967	7
		318			−586			
		338			−460			
		358			−251			

1. Leduc & Desnoyers

2. Kresheck & Hargraves

TABLE 3 (Continued)

3. Musbally, Perron & Desnoyers (1974)

4. Leduc, Fortier & Desnoyers

5. Musbally, Perron & Desnoyers (1976)

6. Jones, Pilcher & Espada

7. Adderson & Taylor

REFERENCES

ADDERSON, J.E., and TAYLOR, R., in *Chem. Phys. Appl. Surface Active Subst.*, *Proc. Int. Congr.*, *4th*, *1964*, Gordon and Breach, New York, 1967, 2, 527

ANACKER, E.W., in *"Cationic Surfactants"*, ed. by JUNGERMANN, E., M. Dekker, New York, 1970, p. 203

ANIANSSON, E.A.G., and WALL, S.N., *J. Phys. Chem.*, 78, 1024 (1974)

ANIANSSON, E.A.G., and WALL, S.N., *J. Phys. Chem.*, 79, 857 (1975)

ANIANSSON, E.A.G., WALL, S.N., ALMGREN, M., HOFFMANN, H., KIELMANN, I., ULBRICHT, W., ZANA, R., LANG, J., and TONDRE, C., *J. Phys. Chem.*, 80, 905 (1976)

BARRY, B.W., and WILSON, R., *Colloid Polymer Sci.*, 256, 251 (1978)

BENJAMIN, L., *J. Phys. Chem.*, 68, 3575 (1964)

BIRCH, B.J., and HALL, D.G., *JCS Faraday I*, 68, 2350 (1972)

BRADY, A.P., and HUFF, H., *J. Colloid Sci.*, 3, 511 (1948)

DANIELSSON, I., ROSENHOLM, J.B., STENIUS, P., and BACKLUND, S., *Progr. Colloid and Polymer Sci.*, 61, 1 (1976)

DÖRFLER, H.-D., and SACKMANN, H., *Z. physik. Chem. Leipzig*, 259, 769 (1978)

EKWALL, P., DANIELSSON, I., and STENIUS, P., in *"Surface Chemistry and Colloids"*, ed. by KERKER, M., MTP Int. Rev. Sci. Phys. Chem. Ser. 1, Vol. 7, p. 97, Butterworths, London, 1972

EKWALL, P., and STENIUS, P., in *"Surface Chemistry and Colloids"*, ed. by KERKER, M., MTP Int. Rev. Sci. Phys. Chem. Ser. 2, p. 215, Butterworths, London, 1975

EMERSON, M.F., and HOLTZER, A., *J. Phys. Chem.*, 71, 1898 (1967)

ESPADA, L., JONES, M.N., and PILCHER, G., *J. Chem. Thermodyn.*, 2, 1 (1970)

FLOCKHART, B.D., *J. Colloid Sci.*, 16, 484 (1961)

FLOCKHART, B.D., and UBBELOHDE, A.R., *J. Colloid Sci.*, 8, 428 (1953)

FORD, W.P., OTTEWILL, R.H., and PARREIRA, H.C., *J. Colloid Interface Sci.*, 21, 522 (1966)

GODDARD, E.D., and BENSON, G.C., *Trans. Faraday Soc.*, 52, 409 (1956)

GODDARD, E.D., and BENSON, G.C., *Can. J. Chem.*, 35, 986 (1957)

GODDARD, E.D., HOEVE, C.A.J., and BENSON, G.C., *J. Phys. Chem.*, 61, 593 (1957)

GODDARD, E.D., and PETHICA, B.A., *J. Chem. Soc.*, 1951, 2659

HUISMAN, H.F., *Proc. Koninkl. Ned. Akad. Wetenschap*, B67, 367 (1964)

JOLICOEUR, C., and PHILIP, P.R., *Can. J. Chem.*, 52, 1834 (1974)

JONES, M.N., AGG, G., and PILCHER, G., *J. Chem. Thermodyn.*, 3, 801 (1971)

JONES, M.N., and PIERCY, J., *JCS Faraday I*, 68, 1839 (1972)

JONES, M.N., PILCHER, G., and ESPADA, L., *J. Chem. Thermodyn.*, 2, 333 (1970)

KLEVENS, H.B., *J. Phys. Colloid Chem.*, <u>52</u>, 130 (1948)

KRESHECK, G.C., in *"Water, A Comprehensive Treatise"*, ed. by FRANKS, F., Plenum Press, New
      York, 1975, <u>4</u>, 95

KRESHECK, G.C., and HARGRAVES, W.A., *J. Colloid Interface Sci.*, <u>48</u>, 481 (1974)

LEDUC, P.-A., and DESNOYERS, J.E., *Can. J. Chem.*, <u>51</u>, 2993 (1973)

LEDUC, P.-A., FORTIER, J.-L., and DESNOYERS, J.E., *J. Phys. Chem.*, <u>78</u>, 1217 (1974)

MANDRU, I., *J. Colloid Interface Sci.*, <u>41</u>, 430 (1972)

MANDRU, I., *Chem., Phys. Chem. Anwendungstech. Grenzflächenaktiven Stoffe*, Ber. Int. Kongr.,
      6., <u>II:2</u>, 1035 (1972)

MARKINA, Z.N., BOVKUN, O.P., TSIKURINA, N.N., and SINYOVA, A.V., in *Proc. Int. Conf. Colloid*
      Surface Sci., ed. by WOLFRAM, E., Akadémiai Kiado, Budapest, <u>1</u>, 465 (1975)

MATIJEVIC, E., and PETHICA, B.A., *Trans. Faraday Soc.*, <u>54</u>, 987 (1958)

MOROI, Y., NISHIKODO, N., UEHARA, H., and MATUURA, R., *J. Colloid Interface Sci.*, <u>50</u>, 254 (1975)

MUKERJEE, P., and MYSELS, K.J., *"Critical Micelle Concentrations of Aqueous Surfactant Systems"*,
      NSRDS-NBS 36, Washington, 1971

MUKERJEE, P., and MYSELS, K.J., *ACS Symp. Ser. 9*, 239 (1975)

MUKERJEE, P., MYSELS, K.J., and KAPAUAN, P., *J. Phys. Chem.*, <u>71</u>, 4166 (1967)

MULLER, N., *J. Magn. Resonance*, <u>28</u>, 203 (1977)

MUSBALLY, G.M., PERRON, G., and DESNOYERS, J.E., *J. Colloid Interface Sci.*, <u>48</u>, 494 (1974)

MUSBALLY, G.M., PERRON, G., and DESNOYERS, J.E., *J. Colloid Interface Sci.*, <u>54</u>, 80 (1976)

PAREDES, S., TRIBOUT, M., FERREIRA, J., and LEONIS, J., *Colloid Polymer Sci.*, <u>254</u>, 637 (1976)

PHILLIPS, J.N., *Trans. Faraday Soc.*, <u>51</u>, 561 (1955)

PILCHER, G.N., JONES, M.N., ESPADA, L., and SKINNER, H.A., *J. Chem. Thermodyn.*, <u>1</u>, 381 (1969)

POWNEY, J., and ADDISON, H., *Trans. Faraday Soc.*, <u>33</u>, 1243 (1937)

ROBINS, D.C., and THOMAS, I.L., *J. Colloid Interface Sci.*, <u>26</u>, 407 (1968)

ROSENHOLM, J.B., STENIUS, P., and DANIELSSON, I., *J. Colloid Interface Sci.*, <u>57</u>, 551 (1976)

SHINODA, K., *Chem., Chem. Phys. Anwendungstech. Grenzflächenaktiven Stoffe, 6.*, <u>II:2</u>, 891 (1972)

STEAD, J.A., and TAYLOR, H., *J. Colloid Interface Sci.*, <u>30</u>, 482 (1969)

STENIUS, P., ROSENHOLM, J.B., and HAKALA, M.-R., in *"Colloid and Interface Science"*, ed. by
      KERKER, M., Academic Press, New York, 1976, <u>2</u>, 397

WASIK, S.P., and ROSCHER, N.M., *J. Phys. Chem.*, <u>74</u>, 2784 (1970)

WEISSLER, A., FITZGERALD, J.W., and RESNICK, I., *J. Appl. Phys.*, <u>18</u>, 434 (1947)

WENNERSTRÖM, H., and LINDMAN, B., *Phys. Reports*, <u>52</u>, 1 (1979)

WHITE, P., and BENSON, G.C., *J. Colloid Sci.*, <u>13</u>, 584 (1958)

WHITE, P., and BENSON, G.C., *Trans. Faraday Soc.*, <u>55</u>, 1025 (1959)

FORMATION OF LYOTROPIC LIQUID CRYSTALS BY ORGANIC SALTS

P. Stenius and P. Ekwall

PHASE EQUILIBRIA OF AMPHIPHILIC SALTS IN BINARY, TERNARY AND QUATERNARY SYSTEMS

THE ISOTROPIC AQUEOUS SOLUTIONS

In aqueous solutions of organic salts having the typical properties of amphiphilic compounds, i.e., a distinct separation of the molecular structure into polar and non-polar parts, appreciable amounts of micelles begin to be formed at the critical micelle concentration, cmc. We will refer to these micelles, whose structure is discussed in the preceding chapter, as being of the "normal" type (type 1): their core consists of the lipophilic chains of the amphiphilic ions and the hydrophilic polar groups face outwards towards the intermicellar aqueous solution. The isotropic aqueous solutions of the organic ions in the following will be designated L1.

Over a range of concentrations above the cmc (up to several times the cmc) the properties of the micelles formed by organic ions in most cases remain more or less constant: the micelles are spherical or close to spherical with a constant size and no drastic changes in counterion binding or the binding of water. At very high concentrations the micellar structure may undergo changes which lead to the formation of rod-like aggregates but also to an increasing polydispersity of the micelles. Two cases that have been studied in particular detail are hexadecyltrimethylammonium bromide (cetyltrimethylammonium bromide, CTAB) (Reiss-Husson & Luzzati; Götz & Heckmann; Scheraga & Backus; Ulmius & Wennerström; Persson & Lindman; Henriksson, Ödberg, Eriksson & Westman; Johansson, Lindblom & Nordén; Lindblom, Lindman & Mandell; Ekwall, Mandell & Solyom) and sodium dodecylsulfate (Mazer, Benedek & Carey; Young, Missel, Mazer, Benedek & Carey).

When the volume fraction of hydrated micelles exceeds a critical value a liquid crystal separates.

It is well known that the micelles are capable of incorporating (solubilizing) added lipophilic or amphiphilic compounds (solubilizates). This capability is best illustrated in three-component phase diagrams. The extent of the solution region may change in various ways depending on the type of solubilizate. Some typical cases are illustrated in Fig. 1.

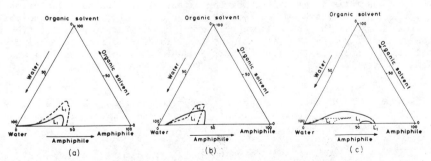

Fig. 1. Schematic diagrams showing the extent of the isotropic aqueous solution L1 in ternary systems of amphiphilic organic salt, water and solubilizates. Each diagram shows two different possibilities for the extent of region L1 (Ekwall, 1975, reproduced by permission from Academic Press, Inc.).

The solubilization of different compounds promotes micelle formation and also leads to changes in the properties of the micelles, e.g., their charge density, binding of counterions and shape. Addition of excess solubilizate in many cases also leads to the formation of lyotropic liquid crystals but in some cases the addition of solubilizate appears to stabilize the L1 solution which then will extend towards higher concentrations; in some systems the solution region may be continually extended to another L1 region in the ternary system (Fig. 1c). In such cases a liquid crystalline phase is formed in the region between the two solutions (Fig. 1c; an example of this behaviour is the system octylmethylammonium chloride/decan-1-ol/water, Ekwall, 1975).

THE ISOTROPIC SOLUTIONS IN ORGANIC SOLVENTS

Solutions of the organic salt and/or water in an organic solvent will be denoted by L2. The structure of the aggregates formed in such solutions has received extensive attention during the last decade due to the possibilities to dissolve large amounts of water in organic solvent by the addition of a surfactant (ionic or non-ionic). In most cases such water solubilization requires that the organic solvent itself is at least weakly amphiphilic; typical examples are the alkanols (C_4–C_{12}) and the aliphatic fatty acids.

The structure and association in these L2 solutions ranges from molecularly disperse solutions of the surfactant over the formation of submicellar aggregates to the formation of micelles of the reversed type (type 2) with a core composed of polar groups and water surrounded by the hydrocarbon chains of organic salt and by those solvent molecules that are included in the aggregate. The mechanisms by which these large aggregates are formed and stabilized is the subject of intensive research, and the regions of existence of such aggregates within the L2 solution region is still a matter of dispute. Some typical extensions of region L2 for different organic solvents in three-component systems are shown in Fig. 2.

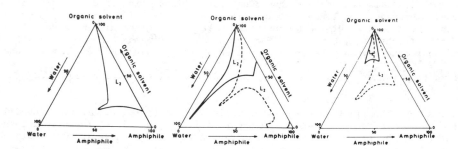

Fig. 2. Schematic diagrams showing the extent of the isotropic solution region L2 in ternary systems of amphiphilic organic salt, water and solubilizates. Diagrams (b) and (c) each represent two different possibilities for the extent of region L2; (a), (b) and (c) illustrate different amphiphiles and organic solvents (Ekwall, 1975, reproduced with permission from Academic Press, Inc.).

A detailed discussion of the various structures occurring in these L2 solutions is beyond the scope of this review. A comprehensive survey of the literature on surfactants in organic solvents is given by Kertes & Gutman.

We note, however, the following features.

(i) It is characteristic of solutions of crystalline surfactants in weakly amphiphilic solvents, of which only the alcohols have been investigated in detail, that the organic salt requires a minimum of water to go into solution; hydrated aggregates containing the solute and probably also the solvent are then formed. When the amount of water and/or organic salt is increased, inverted micelles are formed.

(ii) In other cases the organic salt dissolves in the solvent without the addition of water and then may be hydrated to form a ternary compound containing considerable amounts of water.

(iii) To both types of solution hydrocarbons or chlorinated hydrocarbons may be added. These solvents act as dispersion media between the aggregates formed by the organic salt and the weakly amphiphilic compound.

The ternary L2 solutions containing inverted micelles and quaternary solutions formed by the mechanism described in (iii) may contain large amounts of water in the core of the type 2 micelles. For this reason they are frequently called microemulsions, although this term is somewhat misleading since the L2 region is an isotropic, thermodynamically stable solution.

Region L1 (aqueous solution) and region L2 (solution in an organic solvent) may occur in the same amphiphilic system and in some cases form a continuous region extending from pure water to pure organic solvent. This continuity is often the result of the addition of the surfactant to water and an organic solvent which is not miscible with water in the binary system. Some typical cases are shown in Fig. 3. The extent of such continuous solution regions may be very sensitive to temperature and the nature of the surfactant, in particular in the case of non-ionic surfactants. The sensitivity may be considerably decreased by adding small amounts of an ionic compound (Friberg, Lapczynska & Gillberg, 1976).

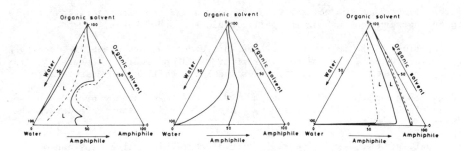

Fig. 3. Some examples of continuous solution regions in ternary systems of amphiphilic compound, organic solvent and water. Several different cases are illustrated in Fig.s (a) and (c), corresponding to different amphiphilic compounds and solvents (Ekwall, 1975, reproduced by permission from Academic Press, Inc.).

LYOTROPIC LIQUID CRYSTALLINE PHASES

In a review of the different types of lyotropic liquid crystalline phases formed by amphiphilic organic salts it is appropriate to distinguish between two types of such salts:

(i) amphiphiles whose solubility exceeds the cmc and form micelles (association colloids);

(ii) amphiphiles with a low solubility in water that are capable of taking up water by swelling.

The distinction between these two types is not a very fixed one: increasing the temperature in many cases shifts the character of the amphiphile towards that of a typical association colloid.

Lyotropic liquid crystalline phases are formed from solid crystalline or liquid association colloids or swelling amphiphiles by the addition of water and, in some cases, also by the addition of an organic solvent. They may also be formed from aqueous solutions of association colloids by three processes:

(i) by separation of a liquid crystalline phase when the solubility of the association colloid in water is exceeded;

(ii) by the addition of an amphiphilic or lipophilic compound that is solubilized in the micelles; when the solubilization limit is exceeded lyotropic liquid crystalline phases may separate at much lower concentrations than in case (i);

(iii) by the formation of aggregates between a solubilizate and the amphiphilic ion below the cmc.

Finally, lyotropic liquid crystalline phases may also be formed from organic solution through an increase in the concentration of organic salt or the organic salt and water above the solubility limit.

In each of the systems the different structures follow each other in a definite sequence, which is governed by the ratio water:amphiphile and also, when there is more than one amphiphilic or lipophilic compound present, by the proportion between these. In general, the amphiphilic molecules arrange themselves together so that the number of water molecules that are separated from each other by the hydrocarbon parts is minimized and the contact between the hydrophilic groups and water is maximized. In addition, the spatial relationships are of great importance to this process of association and aggregation. The most important factors governing the stepwise formation of different lyotropic liquid crystalline structures are:

(i) the interaction between the polar groups of the molecules and the water;

(ii) the structure of the amphiphilic molecules;

(iii) the interaction among the polar groups themselves and the binding of counterions in the boundary regions between aqueous and lipophilic regions of the phases.

A thorough discussion of the importance of these different factors has been given by Ekwall (1975). The general evidence for the conclusions is based on studies of phase equilibria, the thermodynamic properties of the micellar phases and X-ray diffraction measurements.

The detailed evidence on a molecular basis, however, have been largely provided by spectroscopic studies, of which the NMR methods have played a very important role during the last years.

Similarly to micelles, a number of lyotropic liquid crystalline structures can occur in two complementary types, the normal type 1 and the reversed type 2. Type 1 phases are composed of amphiphilic aggregates with a hydrocarbon core and a layer of polar groups in the boundary against the surrounding water which forms a continuum. Type 2 consists of aggregates with a core of unhydrated or hydrated polar groups and water, surrounded by layers of hydrocarbon chains, i.e., the hydrocarbon moiety forms the continuum. In addition to these types there exists lamellar liquid crystalline phases composed of bimolecular layers of amphiphilic molecules with the hydrocarbon chains towards one another and separated by intervening layers of water molecules. The type 1 and type 2 phases may be either optically anisotropic (types E, F; Fig. 4) or isotropic. In the latter case the structure is often not yet definitely established.

LYOTROPIC LIQUID CRYSTALS IN BINARY SYSTEMS

Consider the gradual reduction of the water content of an aqueous L1 solution of an association colloid until a concentration region is reached in which a lyotropic liquid crystal separates. In the case of ionic surfactants this phase is usually the "middle phase" which we will denote by E. This phase is the earliest reported lyotropic crystalline phase and it is well established that its structure (see Fig. 4) consists of parallel amphiphilic rods in a hexagonal array (Husson, Mustacchi & Luzzati; Luzzati & Husson; Luzzati, Mustacchi, Skoulios & Husson).

As the content of H_2O molecules is further decreased another structure is formed, the so-called "soap-boilers' neat phase" (phase D) with a lamellar structure (Husson, Mustacchi & Luzzati; Luzzati & Husson; Luzzati, Mustacchi, Skoulios & Husson). This phase also has been known for a long time and the phase equilibria involving the E and D phases for fatty acid soaps were established many years ago by McBain and coworkers (McBain, Lazarus & Pitter; McBain & Field; McBain & Lee; McBain & Sierichs). It is schematically represented in Fig. 5.

In the concentration region between these two phases D and E other phases have been reported: a "rectangular" phase R and a "complex hexagonal" phase H_c.

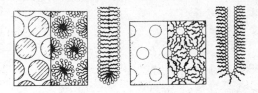

Fig. 4. Schematic representation of the structure of lyotropic liquid crystalline phases with rod-like structures: (a) type 1 (phase E); (b) type 2
(phase F).

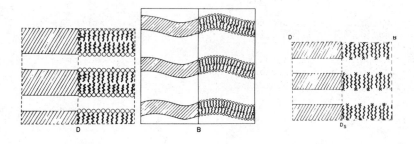

Fig. 5. Schematic representation of the structures of lamellar liquid crystalline phases D and B.

When water is added to swelling amphiphiles concentration regions in which lyotropic liquid
crystals are encountered often occur. In this case, however, the first phase to be formed as
the water content increases often is a hexagonal phase of type 2 (see Fig. 4a) which we denote
by F. The structure of this phase is also well established by X-ray diffraction measurements
(Ekwall, Mandell & Fontell, 1968a, 1969a; Fontell, Mandell, Ekwall & Danielsson, 1962; Fontell,
Mandell, Lehtinen & Ekwall; Luzzati; Luzzati & Husson; Luzzati, Mustacchi, Skoulios & Husson).
At higher concentrations of water the swelling leads to the formation of a phase with structure D.

The phases with structures D, E and F all are optically anisotropic. In addition to these, a
number of optically isotropic liquid crystalline phases occur which may also be of type 1 or
type 2 and are denoted by I_1 and I_2, respectively.

Isotropic liquid crystalline phases of type I_1 occur in the concentration region between mesophases D and E and between E and solution region L1. Their region of existence is usually
rather limited. From the X-ray patterns it is evident that the phases possess cubic symmetry.
For the isotropic phases occurring between the middle and neat phases X-ray investigations have
shown that the structure is composed of short rod-like elements of finite length, all identical
and crystallographically equivalent and joined in threes at each end to form two three-dimensional interlacing but unconnected networks (Luzzati & Spegt; Luzzati, Gulik – Krzywicki & Tardieu; Luzzati, Tardieu & Gulik – Krzywicki; Ekwall, 1975). This structure which is denoted I_1'
is shown in Fig. 6. It is obvious, however, that all isotropic phases of type I_1 cannot have
this structure; this is the case for the I_1 phases occurring in the region between L1 solutions
and the middle (E) phase. Such phases are denoted by I_1'' .

Isotropic liquid crystalline phases of type 2 (denoted by I_2') occur in the region between the
D and F phases. They also evidently have a cubic symmetry. For some systems, at least, their

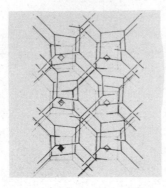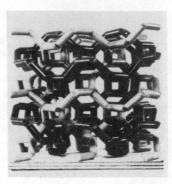

Fig. 6. The structure of the optically isotropic mesophase $I_1^!$. (a) Perspective picture in which the dotted lines show the projections of the solid lines on the basal plane (Luzzati, Gulik-Krzywicki & Tardieu). (b) Visualization of the structure of the phase (Fontell). (Reproduced by permission of Academic Press, Inc.).

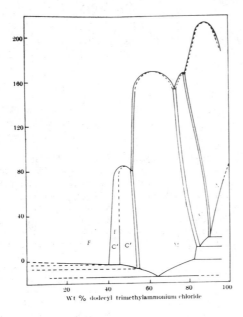

Wt % dodecyl trimethylammonium chloride

Fig. 7. Phase diagram for the binary system dodecyltrimethylammonium chloride/water. F = Ll, aqueous solution. C" = $I_1^{\prime\prime}$, optically isotropic phase of type 1. M = E, type 1, middle phase. C = $I_1^!$, optically isotropic phase of type 1. N = D, lamellar phase (Balmbra, Clunie & Goodman, 1969, reproduced with permission from McMillan & Co.).

structure is similar to that of the phases $I_1^!$, i.e., they would consist of two interwoven networ with the short rods consisting of amphiphilic molecules with their polar end groups and water in the core of the rods (Ekwall, Mandell & Fontell, 1970; Fontell).

The temperature ranges of the most commonly occurring lyotropic liquid crystalline phases in

binary water/amphiphile systems have been examined quite extensively. The studies were often con-
ducted in sealed tubes and hence the total pressure of the system has varied with temperature.
Usually these pressures have not been recorded. The neat phases (D) of the fatty acid alkali
soaps usually are stable up to 250-350 °C and the middle phases (E) up to 130-220 °C. At higher
temperatures they are transformed to an isotropic solution. A typical phase diagram for a binary
mixture of water and an association colloid is shown in Fig. 7.

LYOTROPIC LIQUID CRYSTALS IN TERNARY SYSTEMS

If an organic compound is added to the binary systems considered above they behave quite differently depending on the nature of the organic compound. If the solubilizate is a lipophilic
compound (hydrocarbon or chlorinated hydrocarbon) the L1 solution and the E and D phases usually are able to incorporate small amounts of the solubilizate; the saturated phases are usually
in equilibrium with almost pure solubilizate. In the concentration region between saturated L1
phase and homogeneous E phase the addition of hydrocarbon or chlorinated hydrocarbon frequently
leads to the formation of isotropic liquid crystalline phases of type I_1'' (Ekwall, 1975; Ekwall,
Mandell & Fontell, 1969b). A typical system showing such a phase diagram - sodium octanoate/
p-xylene/water - is shown in Fig. 8.

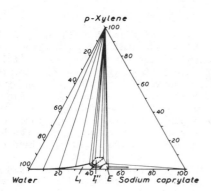

Fig. 8. Phase diagram for the three-component system sodium octanoate/
p-xylene/water at 20 °C (Ekwall, 1975, reproduced with permission from
Academic Press, Inc.).

If the solubilizate is a weakly polar or a non-polar compound (e.g., a fatty acid or an aliphatic alcohol) the L1 solutions as well as the E and D phases are able to incorporate much
larger amounts of the solubilizate. When such a solubilizate is added in excess to a L1 solution at concentrations above the cmc a lamellar phase of type D usually precipitates. The
structure of this phase which may contain very large amounts of water is extremely well established also in the case of three-component systems through X-ray investigations and detailed
through, above all, NMR spectroscopy.

When the solubilizate is added to solutions below the cmc another phase containing amphiphile
aggregates of lamellar structure separates. This phase is denoted B by Ekwall and coworkers.
This phase is slightly cloudy and has a low viscosity; X-ray diffraction studies to establish
its structure are difficult due to the extremely large content of water but indicate a lamellar structure.

In addition to phases B and D the separation of an additional liquid crystalline phase denoted
C from micellar L1 solutions has been reported.This is the case in the system sodium octanoate/
decan-1-ol/water which has been investigated in the greatest detail of all ternary systems

(Fontell, Mandell, Lehtinen & Ekwall). It is shown in Fig. 9.

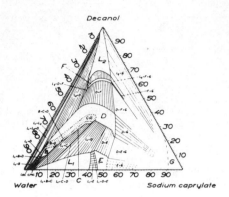

Fig. 9. Phase diagram for the three-component system sodium octanoate/
decan-1-ol/water at 20 °C

In earlier reports, it was suggested that phase C consists of rods of amphiphiles with a rectangular cross section and intercalating water. In subsequent studies of the structure of this
phase by NMR spectroscopy it has been clearly established that the quadrupole splitting of
deuterons in D_2O that replaces H_2O in this phase is incompatible with such a structure. Indeed,
it has been suggested that phase C, which has a milky appearance, is a dispersion of phase D
in an L1 solution (Tiddy; Persson, Fontell, Lindman & Tiddy). Recent detailed studies of the
X-ray diffraction of samples in the D and C regions in the above mentioned system appear to
confirm that the C phase does not exist as a separate phase but also indicate that the phase
boundaries towards high water contents of D phases in systems where the C phase has been reported probably should be shifted towards higher water contents (Danielsson, Friman & Stenius).

The inverted hexagonal phase F that is formed in a few known cases of binary systems of water
and organic salt (it is also formed by the swelling of several uncharged lipids: Lutton; Luzzati; Luzzati,Tardieu, Gulik-Krzywicki, Rivas & Reiss-Husson) may also be formed in ternaries. In
binary systems, this phase is formed by amphiphilic molecules with a bulky hydrocarbon part on
relatively small polar groups {e.g., di(2-ethylhexyl)sulfosuccinate, *Aerosol OT*}. In ternary
systems some cases are known in which a similar result is obtained by adding a hydrocarbon to
a micellar solution of cationic surfactants. Phase F may also be formed by anionic surfactants
when a large amount of an amphiphilic compound (aliphatic alcohol, fatty acid) is added. In
such cases it is observed in equilibrium with a micellar solution of type L2 in analogy with
the formation of phase E from micellar L1 solutions. As is the case with the formation of L2
solutions of water and surface active organic salt, a minimum of water in such cases is required
for the formation of phase F. The H_2O molecules are bound to the polar groups of the surfactant
and the amphiphile. These appear to be able to associate to form the rod-like cores of the amphiphile aggregates in the F phase only if hydrated. Examples are given in Fig. 9 (sodium octan-
oate/ decan-1-ol/ water) and Fig. 10 (potassium oleate/ decan-1-ol/ water).

In addition to these phases, isotropic liquid crystalline phases of the inverted type 2 (I_2)
have been observed in a number of systems.

In the system potassium decanoate/decan-1-ol/water an optically anisotropic phase which is a
stiff, fairly clear gel has been observed in the region between phase D and L2 at low water
contents. This phase exhibits Bragg spacings in the ratio 1:1/2:1/3 in contrast to the ratio
$1:1/\sqrt{3}:1/\sqrt{4}$ for the F phase but can be separated from the D phase. It has been denoted K.

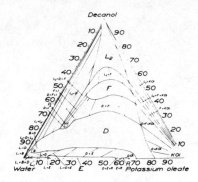

Fig. 10. Phase diagram for the system potassium oleate/decan-1-ol/water at
20 °C (Ekwall, Mandell & Fontell, 1969a, reproduced with permission from
Academic Press, Inc.).

TABLES

Table 1 gives a summary of the phase structures that have been discussed together with notations
used by different authors. The phases that have been observed in different systems are summarized in Tables 2, 3 and 4.

ACKNOWLEDGEMENT

The assistance of Mrs Lena Käll, The Swedish Institute for Surface Chemistry, in the compilation of data for this chapter is gratefully acknowledged.

TABLE 1. Lyotropic liquid crystalline phases occurring in systems containing ionic surfactants

Designation	Type	Basic structure	Description of the supposed structure	Optical properties	Alphabetical notation Luzzati	Winsor	Ekwall, 1975

Structural arrangement displaying Bragg spacing ratio 1:1/2:1/3
  Layer structures

1. Neat phase		Lamellar, double layers	Coherent double layers of amphiphile molecules with intervening layers of water molecules	Anisotropic	L LL L_α	G	D
2. Single layered neat phase		Lamellar, single layers	Coherent single layers of amphiphile molecules oriented with the polar groups towards the intervening layers of water molecules	Anisotropic			D_s
3. Muscous woven phase		Lamellar, double layers	Coherent double layers of amphiphile molecules with intervening layers of water molecules	Slightly anisotropic			B

Other structures

4. White phase		Lamellar	Probably dispersion of D in L1	Anisotropic			C
5. Rectangular phase	1	Two-dimensional orthorhombic	Indefinitely long, mutually parallel rods in orthorhombic array	Anisotropic	R		R
6.			Stiff viscous gel	Anisotropic			K

Structures of indefinitely long rods
Structural arrangement displaying Bragg spacing ratio 1:1/√3:1/√4

7. Middle phase, normal	1	Two-dimensional hexagonal	Indefinitely long, mutually parallel rods in hexagonal array, with a hexagonal or circular cross section; the rods consist of radially arranged amphiphilic molecules	Anisotropic	H_1	M_1	E
8. Middle phase, reversed	2	Two-dimensional hexagonal		Anisotropic	H_2	M_1	F
9. Hexagonal complex phase, normal	1	Two-dimensional hexagonal	Indefinitely long, mutually parallel rods in hexagonal array	Anisotropic	H_c		H_c

TABLE 1 (Continued)

Structural arrangement displaying cubic symmetry

No.	Optical arrangement		Structure				
10.	Optical isotropic mesophase, normal "Viscous isotropic phase", normal	1	Short rod-like elements (axial ratio near 1) joined in threes at each end to a three-dimensional network: two networks interwoven but unconnected	Cubic, body centred Space group $Ia3d$	Isotropic	Q	V_1 I_1
11.	Optical isotropic mesophase, reversed "viscous isotropic phase", reversed	2			Isotropic	Q	V_2 I_2
12	Optical isotropic mesophase, normal	1	Possibly, spherical aggregates caged within a network of short rod-like elements joined in threes at one end and in fours at the other end	Cubic, body centred Possibly, space group $Pm3n$	Isotropic	S_{1c}	I_1''
13.	Optical isotropic mesophase, reversed	2	Structure unknown	Cubic, space group unknown	Isotropic		I_2''

Optically isotropic liquid phases are designated S (by Winsor) and L (by Ekwall); S_1 and L_1 denote aqueous solutions, S_2 and L_2 solutions in organic solvents.

TABLE 2. Liquid crystalline phases occurring at decreasing water content in binary systems of ionic amphiphiles and water [a]

Salt	T / K	L_1 [b]	Liquid crystalline phases					L_1' [b]	Liq. cryst. phases			L_2	Ref.
			I_1''	E	R	H_c	I_1'		D	I_2'	F		
With organic anion													
Na octanoate	293	x	–	x	–	–	–	–	–	–	–	–	1,2
K octanoate	293	x	–	x	–	–	x	–	–	–	–	–	3
NH₄ perfluorooctanoate	298	x	–	–	–	–	–	–	x	–	–	–	1
Na octylsulfate	373	x	–	x	–	–	x	–	–	–	–	–	4,5
Na octylsulfate	293	x	–	x	–	–	–	–	–	–	–	–	2,6
Na 2-ethylhexylsulfate	293	x	–	x	–	–	x	–	x	–	–	–	7
Na octylsulfonate	293	x	–	–	–	–	–	–	–	–	–	–	2,6

TABLE 2 (Continued)

Compound	T/K												Ref.
Na 2-ethylhept-6-enylsulfate	293	x	–	–	–	–	–	–	–	–	–	–	8
K decanoate	293	x	–	x	–	–	–	–	–	–	–	–	3
Na dodecanoate	373	x	–	x	–	–	–	–	–	–	–	–	4,5
K dodecanoate	373	x	–	x	x	x	–	–	–	–	–	–	4,5,9
NH$_4$ dodecanoate	318	x	–	–	–	–	–	–	–	–	–	–	10
Na dodecylsulfate	348	x	–	x	–	x	–	–	–	–	–	–	4,5
Na tetradecanoate	373	x	–	x	–	x	–	–	–	–	–	–	4,5
K tetradecanoate	373	x	–	x	x	x	–	–	–	–	–	–	4,5,9
Rb tetradecanoate	359	x	–	x	–	x	–	–	–	–	–	–	11
Na di(1,3-dimethylbutyl)succinate	293	x	–	x	–	–	–	–	–	–	–	–	4,5
(Aerosol MA)													
Na hexadecanoate	373	x	–	x	–	x	–	–	–	–	–	–	4,5
K hexadecanoate	373	x	–	x	x	x	–	–	–	–	–	–	4,5,9
K di(2-ethylhexyl)acetate	293	x	–	x	–	–	x	–	x	x	x	x ^{c}	12
Na di(2-ethylhexyl)sulfosuccinate	293	x	–	x	–	–	x	x	x	x	x	–	13
(Aerosol OT)													
Na octadecanoate	373	x	–	x	–	x	–	–	–	x	–	–	4,5
K octadecanoate	373	x	–	x	–	x	x	–	–	x	–	–	4,5,9
Na 9.10-octadecenoate *(cis)* (oleate)	338	x	–	x	–	x	x	–	x	x	–	–	4,5
K 9.10-octadecenoate *(cis)*	293	x	–	x	–	x	–	–	x	x	–	–	4,5,6
K eicosanoate	359	x	–	x	–	x	–	–	x	x	–	–	11
K docosanoate	359	x	–	x	–	x	–	–	x	x	–	–	11
Rb docosanoate	359	x	–	x	–	x	–	–	x	x	–	–	11
Cs docosanoate	359	x	–	x	–	x	–	–	–	x	–	–	11
With organic cation													
Octylammonium Cl	293	x	–	x	–	x	–	x	–	x	–	–	2,6
Octylmethylammonium Cl	293	x	–	x	–	x	–	x	–	x	–	–	6

TABLE 2 (Continued)

	T										Ref.
Octyldimethylammonium Cl	293	x	-	x	-	-	-	x	-	-	6
Octyltrimethylammonium Cl	293	x	-	x	-	-	-	-	-	-	6
Dodecylammonium Cl	333	x	-	x	-	x	-	x	-	-	14
Dodecylmethylammonium Cl	333	x	-	x	-	-	-	x	-	-	14
Dodecyldimethylammonium Cl	333	x	x	x	-	-	-	x	-	-	14
Dodecyltrimethylammonium Cl	333	x	-	x	x	-	x	x	-	-	15,16
Tetradecyl-trimethylammonium Cl	298	x	-	-	-	-	-	-	-	-	17
	343	x	-	x	x	-	x	x	-	-	4,5

a: given in order of decreasing water content; x denotes that a phase occurs

b: L_1, L_1' : different L_1 solutions, separated by regions in which liquid crystalline phases occur (see, e.g., Fig. 1c)

c: potassium di(2-ethylhexyl)acetate forms a L_2 solution at higher water content than that of mesophase F

1. Ekwall, Danielsson & Mandell
2. Ekwall (1965)
3. Ekwall, Mandell & Fontell (1969a)
4. Husson, Mustacchi & Luzzati
5. Luzzati, Mustacchi, Skoulios & Husson
6. Ekwall, Mandell & Fontell (1969b)
7. Balmbra, Clunie & Goodman (1965)
8. Balmbra, Clunie & Goodman (1967)
9. Luzzati, Mustacchi & Skoulios
10. Tamamushi
11. Gallot & Skoulios
12. Rogers & Winsor
13. Ekwall, Mandell & Fontell (1970)
14. Brown, Doane & Neff
15. Broome, Hoerr & Harwood
16. Balmbra, Clunie & Goodman (1969)
17. Ekwall, Mandell & Fontell (1969d)

TABLE 3. Lyotropic liquid crystalline phases occurring in ternary systems of an ionic amphiphile (component 1), an organic solubilizate (component 2) and water (component 3) [a]

Components 1 and 2	T /K	Sol=ution L₁			Liq. cryst. phases (type 1)					D	Liq.cr.ph.(type 2)			Sol=ution L₂	Ref.
		B	L₁	C	I₁''	E	R	H_c	I₁'		I₂	F	K		
Component 1 with organic anion															
Non-polar solubilizate															
Na octanoate/ p-xylene	293	x		–	x	x	–	–	–	–	–	–	–	–	1,2
Na octanoate/ octane	293	x		–	x	x	–	–	–	–	–	–	–	–	1,2
Na octanoate/ monochlorooctane	293	x		–	x	x	–	–	–	–	–	–	–	–	1,2
Na octanoate/ tetrachloro= methane	293	x		–	x	x	–	–	–	–	–	–	–	–	1,2
Na tetradecanoate/ ethylbenzene	355	x		–	x	x	–	x	–	x	–	–	–	–	3
K oleate/ p-xylene	293	x		–	–	x	x	–	–	x	–	x	–	–	1
Na di(2-ethylhexyl)sulfosuc= cinate (Aerosol OT)/ p-xylene	293	x		–	x	–	–	–	–	x	x	x	x	–	4,5
Lipophilic solubilizate with weakly hydrophilic group															
Na octanoate/ octylnitrile	293	x		–	–	x	–	–	–	x	–	–	–	–	1,2
Na octanoate/ octylaldehyde	293	x		–	–	x	–	–	–	x	–	–	–	–	1,2
Na octanoate/ methyl octanoate	293	x		–	–	x	–	–	–	x	–	–	–	–	1,2
Na octanoate/ cholesterol	293	x		–	–	x	–	–	–	x	–	–	–	–	6
Na xylenesulfonate/monocapryline	293	x		–	–	–	–	–	–	x	–	–	–	–	7
Na xylenesulfonate/ lecithine	293	x		–	–	–	–	–	–	x	–	–	–	–	7
Na oleate/ lecithine	293	x		–	–	x	x	x	–	x	–	–	–	–	8
Na stearate/ lecithine	293	x		–	–	x	x	x	–	x	–	–	–	–	8
Na cholate/ lecithine	298	x		–	–	x	–	–	x	x	–	–	–	–	9
Na cholate/ monolaurine	278	x		–	–	–	–	–	–	x	–	–	–	–	10
Solubilizate: alcohol															
Na octanoate/ methanol	293	x[b]		–	–	–	–	–	–	–	–	–	–	x[b]	1
Na octanoate/ ethanol	293	x[b]		–	x	x	–	–	–	x	–	–	–	x[b]	1

TABLE 3 (Continued)

Compound	T/K												Ref.
Na octanoate/ propan-1-ol	293	x[b]	–	–	–	x	–	x	–	–	–	x[b]	1
Na octanoate/ butan-1-ol	293	x	–	–	–	x	–	x	–	–	–	x	1
Na octanoate/ pentan-1-ol	293	x	–	–	–	x	–	x	–	–	–	x	1,2
Na octanoate/ hexan-1-ol	293	x	x	–	x	x	–	x	–	–	–	x	1,2
Na octanoate/ heptan-1-ol	293	x	x	–	x	x	–	x	–	–	x	x	1
Na octanoate/ octan-1-ol	293	x	x	–	x	x	–	x	–	–	x	x	1,2
Na octanoate/ nonan-1-ol	293	x	x	–	x	x	–	x	x	–	x	x	1,2
Na octanoate/ decan-1-ol	293	x	x	–	x	x	–	x	x	–	x	x	11,12
Na octanoate/ 1,8-octanediol	293	x	x	–	x	x	–	x	x	–	–	–	13,14
Na octanoate/ 1,2-ethanediol	293	x[b]	–	–	x	–	–	x	–	–	–	x[b]	15
Na octanoate/ tetraethylene glycol	293	x[b]	–	–	x	–	–	x	–	–	–	x[b]	15
Na octanoate/ glycerol	293	x[b]	–	–	x	–	–	x	–	–	–	x[b]	15
Na octanoate/ 2,5-dimethyl=phenol	293	x	–	–	–	x	–	x	–	–	–	–	1
Na octanoate/ 2-naphthol	293	x	–	–	x	–	–	x	–	–	–	–	1
K octanoate/ octan-1-ol	293	x	x	–	x	–	–	x	–	–	x	x	16
K octanoate/ decan-1-ol	293	x	x	–	x	–	x	x	–	–	x	x	16
Na octylsulfate/ decan-1-ol	293	x	x	–	x	–	–	x	–	–	x	x	1,2
Na octylsulfonate/ decan-1-ol	293	x	–	–	–	–	–	x	–	–	–	x	1,2
NH4 perfluorooctanoate/ octan-1-ol	298	x	x	–	–	–	–	–	–	–	x	x	17
K decanoate/ octan-1-ol	293	x	x	–	x	–	–	x	–	–	x	x	16
Na dodecanoate/ o-hydroxy=toluene		x	x	–	–	–	–	x	–	–	–	x	18
NH4 dodecanuoate/ octan-1-ol	313	x	–	–	–	–	–	x	–	–	–	x	19
Na dodecylsulfate/ decan-1-ol	298	x	–	–	–	–	–	x	–	–	–	x	1
K oleate/ decan-1-ol	293	x	x	x	x	–	–	x	–	–	x	x	16
Na di(2-ethylhexyl)sulfosuc=cinate/ decan-1-ol	293	x	–	–	x	–	x	–	–	x	x	x	5

TABLE 3 (Continued)

Component	T (K)											Ref.
Solubilizate: fatty acid												
Na octanoate/ octanoic acid	293	x	x	x	–	–	–	–	x	x	x	1,6,20
K dodecanoate/ dodecanoic acid	373	x	–	–	–	–	–	–	x	x	x	21
	448	x	–	–	–	–	–	–	x	x	x	21
	453	x [b]	–	–	–	–	–	–	x	x	x	21
Na dodecylsulfate/ decanoic acid	298	x	–	–	–	–	–	–	–	x	x	22
K oleate/ oleic acid	298	x	–	–	–	–	–	–	x	x	x	18
Na di(2-ethylhexyl)sulfosuc= cinate/ octanoic acid	293	x	–	–	–	–	–	x	x	x	x	4,5
Component 1 with organic cation												
Non-polar solubilizate												
Octylammonium Cl/ p-xylene	293	x	–	–	–	–	–	–	x	x	x	1
Solubilizate: alcohol												
Octylammonium Cl/ decan-1-ol	293	x	x	–	–	–	–	–	x	x	x	1
Octylmethylammonium Cl/ decan-1-ol	293	x	–	–	–	–	–	–	x	x	x	1
Octyldimethylammonium Cl/ decan-1-ol	293	x	–	–	x	–	–	–	x	x	x	1
Octyltrimethylammonium Cl/ decan-1-ol	293	x	–	–	x	–	–	–	x	x	x	1
Cetyltrimethylammonium Br/ hexan-1-ol	298	x	–	–	–	–	–	–	x	x	x	23

a: given in order of increasing content of or anic solubilizate and decreasing content of water
b: "solubilizate" completely miscible with water

1. Ekwall (1975)
2. Ekwall, Mandell & Fontell (1969b)
3. Spegt, Skoulios & Luzzati
4. Fontell
5. Ekwall, Mandell & Fontell (1970)
6. Ekwall & Mandell
7. Friberg & Rydhag
8. Gulik-Krzywicki, Tardieu & Luzzati

9. Small & Bourges
10. Lawrence, Boffey, Bingham & Talbot
11. Ekwall, Danielsson & Mandell
12. Fontell, Mandell, Lehtinen & Ekwall
13. Ekwall, Mandell & Fontell (1968b)
14. Ekwall, Mandell & Fontell (1969c)
15. Ekwall, Mandell & Fontell (1968c)
16. Ekwall, Mandell & Fontell (1969a)

17. Tiddy & Wheeler
18. McBain & Stewart
19. Tamamushi
20. Ekwall (1965)
21. McBain & Field (1933)
22. Lawrence
23. Ekwall, Mandell & Fontell (1969d)

TABLE 4. Lyotropic liquid crystalline phases occurring in ternary systems of an ionic amphiphile (component 1), an inorganic electrolyte (component 2) and water (component 3) a

Components 1 and 2	$\frac{T}{K}$	Solution	Liquid crystalline phases		References
		L_1	E	D	
Na octanoate/ Na chloride	453	x	x	–	1,2
Na dodecanoate/ Na chloride	363	x	x	x	2,3
K dodecanoate/ K chloride	291	x	x	?	4
	363	x	x	x	4

a: given in order of decreasing water content; x denotes that a phase occurs

1. Ekwall, Mandell & Fontell (1977)
2. McBain, Brock, Vold & Vold
3. McBain, Thornburn & McGee
4. McBain & Field (1926)

REFERENCES

BALMBRA, R.R., CLUNIE, J.S., and GOODMAN, J.F., *Proc. Roy. Soc.*, A285, 534 (1965)

BALMBRA, R.R., CLUNIE, J.S., and GOODMAN, J.F., *Mol. Cryst.*, 3, 281 (1967)

BALMBRA, R.R., CLUNIE, J.S., and GOODMAN, J.F., *Nature*, 222, 1159 (1969)

BROOME, F.K., HOERR, C.W., and HARWOOD, H.J., *J. Amer. Chem. Soc.*, 73, 3350 (1951)

BROWN, G.A., DOANE, J.W., and NEFF, V.D., *"A Review of the Structure and Physical Properties of Liquid Crystals"*, Chem. Rubber Pub. Co., Cleveland, 1971

DANIELSSON, I., FRIMAN, R., and STENIUS, P., *to be published*

EKWALL, P., *Wiss. Z. Friedrich-Schiller-Univ. Jena, Math.-Naturwiss. Reihe*, 14, 181 (1965)

EKWALL, P., *Adv. Liquid Cryst.*, 1, 1 (1975)

EKWALL, P., DANIELSSON, I., and MANDELL, L., *Kolloid-Z.*, 169, 113 (1960)

EKWALL, P., and MANDELL, L., *Acta Chem. Scand.*, 15, 1403 (1961)

EKWALL, P., MANDELL, L., and FONTELL, K., *Acta Chem. Scand.*, 22, 697 (1968)

EKWALL, P., MANDELL, L., and FONTELL, K., *Acta Chem. Scand.*, 22, 365 (1968)

EKWALL, P., MANDELL, L., and FONTELL, K., *J. Colloid Interface Sci.*, 28, 219 (1968)

EKWALL, P., MANDELL, L., and FONTELL, K., *J. Colloid Interface Sci.*, 31, 508 (1969)

EKWALL, P., MANDELL, L., and FONTELL, K., *Mol. Cryst. Liquid Cryst.*, 8, 157 (1969)

EKWALL, P., MANDELL, L., and FONTELL, K., *J. Colloid Interface Sci.*, 29, 542 (1969)

EKWALL, P., MANDELL, L., and FONTELL, K., *J. Colloid Interface Sci.*, 29, 639 (1969)

EKWALL, P., MANDELL, L., and FONTELL, K., *J. Colloid Interface Sci.*, 33, 215 (1970)

EKWALL, P., MANDELL, L., and FONTELL, K., *J. Colloid Interface Sci.*, 61, 519 (1977)

EKWALL, P., MANDELL, L., and SOLYOM, P., *J. Colloid Interface Sci.*, 35, 519 (1971)

FONTELL, K., *J. Colloid Interface Sci.*, 43, 156 (1973)

FONTELL, K., MANDELL, L., EKWALL, P., and DANIELSSON, I., *Acta Chem. Scand.*, 16, 2294 (1962)

FONTELL, K., MANDELL, L., LEHTINEN, H., and EKWALL, P., *Acta Polytech. Scand. Chem.*, 74, III,1, (1968)

FRIBERG, S., LAPCZYNSKA, I., and GILLBERG, G., *J. Colloid Interface Sci.*, 56, 19 (1976)

FRIBERG, S., and RYDHAG, L., *J. Amer. Oil Chem. Soc.*, 48, 113 (1970)

GALLOT, B., and SKOULIOS, A., *Kolloid-Z.*, 208, 37 (1966)

GÖTZ, K.G., and HECKMANN, K., *J. Colloid Sci.*, 13, 266 (1958)

GULIK-KRZYWICKI, T., TARDIEU, A., and LUZZATI, V., *Mol. Cryst. Liquid Cryst.*, 8, 285 (1969)

HENRIKSSON, U., ÖDBERG, L., ERIKSSON, J.C., and WESTMAN, L., *J. Phys. Chem.*, 81, 76 (1977)

HUSSON, F., MUSTACCHI, H., and LUZZATI, V., *Acta Cryst.*, 13, 668 (1960)

JOHANSSON, L., LINDBLOM, G., and NORDEN, B., *Chem. Phys. Lett.*, 39, 128 (1976)

KERTES, A.S., and GUTMAN, H.L., in *Surface and Colloid Science*, ed. by Matijević, J. Wiley, New York, 8, 193 (1975)

LAWRENCE, A.S.C., *Nature*, 183, 1941 (1959)

LAWRENCE, A.S.C., BOFFEY, B., BINGHAM, B., and TALBOT, K., *Chem. Phys. Appl. Surface Active Subst. Proc. Int. Congr. 4th, 1964*, 4, 613

LINDBLOM, G., LINDMAN, B., and MANDELL, L., *J. Colloid Interface Sci.*, 42, 400 (1973)

LUTTON, E.S., *J. Amer. Oil Chem. Soc.*, 42, 1068 (1965)

LUZZATI, V., in *"Biological Membranes"*, ed. by Chapman, D., Academic Press, New York, 1968, 71

LUZZATI, V., GULIK-KRZYWICKI, T., and TARDIEU, A., *Nature*, 218, 1031 (1968)

LUZZATI, V., and HUSSON, F., *J. Cell Biol.*, <u>12</u>, 207 (1962)

LUZZATI, V., MUSTACCHI, H., and SKOULIOS, A., *Disc. Faraday Soc.*, <u>25</u>, 43 (1958)

LUZZATI, V., MUSTACCHI, H., SKOULIOS, A., and HUSSON, F., *Acta Cryst.*, <u>13</u>, 660 (1960)

LUZZATI, V., and SPEGT, A., *Nature*, <u>215</u>, 701 (1967)

LUZZATI, V., TARDIEU, A., and GULIK-KRZYWICKI, T., *Nature*, <u>217</u>, 1028 (1968)

LUZZATI, V., TARDIEU, A., GULIK-KRZYWICKI, T., RIVAS, E., and REISS-HUSSON, F., *Nature*, <u>220</u>, 485 (1968)

MAZER, N.A., BENEDEK, G.B., and CAREY, M.C., *J. Phys. Chem.*, <u>80</u>, 1075 (1976)

McBAIN, J.W., BROCK, G., VOLD, R.D., and VOLD, M.J., *J. Amer. Chem. Soc.*, <u>60</u>, 1870 (1938)

McBAIN, J.W., and FIELD, M.C., *J. Phys. Chem.*, <u>30</u>, 1545 (1926)

McBAIN, J.W., and FIELD, M.C., *J. Amer. Chem. Soc.*, <u>55</u>, 4776 (1933)

McBAIN, J.W., LAZARUS, L.H., and PITTER, A.V., *Z. physik. Chem.*, <u>147</u>, 87 (1930)

McBAIN, J.W., and LEE, W.W., *Oil and Soap*, <u>20</u>, 17 (1943)

McBAIN, J.W., and SIERICHS, W.C., *J. Amer. Oil Chem. Soc.*, <u>25</u>, 221 (1948)

McBAIN, J.W., and STEWART, R., *Disc. Faraday Soc.*, <u>18</u>, 119 (1954)

McBAIN, J.W., THORNBURN, R.C., and McGEE, C.G., *Oil and Soap*, <u>21</u>, 227 (1944)

PERSSON, N.O., FONTELL, K., LINDMAN, B., and TIDDY, G., *J. Colloid Interface Sci.*, <u>53</u>, 471 (1975)

PERSSON, N.O., and LINDMAN, B., *J. Phys. Chem.*, <u>79</u>, 1410 (1975)

REISS-HUSSON, F., and LUZZATI, V., *J. Phys. Chem.*, <u>68</u>, 3504 (1964)

ROGERS, J., and WINSOR, P.A., *Chim. Phys. Appl. Prat. Ag. Surface*, *C.R. Congr. Int. Déterg. 5e*, <u>2</u>, 933 (1968)

SCHERAGA, H.A., and BACKUS, J.K., *J. Amer. Chem. Soc.*, <u>73</u>, 5108 (1951)

SMALL, D.M., and BOURGES, M., *Mol. Cryst.*, <u>1</u>, 541 (1966)

SPEGT, P.A., SKOULIOS, A., and LUZZATI, V., *Acta Cryst.*, <u>14</u>, 866 (1961)

TAMAMUSHI, B., *J. Colloid Polymer Sci.*, <u>254</u>, 571 (1976)

TIDDY, G., *JCS Faraday I*, <u>68</u>, 369 (1972)

TIDDY, G., and WHEELER, B., *J. Colloid Interface Sci.*, <u>47</u>, 59 (1974)

ULMIUS, J., and WENNERSTRÖM, H., *J. Magn. Reson.*, <u>28</u>, 309 (1977)

WINSOR, P.A., *Chem. Rev.*, <u>68</u>, 1 (1968)

YOUNG, C.Y., MISSEL, P.J., MAZER, N.A., BENEDEK, G.B., and CAREY, M.C., *J. Phys. Chem.*, <u>82</u>, 1375 (1978)

Editors' note to Part 3.

The question of the structure of H_2O in multicomponent systems with organic salts should be focussed in future work somewhat more sharply than in the past, e.g., by adopting a proper convention on nomenclature, which might be expressed as follows: (i) when individual molecules of H_2O are considered, without specifying their mutual configuration in the aqueous phase, the reference should always be to "H_2O molecules" never to "water" or "water molecules"; (ii) when an extended region of H_2O molecules is considered, for example within a reverted micelle, or near a solid boundary, the reference should be to "water" with the clear understanding that such water may have tetrahedral packing of its molecules, almost as in pure water, e.g., when the systems are very dilute. But if the region is exposed to strong electrostatic forces, e.g., from the ions, the packing of H_2O molecules in such "water" may be quite remote from that in pure water, since the dipoles are oriented by the local electric field.

APPENDIXES

Appendix 1

ASPECTS OF THE STRUCTURE OF PURE
SOLID ORGANIC SALTS

K. Fontell

INTRODUCTION

The structural information obtainable on pure organic salts is quite diverse. The crystalline structure of organic compounds is, in the main, determined by the geometrical space requirements of the individual atoms building up the molecules. This is in contrast to the situation for the inorganic compounds where the structure is, to a large extent, determined by directional forces between the individual atoms. The organic salts occupy an intermediate position and in consideration of their structures one has to account for these two, sometimes conflicting building principles.

All crystal structure work is continuously abstracted and appraised in "Structure Report" [1]. The compilations by Wyckoff [2a;b] attempt to give a more condensed survey of the published crystal structure data and their reliability. Some recent information may also be found in two standard series published by the International Union of Crystallography [3,4]. However, no comprehensive account exists of the crystalline structures of organic salts.

The crystals of organic molecules are often weak and fragile. Many possess a layer structure and a serious complication from the X-ray structure determination point of view is a tendency to form twin crystals. Therefore, it is difficult to perform a proper single crystal study and one has instead to draw conclusions bases upon polycrystalline powder methods.

As mentioned, in the consideration of the structure of organic crystals in general one has to deal with the problem of bodies of definite shape, the aim being to find the closest packed arrangement which in turn means that one has to find the minimum free energy for the system. Kitagorodskii has especially stressed this geometrical point of view [5]. With organic salts one has in addition to consider the directional forces between the polar parts of the molecules and the counterions.

The number of organic salts is large. The conditions determining their structures are depending on a combination of strong ionic forces and weak Van der Waals forces. The ionic groups have to be in mutual contact but there is an additional requirement that the non-polar often bulky organic parts pack in a reasonable manner. It is clear that in the structure of salts having a small organic part the ionic directional forces will dominate, while when the organic part is large its space requirements will govern the structure. The interplay between these two opposing building principles creates an abundance of different possible structure combinations.

In a short appendix it is impossible to give an account for all structural proposals presented for different groups of organic salts and to appraise their merits. In some rare cases the organic salts will have a very symmetrical shape, as for example the salts of porphyrines, the hemines, but normally the molecules are asymmetrical having one non-polar organic part and one polar ionic part, a structural feature which endows the molecule with surface active properties. A common case for such salts is that the molecule has <u>one</u> ionic group and <u>one</u> organic group, as for instance the soaps, that is the fatty acid salts; in this account we will mainly deal with this group of compounds.

In an appraisal of published structural information on fatty acid salts one must realize that much of the older deductions is of little real value. The work was often performed on polycrystalline material and as the structures were of low symmetry the diffraction patterns offered insurmountable difficulties in indexing. Many of the interpretations were based upon analogy deductions from the few cases where one had been lucky in obtaining crystals suitable for single crystal work. One pitfall is also the mentioned frequent occurrence of twinning in the crystals; such a twinning is not easy to detect and consequently incorrect dimensions for the unit cell could be obtained. Another difficulty is the pleomorphism.

THE SUBCELL CONCEPT

The aliphatic hydrocarbon chains were originally thought always to be in parallel configuration in the solid state; this is not so, in some instances it has been shown that the chains cross each other. That increases the number of structural possibilities. In understanding these phenomena the subcell concept of Vand is useful [6]. The subcell represents the smallest repeti-

343

tion unit in a structure composed of hydrocarbon chains. It has been shown that there exist several modes for packing hydrocarbon chains. This account of the different possibilities follows a recent review ([7]).

The packing modes of the hydrocarbon chains may be triclinic, monoclinic, orthorhombic or hexagonal. The dimensions of the most common subcells are given in Table 1, while Fig. 1 depicts the cross-sections perpendicular to the chain axis for some of them. In the literature one may find other choices of subcell units. There is known just one mode for the triclinic and monoclinic packing, T|| and M||, respectively. The orthorhombic packing has several variants. In two of them, O⊥ and O'⊥, the planes of every second carbon chain row are perpendicular to the planes of the other, while the two others, O|| and O'||, have parallel chains.

TABLE 1. The subcell dimensions of some common close-packing modes of hydrocarbon chains (after Ref.[7])

Triclinic	T||	a_s = 4.41 Å,	b_s = 5.40 Å,	c_s = 2.45 Å,	α = 74.8°, β = 108.5°, γ = 120.5°	
Monoclinic	M||	4.26	4.83	2.57	114.5	
Orthorhombic	O⊥	4.90	7.41	2.54		
Orthorhombic	O'⊥	7.43	5.01	2.50		
Orthorhombic	O||	8.15	9.21	2.55		
Orthorhombic	O'||	7.93	4.74	2.53		
Hexagonal	H	4.9	2.5			

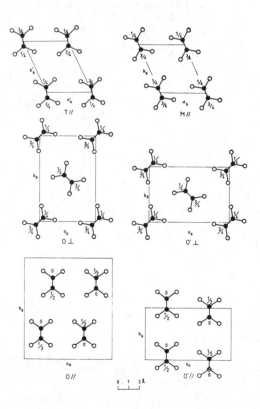

Fig. 1. Subcells corresponding to different close-packing modes of straight-chain hydrocarbon chains (Taken from Ref.[7], by kind permission of *J. Amer. Oil Chem. Soc.*)

The dimensions of the most common orthorhombic structure O⊥ seems to be very sensitive to thermal oscillation. The same holds for the hexagonal chain packing mode. In the latter case the detailed arrangement of the chains in the subcell is relatively unknown.

A displacement along the chain axis by an integral number of c_s units will not affect the regularity of the subcell in the layers. Therefore it is possible to obtain for every chain packing mode different angles of tilt of the chains toward the end group planes. It is not surprising that pleomorphism occurs.

When the chain length increases an increased tendency for parallel packing would be expected. Such a tendency has in fact been observed. Introduction of branched chains, double bonds or other steric hindrances restricts the crystallization capability and affects the structures obtained.

SALTS OF ALIPHATIC CARBOXYLIC ACIDS

ALKALI SALTS OF FATTY ACIDS - THE SOAPS

The monovalent metal salts of the fatty acids constitute the soaps in a restricted sense. The soaps exhibit polymorphism and, since different hydrates as well appear, the nomenclature for the various solid phases has been subject of controversy and confusion. In the literature one may find almost the whole Greek alphabet, primed and unprimed, each notation supposed to define a separate phase. It has been maintained that each such phase is associated with a unique crystal structure, while others have claimed that the different X-ray diffraction patterns do not necessarily represent true crystal structures but instead are merely associated with different types of disorder of the chains. There is also the additional problem of the "descendant" phases, i.e., structures that occur at some elevated temperature and remain unaltered at room temperature in a pseudoequilibrium for a long time.

The early studies of soaps revealed the mentioned abundance of different powder patterns with sharp reflections indicating the existence of well-developed three-dimensional crystal structures ([8-11]). However, the unit cell information obtainable from these data was incomplete and one had to satisfy oneself by including the various powder patterns in the ASTM Card File ([12]).

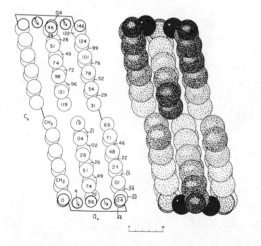
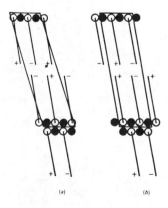

Fig. 2. The monoclinic structure of the A form of potassium decanoate (KC_{10}) projected along its b_0 axis (left-hand axis). *Left*. The original proposal. *Right*. The relation between the original structure proposal (a) and the now established one (b). Chains with plus signs are tilted up from the paper, those with minus signs are tilted down. (Taken from Ref.s [2a] and [16], by kind permission of *John Wiley & Sons* and *Acta Cryst*.)

Crystals suitable for single crystal X-ray work are somewhat easier to obtain for the potassium soaps than for other soaps. The first successful structure determination was performed by Vand, Lomer & Lang in 1949 [13]. They were able to show that the so-called form A of potassium decanoate (KC_{10}) had a monoclinic configuration, belonged to the space group $P2_1/a$ [14] and had four molecules in the unit cell.

This structure has later been refined [15,16]. In the final version the hydrocarbon chains are packed in a crossed-chain structure with an angle of tilt of 57.9±0.5° [16]. The dimensions of the unit cell are given in Table 2. Fig. 2 shows the structure originally proposed as well as the refined one. The packing in the ionic layers is the same in both structures, the difference lies in the packing of the hydrocarbon chains. The wrong alternative of two possible independent values was chosen for the angle β in the original paper.

TABLE 2. The unit cell dimensions of potassium soaps, $C_nH_{2n-1}O_2K$ (abbreviation: KC_n)
       (After page 437 in Vol. 16 of Ref.[1])

$t/°C$	Soap	Phase	Space group	$a/Å$	$b/Å$	$c/Å$	$d_{001}/Å$	$\alpha/°$	$\beta/°$	$\gamma/°$	Ref.
25	KC_4		$P2_1/a$?	8.12	5.69	14.43	14.42	90	92	90	15
	KC_6			8.00	5.74	18.94	18.93	90	91.9	90	15
	KC_8			7.92	5.68	23.01	23.00	90	92.1	90	15
	KC_{10}	A	$P2_1/a$	8.08	5.68	27.58	27.53	90	93.5	90	15
				8.11	5.64	28.87	-	90	108.0	90	13
				8.03	5.69	28.04	27.6	90	101.2	90	16
	KC_{12}			7.97	5.67	31.78	31.62	90	95.7	90	15
	KC_{12}			4.14	5.60	30.04	29.78	97.4	92.0	94	15
	KC_{14}			4.15	5.61	34.07	34.05	91.6	91.5	93	15
				4.13	5.65	34.28	34.26	91.1	91.4	92.2	18
	KC_{16}	B	$P\bar{1}$	4.17	5.61	38.27	38.18	92.7	91.3	93	15
				4.15	5.00	37.87	-	93.06	91.4	92.4	17
	KC_{18}			4.16	5.57	42.11	42.07	91.5	91.6	94	15
75	KC_4	A	$P2_1/a$?	8.11	5.71	14.53	-	90	92.2	90	15
	KC_6			8.07	5.75	19.14	-	90	92.8	90	15
	KC_8			8.09	5.73	23.81	23.80	90	91.7	90	15
	KC_{10}			8.12	5.68	28.07	28.06	90	91.8	90	15
	KC_{12}			8.08	5.67	32.36	32.27	90	94.2	90	15
	KC_{14}	C	$P2_1/a$	8.08	5.69	36.94	36.91	90	92.1	90	15
	KC_{16}			8.05	5.71	41.03	41.03	90	90.1	90	15
	KC_{18}			8.07	5.67	45.46	45.31	90	93.2	90	15

Phase behaviour.
On crystallization from ethanol at room temperature KC_4 to KC_{10} are in form A, KC_{12} to KC_{18} in form B. On heating, KC_4 to KC_6 show only thermal expansion, KC_8 to KC_{18} transform to form C. On cooling, KC_4 to KC_{12} revert to form A, KC_{14} to KC_{18} revert to form B; KC_{12} reverts over a period of months from form A to form B.

The other even-numbered potassium soaps with chain lengths between 4 and 10 are also, in anhydrous state at room temperature, monoclinic with four molecules in the unit cell. With the exception of the C_4 homologue they all belong to the above mentioned space group, $P2_1/a$. The structures of the longer chain ones, C_{12} to C_{18}, are, however, triclinic with two molecules in the unit cell [15]. Originally it was not possible to decide whether the space group for this form B was $P\overline{I}$ or $P\overline{I}$, but later work seems to favour the latter alternative ([17]). The angle of tilt of the chain axis is 52°. Contrary to form A there is no crossing of the chain axis direction in alternate layers, perpendicular to this a axis. The difference in structure is thus that while the packing in the ionic double layers of the B form is the same as in the A form, all the hydrocarbon chains are parallel in the B form.

With the exception of the C_4 and C_6 homologues both these structures transform at elevated temperatures into a third modification, form C, which is monoclinic and belongs, as does form A, to space group $P2_1/a$, but with the difference that the angle of tilt of the crossed-chained hydrocarbon chains is now 55.5°. This monoclinic form C differs from the form A in the positions of the last four carbon atoms ([18]). Their thermal mobility increases markedly; this may have relevance for the mechanism of the phase change between the A and C forms.

The unit cell dimensions for the different potassium soap modifications are given in Table 2. As mentioned the different crystal forms occur in different temperature regions. One exception is potassium dodecanoate (KC_{12}) for which two modifications have been reported at 25 °C. However, it is to be noted that the B form is the equilibrium modification ([15]).

The arrangements of the methyl end groups and the ionic groups for the forms A and B are depicted in Fig.s 3 and 4, respectively.

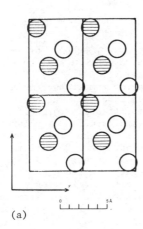

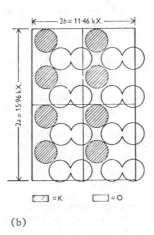

(a)   (b)

Fig. 3. The arrangement of the methyl groups in the crystal (a) and the structure of the ionic layers (b) in the monoclinic form A of potassium soaps (taken from Ref.s [15] and [16], by kind permission of *Acta Cryst.*).

The situation is still more complicated for the sodium soap series. Several definite hydrates exist and experience has shown that it is difficult to obtain suitable single crystals. The reflection intensities may also vary from pattern to pattern of the same recognizable phase modification. Most of the interpretation of the structures is therefore based upon powder X-ray data with all the ambiguities caused by particle size and/or disorder reflection broadening.

Early work by Buerger indicated that the hemihydrates of sodium hexadecanoate (palmitate) and sodium octadecanoate (stearate), $Na\,C_{16}\cdot\frac{1}{2}\,H_2O$ and $NaC_{18}\cdot\frac{1}{2}H_2O$, respectively, had monoclinic structures and belonged either to the space group Aa or $A2/a$ ([20-23]). The unit cell would contain 16 molecules. Later work by Segerman claims that the space group of the sodium soap series should be either $P2_1$ or $P2_1/m$ and that the same structure should exist from C_6 to C_{18} soaps. The unit cell of the C_6 soap would contain twenty molecules and the chain packing may be described by an orthorhombic subcell containing crossed chains ([24]).

Anhydrous soaps give at elevated temperature rise to a series of structures before the transition to isotropic melt. Some of the structures are crystalline and other liquid crystalline;

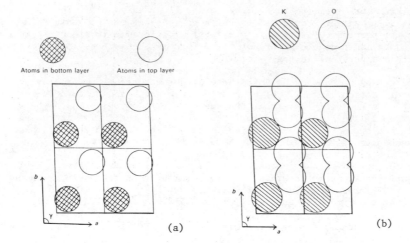

Fig. 4. The arrangement of the methyl groups in the crystal (a) and the structure of the ionic layers (b) in the triclinic form B of potassium soaps (taken from Ref. [17], by kind permission of *Acta Cryst.*)

they go under phenomenological designations as subwaxy, waxy, superwaxy, subneat or neat ([25], [26]). The structures have been studied by various members of the Strasbourg school as functions of temperature, length of the hydrocarbon chains and nature of the cation ([27-30]). Fig. 5 attempts to give an impression of the multitude of transitions. The low temperature modifications are crystalline and their structures may be characterized as lamellar. The angle of tilt of the hydrocarbon chains and their molecular packing vary from modification to modification. However, apart from the few cases already mentioned of single crystal studies of potassium and sodium soaps, all work has been on polycrystalline specimens. It is remarkable that when the temperature is raised the originally lamellar crystalline structures do not go directly over into the lamellar liquid crystalline bilayer structure, the neat soap structure. Instead there are several intermediate structures composed of ribbon-like aggregates in two-dimensional arrangement forming oblique or rectangular (primitive or centered) cells, or forming planar disc-like aggregates in a three-dimensional orthorhombic centered structure. In these structures it is the polar groups and the counterions that form the ribbon-like or disc-like aggregates while the hydrocarbon tails are in a fluid state filling up the intermediate space.

ACID SOAPS

Soaps may associate together with the corresponding acids into crystalline molecular complexes. While the only association complexes reported for the potassium acid soaps have the 1:1 molar ratio, different acid to soap ratios have been reported for sodium acid soaps (1:1, 3:2, 2:1; Ref. [31]). In addition, for the short-chain acid sodium octanoate the ratio 1:2 has been found (1 HC_8: 2 NaC_8 [32]).

The acid soaps exhibit sharp X-ray diffraction powder patterns which are structurally different from those obtained for the corresponding neutral soaps and it has also been possible in some cases to grow suitable single crystals. The structures of the 1:1 acid sodium hexadecanoate and 1:1 acid sodium octadecanoate (HC_{16}:NaC_{16} and HC_{18}:NaC_{18}, respectively) are monoclinic and belong to the space group $P2_1/a$ ([33]). Another study claims that for the acid C_{16} soap the space group is the orthorhombic Pba ([34]). In the monoclinic alternative the unit cell will contain four molecular associates, i.e., eight hydrocarbon chains.

The latter of these compounds (HC_{18}:NaC_{18}) is reported to transform above 50 °C into a triclinic structure modification which belongs to either the space group $P1$ or $P\bar{1}$ ([34]). Its unit cell would contain six molecular associates, i.e., twelve hydrocarbon chains. The same or a similar triclinic structure with six molecular associates in the unit cell is also reported for the

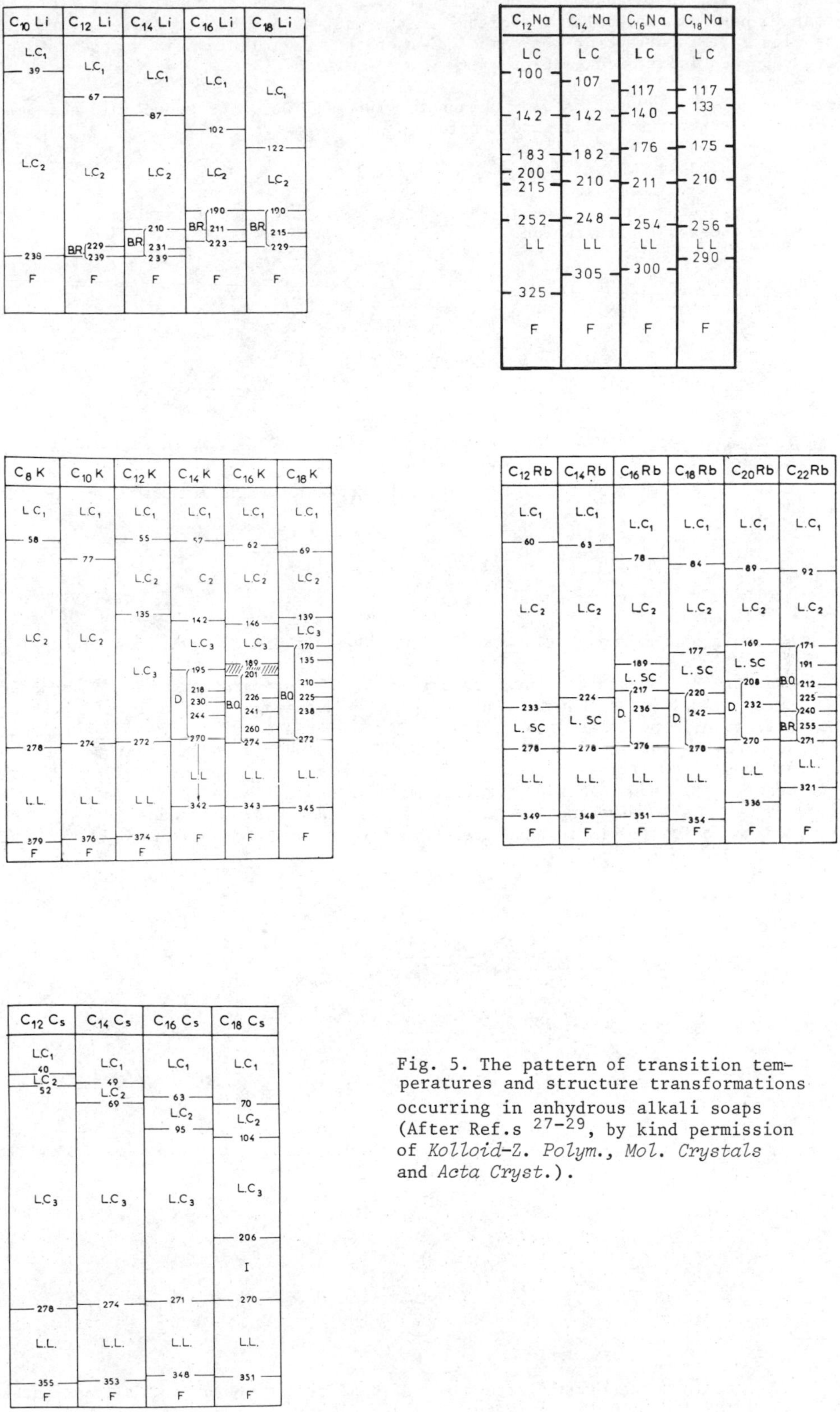

Fig. 5. The pattern of transition temperatures and structure transformations occurring in anhydrous alkali soaps (After Ref.s [27-29], by kind permission of *Kolloid-Z. Polym.*, *Mol. Crystals* and *Acta Cryst.*).

1:1 acid potassium tetradecanoate ($HC_{14}:KC_{14}$ [19]). However, it is rather improbable for a 1:1 acid soap to have six molecular associates in the unit cell. The structure information reported must either be incorrect or the composition of these triclinic acid soaps must be different from the reported 1:1 stoichiometry.

The structure of the 1:2 acid sodium octanoate (1 HC_8: 2 NaC_8) is monoclinic and belongs to the space group Cc ([35]). The unit cell dimensions are

$$a_0 = 8.17 \text{ Å}, \quad b_0 = 15.18 \text{ Å}, \quad c_0 = 23.97 \text{ Å}, \quad \beta = 101°$$

The unit cell contains four molecular associates, i.e., twelve hydrocarbon chains, and the chains are in crossed configuration.

OTHER SALTS OF MONOCARBOXYLIC ACIDS

One example of the effect given by a disturbance in the hydrocarbon chain is the structure of sodium 2-oxooctanoate {$CH_3(CH_2)_5CO\text{-}COONa$ [36]}. The crystals of this substance are orthorhombic and belong to the space group $Pbcn$. The unit cell contains eight molecules and has the dimensions

$$a_0 = 49.57 \text{ Å}, \quad b_0 = 6.05 \text{ Å}, \quad c_0 = 5.97 \text{ Å}$$

and the resulting structure is shown in Fig. 6.

The shorter-chain sodium 2-oxobutanoate {$C_2H_5CO\text{-}COONa$} is practically isostructural ([37]).

There are rather few recent structural reports about the preparation and the structures of other univalent metal soaps. Lawrence has prepared a number of these salts ([38]). Fingerprint patterns are given in a number of articles ([39–41]) and the compilation of Wyckoff has a list of early work (Table 13D 1c in Ref. [2b]). The lithium soaps have been reported to have hexagonal or tetragonal structures whose unit cells are large, containing 24, 48 or 72 molecules while the short-chain homologues are orthorhombic or cubic (Table 13D in Ref. [2b]). Vand and coworkers as well as Matthews and coworkers have studied the silver soaps ([42,43]). These soaps are reported to have a triclinic structure belonging to space groups PI or $P\bar{I}$ and having two molecules per unit cell ([42]).

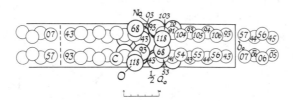

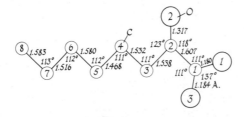

Fig. 6. The orthorhombic structure and bond dimensions of sodium 2-oxooctanoate {$CH_3(CH_2)_5CO\text{-}COONa$} (taken from Ref. [2a], by kind permission of *J. Wiley & Sons*)

The divalent metal salts of fatty acids often form greases but may sometimes give polycrystalline fingerprint patterns showing short and long spacings [39-41, 44-51]. They give a similar
multitude of transitions when the temperature is raised as the univalent soaps [52-55]. Stanley
has tried to evaluate the powder patterns given by calcium octadecanoate monohydrate $\{Ca(C_{18})_2 \cdot H_2O\}$ with the aid of the tentative indexing method of Vand [6]. He concluded that the space
group probably is $P2_1$ and that the unit cell would contain two molecular units, i.e., four hy-
drocarbon chains [56]. The dimensions of the unit cell are

$$a_0 = 5.70 \text{ Å}, \quad b_0 = 6.81 \text{ Å}, \quad c_0 = 50.4 \text{ Å}, \quad \beta = 94°$$

The structure of strontium dodecanoate $\{Sr(C_{12})_2\}$ has been proposed to belong to the space
group $P2_1/n$ [57]. The dimensions of the four-molecule unit cell have been reported as

$$a_0 = 7.819 \text{ Å}, \quad b_0 = 71.00 \text{ Å}, \quad c_0 = 4.76 \text{ Å}, \quad \beta = 102.51°$$

However, Stanley found evidence for a strong tendency to twinning in the strontium soap series
[58]. The crystals obtained for the C_6, C_{10} and C_{12} soaps are triclinic with the additional
feature that β^* is 90°.

In this connection it may be mentioned that the strontium soaps at about 230° have a "viscous
isotropic" modification with a cubic structure built up of two independent interpenetrating
networks [59].

An old report states that the crystals of copper butanoate monohydrate $\{Cu(C_4)_2 \cdot H_2O\}$ are monoclinic and belong to the space group Aa or $A2/a$ [60]. The unit cell of this structure would
contain 32 molecules, i.e., 64 hydrocarbon chains, and would have the dimensions

$$a_0 = 25.17 \text{ Å}, \quad b_0 = 14.65 \text{ Å}, \quad c_0 = 26.85 \text{ Å}, \quad \beta = 109.7°$$

Another old report claims that the crystals of barium dicalcium propanoate $\{BaCa_2(C_3)_6\}$ have a
cubic structure [61].

There is still less to say about the structures of the trivalent metal soaps. For the di- and
tri-aluminium soaps it has been possible to obtain fingerprint patterns but no crystallographic
structure evaluation has been published [62,63]. Similarly, other trivalent metal soaps have
been prepared but no X-ray work has been published [64].

DISOAPS

The salts of the α-ω dicarboxylic acids show a similar pattern of multiple transitions when the
temperature is raised as the salts of monocarboxylic acids [65,66]. The structure information
obtained is just fingerprint patterns. The "bolaform" structure of the molecules leads to a
predomination of lamellar structures.

ORGANIC SULFATES AND SULFONATES

As with the ordinary soaps a large number of different phases exist, each phase differing in
the extent of hydration and the number of hydrocarbon chains in the unit cell. Most of the information obtained is from powder specimens but some single crystal studies heve been performed.
For the 1-alkyl sulfonate series the existence of seven different phases has been reported. The
crystals of this series are less distorted and more stable than those of the sulfate series.
Table 3 lists results obtained by Lingafelter and coworkers [67-76]. The crystal variant with
1/8 mole of water of hydration per mole soap is monoclinic and is claimed to belong to the space
group $A2/a$; its unit cell would contain 32 lipid molecules. The space groups $P2_1/a$, $P1$ or $P\bar{1}$
have been suggested for phases with other hydration stages.

TABLE 3. Data for (A) sodium alkanesulfonates and (B) sodium alkylsulfates after Lingafelter et al. (Ref.s [67-72], [74], [76])

Phase notation	Chain length	Hydration stage	Space group	$a_0/\text{Å}$	$b_0/\text{Å}$	Number of chains in cross section
A.						
α	7, 8, 9, 10, 11, 12, 13, 14, 15, 16, 18	1/8	Aa or $A2/a$	16.8	10.1	8
β	10, 12, 13, 14, 15, 16, 18	1/2	$P2_1/a$	6.85	15.4	4
γ	4, 5, 6, 7, 8, 9,	1	PI or $P\bar{I}$	10.5	5.9	2
δ	10	3.5	?	7.6	19.5	4
ε	12, 13, 14, 16, 18	1/4	PI or $P\bar{I}$	9.9	10.8	4
η	12	1-2	PI or $P\bar{I}$	5.6	8.6	?
ξ	10	(0) ?	?	10.7	12.8	?
B.						
α	6, 7, 8, 9, 10, 11, 12, 14, 16, 18	1/8	Aa or $A2/a$	16.4	10.3	8
ι	11, 12	1	Aa or $A2/a$	9.5	14.0	?
κ	16, 18, 20	1/4	$P2_1/a$	9.5	9.2	4
λ	6, 7	1	Aa or $A2/a$	9.5	14.1	4
μ	6, 7, 8	?	Aa or $A2/a$	8.5	6.1	2
ν	16, 18, 20	1	Cm, $C2$ or $C2/m$	9.9	5.3	2

In the straight chain sodium alkylsulfate series six different phases were observed by the Lingafelter group. Their hydration stages and suggested space groups are also given in Table 3. The structure of the 1/8 hydrate is claimed to be the same as that of the 1/8 hydrate of the sulfonate series ([76]) *viz.* space group $A2/a$ and 32 lipid molecules in the unit cell. However, a complete structure determination of the 1/8 hydrate of sodium dodecylsulfate (C_{12}) has recently been performed by Sundell ([77]). The structure is monoclinic and belongs to the space group $C2/a$. The unit cell contains 32 lipid molecules and four molecules of water, and its dimensions are

$$a_0 = 78.69 \text{ Å}, \quad b_0 = 10.22 \text{ Å}, \quad c_0 = 16.41 \text{ Å}, \quad \beta = 98.28°$$

In the structure the lipid molecules are arranged "tail-to-tail" in a double layer. The chain packing is irregular and can best be described as an intermediate between the orthorhombic O⊥ and the hexagonal H chain packing mode. The molecular arrangement is depicted in Fig. 7. The rigid rippled arrangement of the sulfate head groups allows electrostatic interactions within the polar region as the sodium ions are evenly distributed between the sulfate groups both within each layer and between opposite layers. The arrangement of the polar head groups leaves the hydrocarbon chains with about 10 % more space than otherwise would be required by the solid state, causing disorder and thermal motion in the bilayer hydrocarbon matrix.

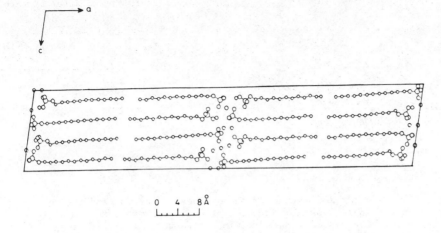

Fig. 7. The molecular arrangement of the hydrocarbon chains and the polar groups in the "double-layered" structure of the 1/8 hydrate of sodium dodecylsulfate (C_{12}). The monoclinic structure is viewed along the b-axis (taken from Ref. 77, by kind permission of *Acta Chem. Scand.*)

ALKYLAMMONIUM HALIDES

Early studies seemed to indicate that all higher monoalkyl-halides were isomorphous with the short-chain monoalkyl-iodides, which possess a bimolecular tetragonal unit cell (p 629 in Ref. 2a; p 6 in Ref. 2b). This was interpreted as evidence that the long aliphatic chains could not occupy fixed positions in the structure but that the zig-zag chains rotated around the c_0-axis.

However later work revealed the existence of structures with lower symmetry, the crystals of the chlorides being monoclinic, having two molecules in the unit cell and the space group being $P2$ or $P2_1/m$ (C_{12}, C_{14}, C_{16}). The structure of the C_{12} bromide was also monoclinic but had in contrast a four-molecule cell (p 629 in Ref. 2a; p 6 in Ref. 2b; [78]). A recent three-dimensional single crystal study confirms the structure of the bromide, whose space group is designed as $P2_1/c$ ([79]).

The dimensions of the unit cells for the C_{12} homologues are:

Chloride $a_0 = 5.68$ Å, $b_0 = 7.17$ Å, $c_0 = 17.86$ Å, $\beta = 91.2°$ (Ref.[78])

Bromide $a_0 = 6.030$ Å, $b_0 = 6.958$ Å, $c_0 = 35.639$ Å, $\beta = 92.09°$ (Ref.[79])

The arrangement of the "head-to-tail" ordered chains in the "single-layered" structure is shown in Fig. 8. The hydrocarbon chain packing is in the orthorhombic O⊥ mode.

When the crystals of the C_{12} bromide are heated there occur phase transitions at 56, 62 and 73 °C before melting under decomposition at about 200 °C ([79]). The structures of these modifications have not been studied.

The structures of n.dodecylammonium bromide and chloride are thus approximately isomorphous in spite of their belonging to different space groups. Interesting is also that these structures resemble closely that for the short-chain propylammonium chloride which in turn belongs to a third space group $C2/m$ ([78]).

One example of the influence of an increase in the bulkiness of the polar part of the molecule is the structure of the N-methyl-n.dodecylammonium chloride, $(C_{12}H_{25})(CH_3)NH_2Cl$, which is triclinic and has two molecules in the unit cell. The space group for this structure is $P\bar{1}$ ([80]).

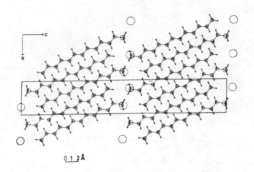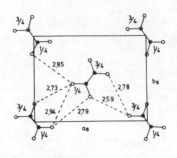

Fig. 8. The molecular arrangement of the hydrocarbon chains and the polar
groups in the "single-layered" structure of n.dodecylammonium bromide (C_{12}),
and its hydrocarbon subcell. The monoclinic structure is viewed along the
b-axis. The orthorhombic subcell O'$\perp$ is viewed along the c_s-axis, the hydrogen Van der Waals contacts are those found in the middle of the chains
(taken from Ref. [79], by kind permission of *Acta Cryst*.)

The odd-chain tridecylammonium chloride (C_{13}) is claimed to have an orthorhombic structure belonging to the space group *C2ca* or *Cmca* and having eight molecules in the unit cell ([80]).

The eighteen carbon homologue octadecylammonium chloride crystallizes according to a rather old
report in an orthorhombic pseudotetragonal structure with the unit cell dimensions

$$a_0 = 5.45 \text{ Å}, \quad b_0 = 5.40 \text{ Å}, \quad c_0 = 69.4 \text{ Å}$$

The unit cell would contain four molecules (p 6 in Ref. [2b]).

There has been no work reported on the structures of the long-chain di- and tri-methylalkylam-
monium halides. It would be interesting to see the influence of the methyl groups. The short-
chain trimethylammonium salts are monoclinic but the triethyl compounds hexagonal (p 7 in Ref.
2b).

BILE ACID SALTS

A special group of surface active salts constitute the bile acid salts. They are crystalline
and give powder fingerprint patterns but it is difficult to obtain suitable single crystals.
The structure of sodium cholate is reported to belong to the space group $P2_1$ ([81]). The bimolecular cell has the dimensions

$$a_0 = 12.593 \text{ Å}, \quad b_0 = 8.215 \text{ Å}, \quad c_0 = 12.196 \text{ Å}, \quad \beta = 107.86°$$

The structure of this and other bile acid salts should conceivably be closely related to the
corresponding parent acids, whose structures are to a high degree determined by the big stiff
hydrocarbon skeleton which consists of three hexaatomic rings and one pentaatomic ring. It may
be expected that the directional forces of the carboxyl group and the hydroxyl groups play a
minor role. The structures of several bile acids have been reported to be orthorhombic ([81-84]).

In this connection one should perhaps also mention the choleic acids. They are inclusion compounds in which the bile acid molecules are helically twisted around guest molecules (for instance long-chain aliphatic molecules). This capability is most prominent for the 3,12-di-
hydroxycholanic acid-24. The structures of such compounds have been determined ([85-91]).

CONCLUSIONS

The above account shows the great variety of structures which occur already in a restricted group of organic salts. The situation is, of course, still more complicated for organic salts in general. As mentioned earlier the structures depend on a combination of the spatial requirements of the organic part of the molecule and the influence of the directional ionic and Van der Waals forces. The organic parts strive to occupy the closest packed arrangement which in the absence of directional forces gives the minimum free energy for the system.

The crystals of the organic salts are in addition often poorly developed and fragile. This fact creates difficulties in performing proper single-crystal studies. One has often to be satisfied with results obtained by polycrystalline powder techniques and to make conclusions based upon analogy deductions from structures of the corresponding parent compounds and/or related salts whose structures one has been lucky to solve. However, due to the variety of structural possibilities and the frequent occurrence of isomorphy caution is indicated.

In the consideration of the structure of a particular organic salt one has always to appraise critically the published information. The number of salts whose structures have been determined by three-dimensional single crystal studies is rather limited. The accessibility of good crystals seems often to have determined the compounds whose structure has been solved. The Cambridge group keeps a continuous databased record of all three-dimensionally solved organic and organometallic compounds [92]. If this collection fails in giving the desired information one may try the Ref.s [1-4] but it is important to be constantly aware that especially the older information given may be incomplete and/or incorrect. An additional obstacle is the frequent occurrence of isomorphous structures and also the fact that structures occurring at one temperature may be apparently stable for a long time at another temperature.

ACKNOWLEDGEMENTS

This work has been supported by a grant from the Swedish Natural Science Research Council. Professor Kåre Larsson is thanked for valuable comments.

REFERENCES

1   *Structure Report* (formerly *Strukturbericht*) 1913-1939; 1940-... International Union of Crystallography; Distr.: Bohn, Scheltema, Holkema, Utrecht, The Netherlands

2   WYCKOFF, R.W.G., *"Crystal Structures"*, (a) book, J. Wiley & Sons, New York, USA, 1966; (b) loose leaf edition, Interscience, New York, USA

3   *"Molecular Structures and Dimensions, A1 Interatomic Distances 1960-1965, Organic and Organometallic Crystal Structures"*, International Union of Crystallography, 1972; Distr.: Oosthoek, A., Utrecht, The Netherlands; Crystallographic Data Centre, Cambridge, U.K.; Polycrystal Book Service, Pittsburgh, Pa, USA

4   *Crystal Structure Communication*, International Union of Crystallography; Distr.: Università degli Studi di Parma, Italy

5   KITAGORODSKII, A.I., *"Organic Chemical Crystallography"*, Consultants Bureau, New York, USA, 1961

6   VAND, V., *Acta Cryst.*, 4, 104 (1951)

7   LARSSON, K., *J. Amer. Oil Chem. Soc.*, 43, 559 (1966)

8   PIPER, S.H., *J. Chem. Soc.*, <u>1929</u>, 234

9   THIESSEN, P.A., and STAUFF, J., *Z. physik. Chem.*, <u>A176</u>, 397 (1936)

10 McBAIN, J.W., de BRETTEVILLE, A., and ROSS, S., *J. Chem. Phys.*, <u>11</u>, 179 (1943)

11 FERGUSON, R.H., ROSEVEAR, F.B., and STILLMAN, R.C., *Ind. Eng. Chem.*, <u>35</u>, 1005 (1943)

12 Powder Diffraction File, Organic Index 1970, Joint Committee on Powder Diffraction Stan
        dards, International Centre for Diffraction Data, Swarthmore, Pa, USA

13 VAND, V., LOMER, T.R., and LANG, A., *Acta Cryst.*, <u>2</u>, 214 (1949)

14 *"International Tables for X-Ray Crystallography"*, Kynoch Press, Birmingham, U.K., 1959

15 LOMER, T.R., *Acta Cryst.*, <u>5</u>, 11, 14 (1952)

16 LEWIS, E.L.V., and LOMER, T.R., *Acta Cryst.*, <u>B25</u>, 702 (1969)

17 DUMBLETON, J.H., and LOMER, T.R., *Acta Cryst.*, <u>19</u>, 301 (1965)

18 DUMBLETON, J.H., Thesis, University of Birmingham, U.K., 1964

19 DUMBLETON, J.H., *Acta Cryst.*, <u>19</u>, 279 (1965)

20 BUERGER, M.J., SMITH, L.B., de BRETTEVILLE, A., and RYER, F.V., *Proc. Nat. Acad. Sci. U.S.*,
        <u>28</u>, 526 (1942)

21 BUERGER, M.J., *Proc. Nat. Acad. Sci. U.S.*, <u>28</u>, 529 (1942)

22 BUERGER, M.J., SMITH, L.B., RYER, F.V., and SPIKE, J.E., *Proc.Nat.Acad.Sci.U.S.*, <u>31</u>, 226 (1945)

23 GARDINER, K.W., BUERGER, M.J., and SMITH, L.B., *J. Phys. Chem.*, <u>49</u>, 417 (1945)

24 SEGERMAN, E., *Acta Cryst.*, 16 Part 13 Suppl (1963) A76

25 NORDSIECK, H., ROSEVEAR, B.F., and FERGUSON, R.H., *J. Chem. Phys.*, <u>16</u>, 175 (1948)

26 LUZZATI, V., Chapter 3 in *"Biological Membranes, Physical Facts and Function"*, Chapman, D.,
        Ed., Academic Press, London, U.K., 1968

27 SKOULIOS, A., and LUZZATI, V., *Acta Cryst.*, <u>14</u>, 278 (1961)

28 GALLOT, B., and SKOULIOS, A., *Mol. Crystals*, <u>1</u>, 263 (1966)

29 GALLOT, B., and SKOULIOS, A., *Kolloid-Z. Polym.*, <u>209</u>, 164 (1966); <u>210</u>, 143 (1966); <u>213</u>,
        143 (1966); <u>222</u>, 51 (1967)

30 GALLOT, B., and SKOULIOS, A., *Acta Cryst.*, <u>15</u>, 826 (1962)

31 RYER, F.V., *Oil & Soap*, <u>23</u>, 310 (1946)

32 MANDELL, L., *Finska Kem. Samf. Medd.*, <u>72</u>, 49 (1963); EKWALL, P., and MANDELL, L., *Kolloid-*
        Z. Polym., <u>233</u>, 638 (1969)

33 BUERGER, M.J., *Amer. Min.*, <u>30</u>, 551 (1945)

34 KOHLHAAS, R., *Chem. Ber.*, <u>82</u>, 487 (1949)

35 FONTELL, K., unpublished results, 1963

36 TAVALE, S.S., PANT, L.M., and BISWAS, A.B., *Acta Cryst.*, <u>17</u>, 215 (1964)

37 TAVALE, S.S., PANT, L.M., and BISWAS, A.B., *Acta Cryst.*, <u>16</u>, 566 (1963)

38 LAWRENCE, A.S.C., *Trans. Faraday Soc.*, <u>34</u>, 660 (1938)

39 VOLD, R.D., and HATTIANGDI, G.S., *Ind. Eng. Chem.*, <u>41</u>, 2311 (1949)

40 HATTIANGDI, G.S., VOLD, M.J., and VOLD, R.D., *Ind. Eng. Chem.*, <u>41</u>, 2320 (1949)

41 VOLD, M.J., and VOLD, R.D., *J. Amer. Oil Chem. Soc.*, <u>26</u>, 520 (1949)

42 VAND, V., AITKENS, A., and CAMPBELL, R.K., *Acta Cryst.*, <u>2</u>, 398 (1949)

43 MATTHEWS, F.W., WARREN, G.G., and MICHELL, J.H., *Anal. Chem.*, <u>22</u>, 514 (1950)

44 VOLD, R.D., and coworkers, *J. Phys. Coll. Sci.*, <u>52</u>, 1421 (1948); *J. Coll. Sci.*, <u>3</u>, 339(1948);
        <u>4</u>, 93 (1949); *Ind. Eng. Chem.*, <u>41</u>, 2539 (1949); *J. Amer. Chem. Soc.*, <u>73</u>, 2006 (1951)

45 SMITH, G.H., and ROSS, S., *Oil & Soap*, <u>23</u>, 77 (1946)

46 SMITH, T.D., *Petroleum* (London), <u>15</u>, 5 (1952); Structure Reports, <u>16</u>, 439 (1952)

47 BASSET, D.C., and IBALL, J., *Trans. Faraday Soc.*, <u>50</u>, 411 (1954)

48   SPINK, J.A., *Nature*, <u>163</u>, 441 (1949); see also: BRUMMAGE, K.G., *Nature*, <u>164</u>, 244 (1949)

49   DAVIS, J.T., and PHILIPPOFF, W., *Nature*, <u>164</u>, 1087 (1949)

50   TRILLAT, J.J., and BARBEZAT, S., *J. Recherche CNRS*, <u>16</u>, 18 (1951)

51   TRILLAT, J.J., BARBEZAT, S., and JAQUOT, A., *Compt. Rend.*, <u>243</u>, 77 (1956)

52   SPEGT, P.A., and SKOULIOS, A., *Acta Cryst.*, <u>16</u>, 301 (1963)

53   SPEGT, P.A., and SKOULIOS, A., *Acta Cryst.*, <u>17</u>, 198 (1964)

54   SPEGT, P.A., and SKOULIOS, A., *Acta Cryst.*, <u>21</u>, 892 (1966)

55   LUZZATI, V., TARDIEU, A., and GULIK-KRZYWICKI, T., *Nature*, <u>217</u>, 1028 (1968)

56   STANLEY, E., *Nature*, <u>175</u>, 165 (1955)

57   MORLEY, W.M., and VAND, V., *Nature*, <u>163</u>, 285 (1949)

58   STANLEY, E., and SHAIKH, M.S., *Nature*, <u>209</u>, 498 (1966)

59   LUZZATI, V., and SPEGT, P.A., *Nature*, <u>215</u>, 701 (1967)

60   IBALL, J., *Nature*, <u>159</u>, 95 (1947)

61   SEKI, S., MOMOTAMI, M., and NAKATSU, K., *J. Chem. Phys.*, <u>19</u>, 1061 (1951)

62   BAUER, W.H., FISHER, J., SCOTT, F.A., and WIBERLEY, S.E., *J. Phys. Chem.*, <u>59</u>, 30 (1955)

63   GILMOUR, A., JOBLING, A., and NELSON, S.M., *J. Chem. Soc.*, <u>1956</u>, 1972

64   WOOD, J.A., and RYCROFT, C.P., *Colloid & Polymer Sci.*, <u>253</u>, 311 (1975)

65   GALLOT, B., and SKOULIOS, A., *Kolloid-Z. Polym.*, <u>222</u>, 51 (1968)

66   GALLOT, B., *Mol. Crystals and Liquid Crystals*, <u>13</u>, 323 (1971)

67   JENSEN, L.H., and LINGAFELTER, E.C., *J. Amer. Chem. Soc.*, <u>66</u>, 1946 (1944)

68   JENSEN, L.H., and LINGAFELTER, E.C., *J. Amer. Chem. Soc.*, <u>68</u>, 1729 (1946)

69   LINGAFELTER, E.C., and JENSEN, L.H., *Acta Cryst.*, <u>3</u>, 257 (1950)

70   MINOR, J.E., and LINGAFELTER, E.C., *Acta Cryst.*, <u>4</u>, 183 (1951)

71   WILCOX, L.A., and LINGAFELTER, E.C., *J. Amer. Chem. Soc.*, <u>75</u>, 5761 (1953)

72   RAWLINGS, F., and LINGAFELTER, E.C., *J. Amer. Chem. Soc.*, <u>77</u>, 870 (1955)

73   PRINS, J.A., and PRINS, W., *Nature*, <u>177</u>, 535 (1956)

74   JENSEN, L.H., and LINGAFELTER, E.C., *J. Amer. Chem. Soc.*, <u>68</u>, 2730 (1946)

75   LINGAFELTER, E.C., JENSEN, L.H., and MARKHAM, A.E., *J. Phys. Chem.*, <u>57</u>, 428 (1953)

76   RAWLINGS, F., and LINGAFELTER, E.C., *J. Amer. Chem. Soc.*, <u>72</u>, 1852 (1950)

77   SUNDELL, S., *Acta Chem. Scand.*, <u>A31</u>, 799 (1977)

78   GORDON, M., STENHAGEN, E., and VAND, V., *Acta Cryst.*, <u>6</u>, 739 (1953)

79   LUNDEN, B.-M., *Acta Cryst.*, <u>B30</u>, 1756 (1974)

80   CLARK, G.L., and HUDGENS, C.R., *Science*, <u>112</u>, 309 (1950)

81   NORTON, D.A., and HANER, B., *Acta Cryst.*, <u>19</u>, 477 (1965)

82   HALL, S.R., MASLEN, E.N., and COOPER, A., *Acta Cryst.*, <u>B30</u>, 1441 (1974)

83   GIGLIO, E., and QUAGLIATA, C., *Acta Cryst.*, <u>B31</u>, 743 (1975)

84   ARORA, S.K., GERMAIN, G., and DECLERQ, J.P., *Acta Cryst.*, <u>B32</u>, 415 (1976)

85   CAGLIOTI, V., and GIACOMELLO, G., *Gazz. Chim. Ital.*, <u>60</u>, 245 (1939)

86   FISCHMEISTER, H., *Monatsh.*, <u>85</u>, 182 (1954)

87   SCHAEFER, J.P., and REED, L.L., *Acta Cryst.*, <u>B28</u>, 1743 (1972)

88   JOHNSON, P.L., and SCHAEFER, J.P., *Acta Cryst.*, <u>B28</u>, 3083 (1972)

89   CANDELORO DE SANCTIS, S., GIGLIO, E., PAVEL, V., and QUAGLIATA, C., *Acta Cryst.*, <u>B28</u>, 3656
       (1972)

90   CANDELORO DE SANCTIS, S., COIRO, V.M., GIGLIO, E., PAGLIUCA, S., PAVEL, V., and QUAGLIATA,
       C., *Acta Cryst.*, <u>B34</u>, 1928 (1978)

91   CANDELORO DE SANCTIS, S., GIGLIO, E., PETRI, F., and QUAGLIATA, C., *Acta Cryst.*, <u>B35</u>, 226 (1979)

92   Cambridge Crystallographic Data Centre, Cambridge, U.K.

Appendix 2

ELECTROCHEMICAL STUDIES IN MOLTEN ORGANIC SALTS

M. Fiorani

INTRODUCTION

Molten organic salts have been the object of systematic investigations since comparatively recent times, in spite of their theoretical and technological importance. Such investigations indeed represent only a small percentage of all the studies performed on fused salt systems, the main interest having been devoted to the inorganic ones, in particular the alkali chlorides. On the contrary, greater attention to salts with organic anion (e.g., alkanoate, benzenesulfonate) and/or cation (e.g., alkylammonium, pyridinium) should arise from some advantages they offer in respect to inorganic melts: a generally much lower melting point; a certain possibility of gradually changing some of their physico-chemical properties; and the ability to be employed as useful media for organic reactions [1].

Most probably, the comparatively scanty interest given to organic fused systems may be ascribed in part to the erroneous opinion that such salts have poor thermal stability. It has been found, on the contrary, that (e.g., with the alkali alkanoates) the elimination of impurities such as oxygen, transition metal ions and hydroxide ions, allows to obtain melts of pronounced stability (in several cases up to about 300 °C). At higher temperatures a true instability takes place, caused by spontaneous breaking of the C-C linkages [2].

The instability caused by the presence of oxygen can be explained through a process of charge transfer onto molecular oxygen; in turn, that caused by ions of transition metals with variable oxidation number can be referred to the transformation of the carboxylate ion into a neutral free radical. Finally, the reaction

$$OH^- + R\text{-}COO^- \rightarrow RH + CO_3^{2-} \tag{1}$$

may account for the instability in the presence of hydroxide ions.

The spontaneous decomposition occurring at a certain temperature as a consequence of the reaction

$$R\text{-}CH_2COO^- \rightarrow OH^- + R\text{-}CH=CO \tag{2}$$

can be further sustained , through reaction (1), by the OH^- ions formed in reaction (2) [2-4].

For what concerns, on the other hand, the thermal instability of quaternary ammonium salts it was found [5] that two decomposition types are possible, one represented by the Menschutkin reaction (3) and the other by the Hofmann reaction (4):

$$R_4N^+X^- \rightarrow RX + R_3N \tag{3}$$

$$\left[-\overset{|}{\underset{H}{C}}-\overset{|}{\underset{|}{C}}- \right]_4 N^+X^- \rightarrow ^{\diagdown}C=C^{\diagup} + \left[\overset{|}{\underset{|}{C}}H-\overset{|}{\underset{|}{C}}- \right]_3 NH^+X^- \tag{4}$$

where X = Br, I, SCN, NO_3, picrate.

The present review deals with electroanalytical studies carried out in molten organic salts as solvents, whereas investigations concerning salt solutions in organic solvents will be omitted.

As for the latter, only a brief mention might be made here of the researches performed, e.g., in molten dimethylsulfone (T_F= 400 K), where it was possible by polarography and with a tetraalkylammonium perchlorate as supporting electrolyte, to determine the half-wave potentials of several redox couples [6] and to study Cu(I) and Cu(II) complexes with Cl^-, Br^- and I^- [7]. Worthy of particular mention is acetamide, which behaves as an interesting solvent (as well as other amides: formamide, N-methylformamide, N,N-dimethylformamide, etc.) for its dissociating properties. In certain respects acetamide is similar to water (its dielectric constant is 59.2 at 373 K, compared to 55.3 for water at the same temperature), since for the autodissociation reaction

$$2\ AcNH_2 \rightleftarrows AcNH_3^+ + AcNH^- \tag{5}$$

the ionic product value was found to be 10^{-14} at 371 K [8]. In this solvent several electrochemical measurements were carried out, among which, as examples, the researches on thermal cells ([9]) and the voltammetric investigations on solutions of U(VI) ([10]) and Hg(II) and Bi(III) ions ([11]) can be mentioned.

A survey of the information reported in the following sections will make apparent that an increase in the number of electrochemical researches performed in fused organic salts is desirable. Besides other considerations, the possibility of substantially reducing experimental difficulties (as a consequence of the fact that working temperatures lower than in the case of inorganic melts are needed) is surely a topic of not minor interest.

MOLTEN SALTS WITH ORGANIC ANION

Salts with organic anion (particularly carboxylates and in a lesser degree sulfonates) have been rather extensively investigated for what concerns their thermal behaviour and their structural features: comprehensive reviews on these subjects have been given in previous chapters of this book.

It seems rather surprising, on the contrary, that molten salts with organic anion have been scarcely employed as solvents for electroanalytical researches, in spite of their promising features: apart from electrical conductivity measurements on pure and/or mixed alkali formates, acetates and benzenesulfonates (for which the reader is referred to Chapters 1.4 and 2.2) and on several Pb, Zn, Cd and Tl carboxylates ([12-15]), investigations in the electroanalytical area are not numerous and are mainly due to Italian researchers ([16]).

The earliest polarographic measurements in molten ammonium formate at 398 K on a series of inorganic compounds, such as salts of uranium, thorium, plutonium, zirconium and rare earths, allowed to put into evidence the potentialities of this medium for both qualitative and quantitative analysis ([17]). Subsequently, the use of oscillographic polarography made possible the determination, in the same solvent and after electrolytic enrichment, of the species Pb, Zn and Cd at concentration levels from 10^{-4} down to 10^{-6} % ([18]). More recently ([19]), attention was given to potassium formate (T_F= 440 K), the lowest melting among the alkali formates; the drawback encountered with ammonium formate, i.e., the fact that the latter salt, although melting at a lower temperature (389 K), undergoes decomposition from about 403 K, can thus be substantially obviated. In this solvent the potentials of some redox couples were measured with amalgam electrodes, in order to prevent oxidation of the metals. Also in amalgams the anodic polarography of the metals Cd, In, Pb, Sn and Zn was performed. From these measurements, as well as from potentiometric ones, it has been deduced that all the studied elements show a remarkable degree of irreversibility. Other investigations concerned the coulometric production of sulfide ions from the second kind electrode Ag/Ag_2S for the quantitative determination of cations, the sulfides of which are sparingly soluble in the same solvent.

Molten alkali acetates have been much more widely studied than formates, in view also of their acid-base properties and complexing ability. In the ternary mixture K,Li,Na /$C_2H_3O_2$ (50, 20, 30 mol %, respectively) the standard potentials of the couples Cd/Cd(II), In/In(III), Pb/Pb(II), Tl/Tl(I) and Zn/Zn(II) were measured over the temperature range 470÷520 K, by means of amalgam electrodes (whose molality, m, was about 10^{-3} mol kg^{-1}) ([20]). The activity of the metal was evaluated by extrapolation of the existing thermodynamic data ([21]) and the number of electrons involved, n, derived from the plot pE vs $\log m_{M^{n+}}$ (where $pE = EF/2.303\,RT$) had values very close to the theoretical ones, except for lead.

Concerning polarography in the same ternary mixture (only slightly differing in composition), some preliminary investigations dealt with the reduction of Pb(II), Cd(II) and Zn(II) (concentration range: $10^{-4} \leq m \leq 2\;10^{-3}$ mol kg^{-1}; T= 470 K) ([22]), while by employing both conventional and single sweep oscillographic polarography it was aimed at determining the diffusion coefficients of Tl(I), Cd(II), Pb(II) and Zn(II) ([23]). Other polarographic studies were performed on the anodic oxidation of Cd, In, Pb, Tl and Zn amalgams at 495 K. By comparing the half-wave potentials of the oxidation, reduction and composite waves it was shown that the couples studied, with the exception of Tl/Tl(I), are affected by a certain degree of irreversibility ([24]).

The problem of water determination and its possible removal was studied in the molten binary K,Na/$C_2H_3O_2$ (46 mol % Na; T = 523÷573 K), also in connection with the behaviour of the couple H_2/H^+, to which comparatively little attention had been given in fused salt media, in spite of its potential employment in fuel cells ([25-27]). The electro-oxidation of hydrogen and the reduction of water and of acetic acid were investigated on smooth platinum electrodes by means of several electroanalytical methods including chronopotentiometry: the latter, as known, can certainly compete in fused salt media with other techniques, such as polarography and linear sweep voltammetry ([28]). Of the two pairs of peaks exhibited by a cyclic voltammogram (obtained from a partially moist melt saturated with hydrogen at 90 kPa), the former can be ascribed to the reaction

$$H_2 + 2 CH_3COO^- = 2 CH_3COOH + 2 e \tag{6}$$

and the latter to the reaction

$$H_2O + e = 1/2 H_2 + OH^- \tag{7}$$

the diffusive character of which was established by means of the Sand equation from chrono-potentiometric measurements on smooth and carefully cleaned platinum electrodes.

Good results were also obtained when the voltammograms were derived from current vs time curves plotted at different potentials and the current values measured at fixed times (usually 5 s after the application of voltage). In the present of acetic acid the occurrence of the reaction

$$2 CH_3COOH + 2 e = H_2 + 2 CH_3COO^- \tag{8}$$

was proved.

The interest devoted in recent years to fused hydrated salts ([29-32]) gave rise, besides spectroscopic investigations ([33]), also to electroanalytical studies in sodium acetate trihydrate. In particular, polarographic researches were performed ([34]), as well as potentiometric and biamperometric titrations of halides with silver ions, both produced coulometrically ([35]). In these titrations the error was always below 2 %. Further, it was possible to evaluate from potentiometric data the solubility products of silver halides. In the field of polarography the reduction processes concerning Co(II), Ni(II), Tl(I), Zn(II), Cd(II) and Pb(II) were investigated: in general the results did not show any significant difference in comparison with those obtained in anhydrous molten acetate.

MOLTEN SALTS WITH ORGANIC CATION

As in the case of molten salt systems with organic anion, also those with organic cation form a class of useful and proper solvents for the study of chemical and electrochemical properties and in several instances are suitable means for organic reactions, as a consequence of their low fusion temperature, sometimes even markedly lower than 373 K (e.g., tetraamylammonium thiocyanate melts at 322.7 K ([36])). A number of investigations aimed at interpreting the melting mechanism and studying physico-chemical properties, particularly in the case of alkylammonium salts with various anions (chlorides, bromides, iodides, thiocyanates, perchlorates, nitrates, fluoborates) (see [37-47]). Molten quaternary ammonium salts found an employment, among others, as stationary liquid phases in gaschromatography ([48]).

In general, these salts exhibit a comparatively wide range of electrostability; thus, they can be considered as a class of solvents especially suitable for electrochemical investigations: such studies are mainly due to French and Polish authors. Among alkylammonium chlorides, that of ethylammonium was most frequently used, since it has the lowest melting point and the widest temperature range over which it can be employed without decomposition.

Table 1 reports a list of the most commonly used alkylammonium salts, along with their fusion temperatures and the abbreviations adopted hereafter for the sake of simplicity.

ELECTROSTABILITY OF THE SOLVENTS

The knowledge of the electrostability range of the solvents is of basic importance in electrochemical studies, obviously in relation also with the nature of the electrodes employed.

In this regard, the investigations on TBABr, TEABr, TEAI and TPABF$_4$ ([49]) and those on EPBr ([50]) can be mentioned: in the latter solvent, independently of the electrode nature (gold, platinum, graphite, mercury) the reduction process was found to be

$$2 \; E\text{-}\overset{+}{N}\hspace{-0.5em}\bigcirc + 2 e \rightarrow E\text{-}N\hspace{-0.3em}\bigcirc\hspace{-0.3em}\bigcirc\hspace{-0.3em}N\text{-}E \tag{9}$$

through intermediate formation of the blue radical $E\text{-}N\bigcirc\cdot$, which slowly dimerizes to

N,N'-diethyl-bipyridyl-4,4'.

In the anodic process on platinum and graphite the bromide ion is oxidized to bromine with a total electrostability range of about 1.3 V.

The evolution of hydrogen on mercury and platinum electrodes was studied in EACl, MACl, DMACl, DEACl (see [51,52]), while the oxygen evolution in DMAS (strictly speaking, a mixture of $\{(CH_3)_2NH_2\}_2SO_4$ and $(CH_3)_2NH_2HSO_4$ in the weight ratio 70:30), according to the reaction

$$HSO_4^- + SO_4^{2-} \rightarrow HS_2O_7^- + O_{ads} + 2\ e \qquad (10)$$

was studied on platinum and gold electrodes ([54,55]): in the latter case the evolution of oxygen proved to be an irreversible process, although faster than that found on platinum.

ELECTRODE POTENTIALS

The availability of a scale of electrode potentials (in molten salt solvents) for several redox couples is of course remarkably important in order to make useful comparisons.

The potentials determined in some salts with organic cation are reported in Table 2. It can be noted that the potentials observed in the bromide solvent (EPBr) are more reducing than those in chloride, whereas those in sulfate (DMAS) are much more positive: this can be attributed to a strong complexing tendency of the metal ions with Br^-, and to a much less extensive one with the SO_4^{2-} ions (although the number of available measurements is smaller in this latter case), always taking the Cl^- ion as the reference.

Considering DMAS as an example, the cells employed were of the type

$$Pt,\ H_2\ (1\ atm)\ /\ DMAS\ //\ DMAS + MeSO_4\ (x)\ /\ Me$$

where Me is the studied species and x the molar fraction ([56]).

With metals which undergo spontaneous dissolution in the solvents employed, as it is the case, e.g., with cadmium in DMACl ([57]), amalgams can be advantageously used, after determination of the metal activity through e.m.f. measurements of concentration cells.

ACID-BASE EQUILIBRIA

Among acid-base equilibria, besides the evaluation from e.m.f. measurements of the dissociation constants of alkyl- and arylammonium ions dissolved in alkylammonium chlorides ([58,59]), one can mention, e.g., the pK determination of ethanolamine, tributylamine and pyridine obtained by means of potentiometric titrations with HBr in TBABr at 411 K ([49]).

Concerning the solvents EACl and EPBr, an interesting comparison was made between the two pertinent acidity scales at 400 K. On the assumption that the potential of the couple ferrocene/ferricinium is independent of the solvent, the origin of the acidity scale in EACl was found to be at + 2.7 units on the EPBr scale: this means that hydrochloric acid is weaker in EACl than hydrobromic is in EPbr ([60]).

When dissolved in EACl the strong bases ammonia and acetate cause decomposition of the solvent itself, with development of ammoniacal smell; the bases piperidine, fluoride and ethanolamine are capable to decompose also EPBr.

CHRONOPOTENTIOMETRIC INVESTIGATIONS

As already said, molten salts generally satisfy the conditions for chronopotentiometric applications better than water; in this respect, the main feature is a tendency to show ideal behaviour over a larger concentration range of the solute ([28]).

Most measurements have been carried out in EACl: in Table 3 are reported the values of the diffusion coefficients, D, and in a few cases also of the activation energies, ΔH, for several cations studied in this solvent.

In the case of Pt(II) it was shown that reduction occurs in two steps, through the intermediate species Pt(I). Concerning the sequence Ag(I), Au(I), Pt(II), the observed decrease in the cationic mobility was related to the correspondingly increasing tendency to complexation. It is to be mentioned that in the study of Au(I) also chronopotentiometry with current reversal was used in addition to the usual chronopotentiometric technique.

In the same solvent and at the same temperature chronoamperometric measurements [61] led for Au(I) to a D value of 1.51 10^{-6} cm^2 s^{-1}, in good agreement with the chronopotentiometric one, with an activation energy of 8.7±2.1 kJ mol^{-1}.

Still in EACl, the application of chronopotentiometry with current reversal at 413 K to indium amalgams [62] led for the diffusion coefficient of In in Hg to the value 2.87 10^{-6} cm^2 s^{-1}. From these measurements clear evidence was gained of the existence of the species In(I) in solution.

Finally, investigations on the process

$$Cd^{2+} + 2 e \; \underset{\rightarrow}{\leftarrow} \; Cd \; (in \; Hg) \tag{11}$$

which is chronopotentiometrically reversible in DMAS gave the value of 5.48 10^{-6} cm^2 s^{-1} for the diffusion coefficient of the species Cd(II) [63].

COMPLEX FORMATION

Several studies have been devoted to complex formation in EACl.

Particular attention was given to the complexes of Hg(II), Cu(I) and Au(I) with thiourea, iodides and thiocyanates [64], of Ag(I) with bromides and iodides [65] and of Ni(II) with pyridine [66].

Several techniques have been employed in the different cases, i.e., potentiometry for Hg(II) and Au(I); polarography for Cu(I); chronopotentiometry with hanging drop mercury electrode for Ni(II); e.m.f. measurement of concentration cells for Ag(I).

Table 4 reports data on the overall constants of formation, β, for the different complexes studied in EACl at 400 K.

Concerning the solvent EPBr, attention was given to Ti(III) and Ti(IV) complexes with nitrogen organic compounds (such as pyridine, 2,2'-bipyridyl, 1,10-phenanthroline and 1-(2-pyridylazo)-2-naphthol (PAN) [67] and Ti(IV) oxocomplexes [68]. The Ti(III) complexes were generally found to be more stable than the corresponding Ti(IV) complexes. All these investigations were performed by voltammetry with rotating platinum electrode, and as regards the oxocomplexes they allowed to state an order of decreasing stability according to the sequence TiO^{2+} > ZrO^{2+} > AlO$^+$.

Also the weakly complexing solvent DMAS has been studied with regard to Hg(II) complexes with chloride, bromide and iodide ions [69]. The constants for complex formation were evaluated through e.m.f. measurements of concentration cells in the temperature range 363÷388 K: it was proved that in these cases the complexation proceeds *via* formation of precipitated HgX$_2$ and subsequent dissolution of the same. The overall constants, β, at T = 363 K are reported in Table 5.

THERMODYNAMIC PROPERTIES OF SOLUTIONS OF INORGANIC SALTS

A number of measurements concerning the study of thermodynamic properties of mixing have been performed for solutions of inorganic salts in alkylammonium chlorides.

Ideal behaviour was first put into evidence for AgCl in DMACl [57]; subsequently the thermodynamic behaviour of silver halides in ethylammonium halides [70,71] was investigated through e.m.f. measurements of formation cells of the type

$$\overset{-}{Ag} \; / \; AgX \; (x_2) \; + \; EAX \; (x_1) \; / \; X_2 \; (1 \; atm), \; \overset{+}{C}$$

where x_1, x_2 are molar fractions. Direct employment of the halogen X at 1 atm pressure is not possible, since it would react with the melt; this difficulty was overcome by means of the SPI (Single Polarization Impulse) method, which allowed instantaneous electrolytical generation of the halogen at unit activity. Besides AgCl, also AgI exhibited ideal behaviour, whereas negative deviations were shown by AgBr solutions. Negative deviations were also found in dilute solutions of PtCl$_2$, AuCl, CuCl, PbCl$_2$ in EACl [72].

As already said, alkylammonium salts can be obtained as series of cations with regularly varying size: thus they can be advantageously employed to study the effect produced by the steric hindrance of the solvent cation on the thermodynamic properties of solutions of transition metal salts. In this connection, $HgCl_2$ was studied (once more by means of e.m.f. measurements of formation cells) in the solvents EACl, DMACl, i.PACl, MACl, BACl and DEACl. By plotting the partial molar excess free energy of $HgCl_2$ vs $(1/d_1 + 1/d_2)$, where d_1 is the sum of the ionic radii of the alkylammonium cation and of the chloride anion and d_2 is the corresponding sum for the Hg(II) cation and the chloride anion, a straightline trend was observed in the primary ammonium salts sequence MACl $\rightarrow$ EACl $\rightarrow$ i.PACl $\rightarrow$ BACl; this linear dependence, however, was different from that observed with the secondary ammonium salts DMACl and DEACl [73]. The effect of the nature of the inorganic anion (of the organic salt solvent) was also investigated [74]: ethylammonium was chosen as the common cation of the solvent and silver and mercury halides as solutes. In the case of AgX (X = Cl^-, Br^- and I^-) small deviations (of the order of 4.2 kJ mol^{-1}) from ideal behaviour were detected: they were positive for chloride solutions, negative for both bromide and iodide solutions. In the case of HgX_2 the deviations, always negative, decreased (in absolute value) according to the sequence $Br^- > Cl^- > I^-$.

Further, the effect of the transition metal cationic radius on the thermodynamic properties of chlorides ($CoCl_2$, $NiCl_2$, CuCl, AgCl, $SnCl_2$, $PbCl_2$, $CdCl_2$, $HgCl_2$ and $BiCl_3$) was investigated in fused EACl and DMACl [75]. By comparing the values of the partial molar excess enthalpy in pairs of metal halides having a similar electronic structure, it was observed that the deviations from ideal behaviour decreased with increasing ionic radius.

The thermodynamic quantities of $CoBr_2$ and $NiBr_2$ were measured in EABr over the temperature range 443÷503 K [76]. Small negative and positive deviations were found for the $CoBr_2$ and the $NiBr_2$ solutions respectively.

VELOCITY OF ELECTRODE PROCESSES

In order to evaluate the velocity of electrode processes through measurements of the faradaic impedance (which was expected to be particularly suitable in the case of fused salts) the following processes were taken into account for investigation in the solvents EACl and DMAS [77]:

$$Cd^{2+} + 2\ e \rightleftarrows Cd \text{ (in Hg)} \tag{11}$$

$$Cd^{2+} + 2\ e \rightleftarrows Cd \tag{12}$$

For the exchange current $i°$ in the case of reaction (11) the values $1.6\ 10^{-2}$ A cm^{-2} (at T = 397 K) and $8\ 10^{-2}$ A cm^{-2} (at T = 363 K) were found in EACl and DMAS, respectively. These results are a clear indication of poor reversibility of the process in EACl, whereas it is reversible in DMAS. Reaction (12) was studied in DMAS only: the reversibility of the process had been also verified, as said in a previous section, by chronopotentiometry [63].

THE SYSTEM ALUMINIUM CHLORIDE - ETHYLPYRIDINIUM BROMIDE

It may often be useful to have at disposal a room-temperature-fused solvent with a wide electrostability range and acidic properties: the mixture $AlCl_3$-EPBr in the molar ratio 2:1, which was first studied by Wier and coworkers [78] and combines the best features of fused $AlCl_3$ with those of low melting organic salts, completely meets the above requirements.

In particular, the employment of cyclic and pulse voltammetry allowed to study the complexes of Fe(II) with aliphatic amines, ferrocene and hexamethylbenzene [79]. Moreover, taking advantage of the complete miscibility of the same mixture with benzene, it was also possible to employ as solvent medium for electrochemical investigations a system composed with the above fused salts mixture and benzene in the volume ratio 1:1. It was found that the cathodic and anodic limits (electrostability range: $\simeq$ 2 V), respectively bound to aluminium deposition and bromine evolution, are not altered by the presence of benzene. Through cyclic voltammetry and with a rotating disc electrode it was found that, both in the absence and in the presence of benzene, the hexamethylbenzene radical cation was stabilized [80].

Finally, the mixture $AlCl_3$-EPBr was employed at 296 K in experiments of electrolytical attack of aluminium in the two-phase matrix $Al-Al_3Ni$. By operating under controlled potential it was possible to prove that etching of aluminium proceeds without involving the Al_3Ni phase [81].

MOLTEN SALTS WITH ORGANIC ANION AND CATION

Salts with both organic anion and cation generally melt at substantially low temperatures, so that they should deserve more attention than was actually given so far to their study and to their employment in the electrochemical area. Besides the viscosity and electrical conductivity measurements on several tetraalkylammonium picrates (for which reference is made to Chapters 1.4 and 2.2), mention can be made here, as an example, of polarographic investigations performed at 298 K on molten tetrahexylammonium benzoate [82], a salt particularly suitable for electrochemical work, owing to its good solubility in most organic solvents and wide electrostability range. From these measurements the half-wave potentials of oxygen, fumaric acid, benzophenone, anthracene and β-naphthol were obtained.

TABLES

TABLE 1. A few alkylammonium salts used in electroanalytical work

Salt	abbreviation	T_F/K	Ref.
Methylammonium chloride	MACl	498	[73]
Ethylammonium chloride	EACl	382	[73]
iso.Propylammonium chloride	*i*.PACl	426	[73]
Butylammonium chloride	BACl	489	[73]
Dimethylammonium chloride	DMACl	440	[73]
Diethylammonium chloride	DEACl	497	[73]
Ethylpyridinium bromide	EPBr	394	[83]
Dimethylammonium sulfate	DMAS	339	[53]
Tetrabutylammonium bromide	TBABr	388	[49]
Tetraheptylammonium bromide	TEABr	362.4	[49]
Tetraheptylammonium iodide	TEAI	395	[49]
Tetrapropylammonium fluoborate	TPABF$_4$	521	[49]

TABLE 2. Electrode potentials in some fused salts with organic cation

System	Electrode potentials (E/V) in the indicated solvents				
	DMACl [57] (T = 453 K)	EACl [56] (T = 403 K)	EACl [64] (T = 400 K)	DMAS [56] (T = 376 K)	EPBr [84] (T = 400 K)
Cd(II)/Cd	− 0.906			− 0.294	
Sn(II)/Sn	− 0.657		− 0.529	− 0.255	− 0.821
Pb(II)/Pb	− 0.628	− 0.635			
Co(II)/Co	− 0.461				
V(III)/V(II) (Hg)			− 0.420		
Cu(I)/Cu (Hg)			− 0.375		
Cu(I)/Cu	− 0.404	− 0.393	− 0.278		− 0.723
Ni(II)/Ni	− 0.372				
Bi(III)/Bi	− 0.311		− 0.189	+ 0.205	− 0.534
Ag(I)/Ag	− 0.116	− 0.148	− 0.036		− 0.558
Hg(II)/Hg	− 0.108	− 0.288			− 0.498
H(I)/H$_2$	0.000	0.000	0.000	0.000	0.000
Fe(III)/Fe(II)		+ 0.163	+ 0.388		+ 0.227
Ferrocene/ Ferricinium			+ 0.415		
Pt(II)/Pt	+ 0.223	+ 0.377			
Cu(II)/Cu			+ 0.486		
Au(I)/Au		+ 0.479	+ 0.691		+ 0.110

TABLE 3. Diffusion coefficients and activation energies of some cations in EACl

Cation	T/K	$\dfrac{D\ 10^6}{\mathrm{cm^2\ s^{-1}}}$	$\dfrac{\Delta H}{\mathrm{kJ\ mol^{-1}}}$	Ref.
Pt(II)	403	0.956 ± 0.036	17.6	[85]
Au(I)	403	1.18 ± 0.01	13.4 ± 2.1	[86]
Ag(I)	403	3.21 ± 0.13	16.6	[87]
Ag(I)	423	3.97 ± 0.16		[87]
Ag(I)	443	5.03 ± 0.13		[87]
Fe(III)	400	1.82 ± 0.05		[64]
Sn(II)	400	1.24 ± 0.18		[64]
Sn(IV)	400	1.27 ± 0.06		[64]
Cd(II)	396	0.69		[63]

TABLE 4. Formation constants of complexes in EACl at 400 K ([64], [65], [66])
(concentrations in mol kg^{-1})

Cation	Ligand				
	Thiourea	Iodide	Thiocyanate	Bromide	Pyridine
Hg(II)	$\beta_1 = 17.4\pm3.9$ $\beta_2 = 144\pm9$	$\beta_1 = 100\pm40$ $\beta_2 = (89\pm23)\ 10^2$ $\beta_3 = (90\pm32)\ 10^3$ $\beta_4 = (122\pm83)\ 10^3$			
Cu(I)	$\beta_1 = 13.4\pm9.4$ $\beta_2 = 80\pm23$				
Au(I)	$\beta_2 = (56\pm3)\ 10^4$	$\beta_1 = 118\pm81$ $\beta_2 = (170\pm22)\ 10^2$	$\beta_1 = 15.6\pm4.7$ $\beta_2 = 157\pm15$		
Ag(I) (†)		$\beta_1 = 139\pm1$		$\beta_1 = 12.6\pm0.5$	
Ni(II)					$\beta_1 = 3.2\pm5.6$ $\beta_3 = 23.9\pm13.6$

(†) T = 423 K

TABLE 5. Formation constant of complexes in DMAS at 363 K ([69])
(concentrations in mol kg^{-1})

Cation	Ligand		
	Chloride	Bromide	Iodide
Hg(II)	$\beta_2 = (1.6\pm0.9)\ 10^{12}$ $\beta_4 = (5.01\pm0.01)\ 10^{14}$	$\beta_2 = (1.0\pm0.5)\ 10^{14}$ $\beta_4 = (1.5\pm0.2)\ 10^{18}$	$\beta_2 = (2.9\pm1.5)\ 10^{19}$ $\beta_4 = (2.9\pm0.6)\ 10^{22}$

REFERENCES

1   GORDON, J.E., *"Applications of Fused Salts in Organic Chemistry"* in *"Techniques and Methods of Organic and Organometallic Chemistry"*, D.B.DENNEY Edit., New York, 1969.

2   UBBELOHDE, A.R., *Rev. Int. Htes. Temp. Réfract.*, 13, 5 (1976)

3   DURUZ, J.J., MICHELS, H.J., and UBBELOHDE, A.R., *Proc. R. Soc. London*, A 322, 281 (1971)

4   UBBELOHDE, A.R., *Pontif. Acad. Sci. Commentarii*, II, n.56, 1 (1972)

5   GORDON, J.E., *J. Org. Chem.*, 30, 2760 (1965)

6   BRY, B., and TREMILLON, B., *J. Electroanal. Chem.*, 46, 71 (1973)

7   MACHTINGER, M., VUAILLE, M.J., and TREMILLON, B., *J. Electroanal. Chem.*, 83, 273 (1977)

8   POURNAGHI, M., DEVYNCK, J., and TREMILLON, B., *Anal. Chim. Acta*, 89, 321 (1977)

9   WALLACE, R.A., and BRUINS, P.F., *J. Electrochem. Soc.*, 114, 209 (1967)

10 PETIT, N., *J. Electroanal. Chem.*, 31, 375 (1971)

11 POURNAGHI, M., DEVYNCK, J., and TREMILLON, B., *Anal. Chim. Acta*, 97, 365 (1978)

12 HALMOS, Z., SEYBOLD, K., and MEISEL, T., *Therm. Anal. Proc. Int. Conf. 4th*, 2, 429 (1974)

13 EKWUNIFE, M.E., NWACHUKWU, M.U., RINEHART, F.P., and SIME, S.J., *J.C.S. Faraday I*, 71, 1432 (1975)

14 ADEOSUN, S.O., SIME, W.J., and SIME, S.J., *J.C.S. Faraday I*, 72, 2470 (1976)

15 MEISEL, T., SEYBOLD, K., and ROTH, J., *J.Thermal Anal.*, 12, 361 (1977)

16 BARD, A.J., Edit., *"Encyclopedia of Electrochemistry of the Elements"*, Vol. X, M.Dekker Inc., New York, 1976.

17 COLICHMAN, E.L., *Anal. Chem.*, 27, 1559 (1955)

18 CUAN, S.P., DOLEZAL, J., KALVODA, R., and ZYKA, J., *Coll. Czech. Chem. Comm.*, 30, 4111 (1965)

19 BARTOCCI, V., CESCON, P., MARASSI, R., and FIORANI, M., *Ric. Sci.*, 39, 585 (1969)

20 MARASSI, R., BARTOCCI, V., CESCON, P., and FIORANI, M., *J. Electroanal. Chem.*, 22, 215 (1969)

21 HULTGREN, R., ORR, R., ANDERSON, P.D., and KELLEY, K.K., *"Selected Values of Thermodynamic Properties of Metals and Alloys"*, J. Wiley & Sons, New York, 1963.

22 BOMBI, G.G., FIORANI, M., and MACCA', C., *Chem. Comm.*, 1966, 455

23 GIESS, H., FRANCINI, M., and MARTINI, S., *Electrochim. Metall.*, 4, 17 (1969)

24 MARASSI, R., BARTOCCI, V., and FIORANI, M., *Chim. Ind.*, 52, 365 (1970)

25 MARASSI, R., BARTOCCI, V., and PUCCIARELLI, F., *Talanta*, 19, 203 (1972)

26 MARASSI, R., BARTOCCI, V., PUCCIARELLI, F., and CESCON, P., *J. Electroanal. Chem.*, 47, 509 (1973)

27 MARASSI, R., BARTOCCI, V., GUSTERI, M., and CESCON, P., *J. Appl. Electrochem.*, 9, 81 (1979)

28 JAIN, R.K., GAUR, H.C., FRAZER, E.J., and WELCH, B.J., *J. Electroanal. Chem.*, 78, 1 (1977)

29 BRAUNSTEIN, J., in *"Ionic Interaction"*, PETRUCCI, S., Edit., Vol. I, Academic Press, New York, 1971.

30 BRAUNSTEIN, H., BRAUNSTEIN, J., and HARDESTY, P.T., *J. Phys. Chem.*, 77, 1907 (1973)

31 BRAUNSTEIN, J., BACARELLA, A.L., BENJAMIN, B.M., BROWN, L.L., and GIRARD, C.A., *J. Electrochem. Soc.*, 124, 844 (1977)

32 BHATIA, K., SHARMA, R.C., and GAUR, H.C., *Electrochim. Acta*, 23, 1367 (1978)

33 BARTOCCI, V., CESCON, P., PUCCIARELLI, F., and MARASSI, R., *Ann. Chim. (Rome)*, 63, 581 (1973)

34 BARTOCCI, V., MARASSI, R., CESCON, P., and PUCCIARELLI, F., *Gazz. Chim. Ital.*, 104, 509 (1974)

35 BARTOCCI, V., MARASSI, R. PUCCIARELLI, F. and CESCON, P., *J. Electroanal. Chem.*, 42, 139
        (1973)

36 COKER, T.G., WUNDERLICH, B., and JANZ, G.J., *Trans. Faraday Soc.*, 65, 3361 (1969)

37 TAFT, R.W., and RAKSHYS, J.W., *J. Amer. Chem. Soc.*, 87, 4387 (1965)

38 KISZA, A., and HAWRANEK, J., *Z. Phys. Chem.*, 237, 210 (1968)

39 KISZA, A., *Acta Phys. Pol.*, 34, 1063 (1968)

40 GATNER, K., and KISZA, A., *Z. Phys. Chem.*, 241, 1 (1969)

41 KELLER, P., and HARRINGTON, G.W., *Anal. Chem.*, 41, 523 (1969)

42 COKER, T.G., AMBROSE, J., and JANZ,G.J., *J. Amer. Chem. Soc.*, 92, 5293 (1970)

43 LEVKOV,J., KOHR, W., and MACKAY, R.A., *J. Phys. Chem.*, 75, 2066 (1971)

44 GORDON, J.E., and VARUGHESE, P., *Chem. Comm.*, 1971, 1160

45 KOWALSKI, M.A., and HARRINGTON, G.W., *Anal. Chem.*, 44, 479 (1972)

46 ANDREWS, J.T.S., and GORDON, J.E., *J.C.S. Faraday I*, 69, 546 (1973)

47 GORDON, J.E., and VARUGHESE, P., *J. Org. Chem.*, 38, 3726 (1973)

48 GORDON, J.E., SELWYN, J.E., and THORNE, H.L., *J. Org. Chem.*, 31, 1925 (1966)

49 TEXIER, P., BADOZ-LAMBLING, J., *Bull. Soc. Chim.*, 1968, 1273

50 VEDEL, J., and TREMILLON, B., *Bull. Soc. Chim.*, 1966, 220

51 KISZA, A., and TWARDOCH, U., *Roczniki Chem.*, 44, 1503 (1970)

52 KISZA, A., and ROL, B., *Bull. Acad. Pol. Sci., Ser. Sci. Chim.*, 22, 801 (1974)

53 KISZA, A, and KRAJEWSKA, A., *Roczniki Chem.*, 47, 1751 (1973)

54 KISZA, A., and ZABINSKA, G., *Bull. Acad. Pol. Sci., Ser. Sci. Chim.*, 23, 857 (1975)

55 KISZA, A., and ZABINSKA, G., *Bull. Acad. Pol. Sci., Ser. Sci. Chim.*, 25, 473 (1977)

56 KISZA, A., and GRZESZCZUK, M., *Bull. Acad. Pol. Sci., Ser. Sci. Chim.*, 23, 765 (1975)

57 KISZA, A., *Z. Phys. Chem*, 237, 97 (1968)

58 KISZA, A., *Bull. Acad. Pol. Sci., Ser. Sci. Chim.*, 24, 839 (1976)

59 KISZA, A., and SEKOWSKA, A., *Bull. Acad. Pol. Sci., Ser. Sci. Chim.*, 25, 479 (1977)

60 VEDEL, J., *Bull. Soc. Chim.*, 1968, 3069

61 KISZA, A., and ROL, B., *Bull. Acad. Pol. Sci., Ser. Sci. Chim.*, 21, 37 (1973)

62 KISZA, A., and KAZMIERCZAK, J., *Bull. Acad. Pol. Sci., Ser. Sci. Chim.*, 23, 345 (1975)

63 KISZA, A., and GRZESZCZUK, M., *Bull. Acad. Pol. Sci., Ser. Sci. Chim.*, 25, 743 (1977)

64 PICARD, G., and VEDEL, J., *Bull. Soc. Chim.*, 1969, 2557

65 KISZA, A., *Bull. Acad. Pol. Sci., Ser. Sci. Chim.*, 19, 109 (1971)

66 PICARD, G., and VEDEL, J., *J. Electroanal. Chem.*, 40, 81 (1972)

67 LAPIDUS, M., and TREMILLON, B., *J. Electroanal. Chem.*, 15, 359 (1967)

68 LAPIDUS, M., and TREMILLON, B., *J. Electroanal. Chem.*, 15, 371 (1967)

69 KISZA, A., and ROL, B., *Bull. Acad. Pol. Sci., Ser. Sci. Chim.*, 20, 473 (1972)

70 GATNER, K., and KISZA, A., *Bull. Acad. Pol. Sci., Ser. Sci. Chim.*, 18, 199 (1970)

71 GATNER, K., KAZMIERCZAK, J., and KISZA, A., *Electrochim. Acta*, 17, 2055 (1972)

72 TWARDOCH, U., and KISZA, A., *Roczniki Chem.*, 51, 1449 (1977)

73 GATNER, K., and KISZA, A., *Roczniki Chem.*, 47, 1007 (1973)

74 GATNER, K., and KISZA, A., *Roczniki Chem.*, 48, 105 (1974)

75 GATNER, K., and KISZA, A., *Roczniki Chem.*, 48, 1369 (1974)

76 KISZA, A., and KAZMIERCZAK, J., *Bull. Acad. Pol. Sci., Ser. Sci. Chim.*, 22, 815 (1974)

77 KISZA, A., and GRZESZCZUK, M., *J. Electroanal. Chem.*, 91, 115 (1978)

78 HURLEY, F.H., and WIER, T.P., *J. Electrochem. Soc.*, 98, 203 (1951)

79 CHUM, H.L., KOCH, V.R., MILLER, L.L., and OSTERYOUNG, R.A., *J. Amer. Chem. Soc.*, <u>97</u>, 3264 (1975)

80 KOCH, V.R., MILLER, L.L., and OSTERYOUNG, R.A., *J. Amer. Chem. Soc.*, <u>98</u>, 5277 (1976)

81 HUSSEY, C.L., NARDI, J.C., KING, L.A., and ERBACHER, J.K., *J. Electrochem. Soc.*, <u>124</u>, 1451 (1977)

82 GARDNER SWAIN, C., OHNO, A., ROE, D.K., BROWN, R., and MAUGH, II, T., *J. Amer. Chem. Soc.*, <u>89</u>, 2648 (1967)

83 WILLEMS, J., and NYS, J., *Bull. Soc. Chim. Belg.*, <u>66</u>, 502 (1957)

84 PICARD, G., and VEDEL, J., *Electrochim. Acta*, <u>17</u>, 1 (1972)

85 KISZA, A., and TWARDOCH, U., *Bull. Acad. Pol. Sci., Ser. Sci. Chim.*, <u>20</u>, 481 (1972)

86 TWARDOCH, U., and KISZA, A., *Bull. Acad. Pol. Sci., Ser. Sci. Chim.*, <u>20</u>, 1069 (1972)

87 KISZA, A., and TWARDOCH, U., *J. Electroanal. Chem.*, <u>46</u>, 427 (1973)